Springer Series in
Surface Sciences

Guest Editor: J. Peter Toennies

Springer Series in **Surface Sciences**

Editors: Gerhard Ertl and Robert Gomer

Volume 1: **Physisorption Kinetics**
By H. J. Kreuzer, Z. W. Gortel

Volume 2: **The Structure of Surfaces**
Editors: M. A. Van Hove, S. Y. Tong

Volume 3: **Dynamical Phenomena at Surfaces, Interfaces and Superlattices**
Editors: F. Nizzoli, K.-H. Rieder, R. F. Willis

Volume 4: **Desorption Induced by Electronic Transitions, DIET II**
Editors: W. Brenig, D. Menzel

Volume 5: **Chemistry and Physics of Solid Surfaces, VI**
Editors: R. Vanselow, R. Howe

Volume 6: **Secondary Ion Mass Spectrometry SIMS V**
Editors: R. J. Colton, D. S. Simons, A. Benninghoven, H. W. Werner

H. J. Kreuzer
Z. W. Gortel

Physisorption Kinetics

With 133 Figures

Springer-Verlag
Berlin Heidelberg New York Tokyo

Professor Dr. Hans Jürgen Kreuzer

Department of Physics, Dalhousie University
Halifax, Nova Scotia, Canada B3H 3J5

Professor Dr. Zbigniew Wojciech Gortel

Department of Physics, University of Alberta
Edmonton, Alberta, Canada T6G 2J1

Guest Editor

Professor Dr. J. Peter Toennies

Max-Planck-Institut für Strömungsforschung, Böttingerstraße 6–8
D-3400 Göttingen, Fed. Rep. of Germany

Series Editors

Professor Dr. Gerhard Ertl

Institut für Psysikalische Chemie, Universität München, Sophienstraße 11
D-8000 München, Fed. Rep. of Germany

Professor Robert Gomer

The James Franck Institute, The University of Chicago, 5640 Ellis Avenue,
Chicago, IL 60637, USA

ISBN-13:978-3-642-82697-9   e-ISBN-13:978-3-642-82695-5
DOI: 10.1007/978-3-642-82695-5

Library of Congress Cataloging-in-Publication Data. Kreuzer, H.J. (Hans J.) Physisorption kinetics. (Springer series in surface sciences ; v. 1) Bibliography: p. Includes index. 1. Adsorption. I. Gortel, Z.W. (Zbigniew Wojciech), 1944-. II. Title. III. Series: Springer series in surface sciences ; 1. QD547.K74 1986 541.3'453 85-30307

© Springer-Verlag Berlin Heidelberg 1986
Softcover reprint of the hardcover 1st edition 1986

Offsetprinting: Beltz Offsetdruck, 6944 Hemsbach/Bergstr. Bookbinding: J. Schäffer OHG, 6718 Grünstadt
2153/3150-543210

Preface

This monograph deals with the kinetics of adsorption and desorption of molecules physisorbed on solid surfaces. Although frequent and detailed reference is made to experiment, it is mainly concerned with the theory of the subject. In this, we have attempted to present a unified picture based on the master equation approach. Physisorption kinetics is by no means a closed and mature subject; rather, in writing this monograph we intended to survey a field very much in flux, to assess its achievements so far, and to give a reasonable basis from which further developments can take off. For this reason we have included many papers in the bibliography that are not referred to in the text but are of relevance to physisorption.

To keep this monograph to a reasonable size, and also to allow for some unity in the presentation of the material, we had to omit a number of topics related to physisorption kinetics. We have not covered to any extent the equilibrium properties of physisorbed layers such as structures, phase transitions and thermodynamic properties in general. A number of excellent review articles, listed in the bibliography, cover this material. Likewise, little is said about scattering off solid surfaces; this subject is again covered in several books and many review articles. Lastly, little is said about chemisorption kinetics, for which microscopic theories and models have not been fully developed but are still at a rather early exploratory stage. It is the hope of the eternal optimists that we might learn something from physisorption kinetics to continue the task.

Halifax and Edmonton,   *H.J. Kreuzer*
Canada, 1985   *Z.W. Gortel*

V

Contents

1. **Introduction** 1
   1.1 Adsorption Phenomena: a Brief Survey 1
   1.2 Survey of Experimental Methods 17

2. **Gas–Solid Interaction** 23
   2.1 The Static Surface Potential $V_S(r)$ 23
   2.2 Mean–Field Surface Potential at Finite Coverage 43
   2.3 The Dynamic Atom–Solid Interaction 55
   2.4 Phonon Dynamics of a Solid 64
   2.5 The Gas–Solid Hamiltonian 86

3. **The Master Equation** 89
   3.1 The Mesoscopic Approach to the Master Equation 89
   3.2 The Master Equation for Physisorption Kinetics 96
   3.3 The Microscopic Approach to the Master Equation 103
   3.4 The Master Equation at Finite Coverage 107

4. **Transition Probabilities in the Master Equation** 112
   4.1 One–phonon Processes at Low Coverage 112
   4.2 Multi–phonon Processes and Correlation Functions 122
   4.3 Soft Cube Model and Phenomenological Models at Low Coverage .. 133
   4.4 Mean–Field Theory at Finite Coverage 141

5. **Desorption Times** 142
   5.1 Model System with Two Bound States 143
   5.2 Systems with a Few Shallow Bound States 146
   5.3 Perturbation Theory of the Master Equation 150
   5.4 Desorption Kinetics Mediated by Surface Phonons 157
   5.5 Inclusion of Parallel Phonon Momentum for Mobile Desorption ... 159
   5.6 Desorption from a Localized Adsorbate 164

5.7 Cole-Toigo Corrections 173

5.8 Multiphonon Contributions 176

5.9 Multilayer Desorption 181

5.10 Adsorbent Cooling in Thermal Desorption 186

5.11 Flash Desorption and Thermalization 190

6. Time of Flight Spectra 198

6.1 Experiments 198

6.2 Flux of Desorbed Particles 207

6.3 One-phonon Processes 213

6.4 Classical Models 220

7. Sticking and Accommodation 223

7.1 Experimental Results 223

    7.1.1 Sticking Coefficient 223

    7.1.2 Accommodation Coefficient 227

7.2 Theory 229

    7.2.1 Sticking 229

        7.2.1(a) Quantum Theories of Sticking 229

        7.2.1(b) Classical Theories of Sticking 242

    7.2.2 Energy Accommodation 251

        7.2.2(a) Quantum Theories of Accommodation 251

        7.2.2(b) Classical Theories of Accommodation 256

8. Kramers Equation 259

8.1 Derivation from the Master Equation 261

8.2 Macroscopic Laws 269

8.3 Friction Coefficient 276

8.4 Hydrodynamics of Adsorption 277

9. Summary and Outlook 282

9.1 Progress and Problems 282

9.2 Related Topics I: Photodesorption 283

9.3 Related Topics II: Electron-Stimulated Desorption 285

9.4 Approaches to Chemisorption Kinetics 286

References 293

Subject Index 317

1. Introduction

1.1 Adsorption Phenomena: a Brief Survey

Adsorption is the result of the interaction of a gas with a solid. Gas particles approaching the surface of a solid experience, at distances of a few
angstroms above the surface, a net attraction that changes over to a strong
repulsion closer in. If a gas particle can rid itself of enough energy, it
will be bound to the surface: it is said to be part of the adsorbate. To
understand the structure and the dynamics of an adsorbate, one must con-
ceive the gas phase and the adsorbate as two coupled systems open to
exchange of energy and particles, coupled in addition to the solid for
energy exchange, and sometimes particle exchange, e.g., when gas particles
or products thereof diffuse into the solid, or when surface reactions like
oxidation of the solid take place. The term adsorption was, according to
Kayser (1881), introduced by du Bois-Reymond.

The interaction between a gas and a solid is, of course, electromagnetic
in nature. To develop a meaningful picture that incorporates its salient
features, we follow Lennard-Jones (1932) and start with an isolated solid
and an atom far away from its surface. For the present discussion, we consider a metal for which we adopt the Sommerfeld model of free electrons in
which the lattice of ionic cores is smeared out into a uniform positively
charged "jellium" background. The conduction band is filled with electrons
up to the Fermi level E_F as indicated on the left of Fig. 1.1. The work
function ϕ of the metal is the energy required to remove an electron at the
Fermi level from inside the metal to infinity. On the far right of Fig. 1.1
we depict the electronic potential well of a free atom far from the surface. Its ionization energy E_I is the energy required to remove the most
weakly bound electron from the atom or molecule; the affinity level E_A is
the energy gained by attaching an electron to the particle to create a negative ion; typical values are given in Table 1.1. Let us now adiabatically
bring the atom or molecule close to the surface and eventually into an
adsorption site. As a result of the interaction between the electrons of
the particle and those of the metal, the ionization and affinity levels will

1

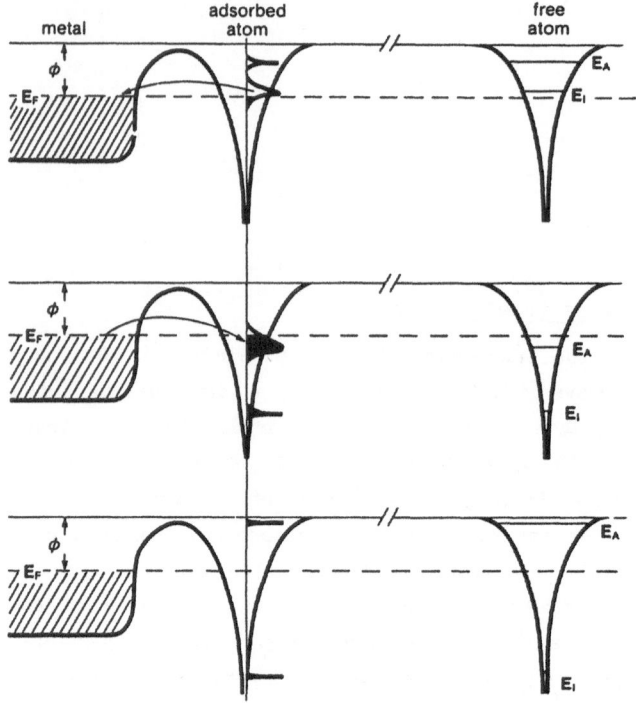

Fig. 1.1. The Lennard-Jones model of atomic adsorption on a metal. E_F = Fermi energy; ϕ = work function; E_I = ionisation energy; E_A = affinity level. Also indicated is the electronic density of states on the adsorbed atom. Arrows indicate charge transfer.

Table 1.1. Ionization energies E_I and electron affinities E_A for isolated atoms and molecules (Franklin and Harland 1974) and work functions ϕ of some metals (Hölzl and Schulte 1979).

	E_I[eV]	E_A[eV]	Metal	ϕ[eV]
H	13.59	0.76	Al	4.06-4.41
He	24.48	0.08	Mg	3.66
Li	5.39	0.65	Na	2.36-2.46
C	11.27	1.2	K	2.01-2.55
O	13.61	1.5	Cu	4.48-5.10
F	17.42	3.4	Ag	4.0 -4.74
O_2	12.06	≥0.45	Au	5.1 -5.47
S	10.36	2.07	W	4.3 -5.22
Cl	13.01	3.6	Pt	5.12-5.93
W	7.98	0.5	Ru	4.71
N_2O	12.89	≤1.5	Fe	4.5 -4.81

shift and broaden into resonances as Gurney pointed out in 1935. Depending on the relative positions of E_I and E_A with respect to E_F, we can distinguish several broad areas of adsorption phenomena, depending on the charge transfer and the kind of bonds established.

If, in the adsorbed particle, the ionization energy E_I is above the Fermi energy, one typically encounters a positively charged adsorbate resulting from electron transfer to the metal. This is the case, for example, for alkalis on most metals. If E_A, on the other hand, is below the Fermi level, electrons are transferred to the adsorbate as happens for halogens on transition metals. For E_A above and E_I below the Fermi level, neutral adsorbates are encountered that for $|E_A-E_F|$ and $|E_I-E_F|$ small, as in the case of Si on metals, establish covalent bonds. Gurney (1935) emphasized that, whereas E_I and E_A correspond to sharp energy levels in the isolated molecule, they broaden into resonances upon adsorption, indicated in Fig. 1.1 by the density of states localized on the adparticle. Depending on the Fermi energy E_F of the metal, these resonances might be partially occupied, leading to a fractionally charged adsorbate. As the latter now makes up an electrical double layer, the work function of the metal is effectively changed as a function of the amount of gas adsorbed. In all three of the above situations, we speak of chemisorption: rearrangement of electronic orbitals possibly accompanied by charge transfer, results in net energy gains upon adsorption, i.e., in heats of adsorption, of the order of electron volts. Because bonding orbitals are established between the adsorbate and the solid, chemisorption is usually restricted to less than a monolayer, with often several adsorption sites present on the various surface planes of a crystal that may change their characteristics as a monolayer fills up. That adsorption can be likened to chemical reactions, was emphasized by Langmuir (1916); earlier speculations are due to Bone and Wheeler (1906) (see also Bone 1922) and Haber (1914a and b). The term chemisorption was, according to Lennard-Jones (1932), coined by Benton and White (1931), although in that reference they refer to primary and secondary adsorption. It took another 40 years until the picture of chemisorption sketched in Fig. 1.1 was quantitatively worked out in the functional density approach (e.g., Lang and Williams 1978) and in Anderson-type models (e.g., Newns 1969; Einstein 1975).

Returning to the discussion of Fig. 1.1, we lastly look at a situation where the electron affinity of the adsorbing molecule is very small or even negative and its ionization energy E_I is large compared to the work function ϕ of the metal. In such systems, the electronic configuration of the

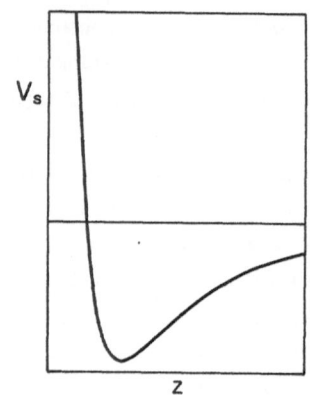

Fig.1.2. The surface potential of physisorption.

V_s

z

adsorbing particle is but slightly changed, i.e., only to the extend of an overall polarization producing an induced dipole that interacts with its image in the metal. This interaction can be modeled by an attractive van der Waals-type potential, acting on the (pointlike) adsorbate particle. Assymptotically far from the surface, it decreases as z^{-3} where z is the distance of the molecule from the surface. Closer to the surface, electron overlap produces strong repulsion. The resulting surface potential is sketched in Fig. 1.2. Particles adsorbed by this van der Waals mechanism are said to be physically adsorbed or physisorbed. Most typical are rare gases on metals, on alkali halides, and on graphite. A list of physisorbed systems, for which at least some kinetic data are available, is given in Table 1.2 where we also give the heat of adsorption (which is roughly the depth of the surface potential) and various kinetic data to be discussed later. In addition to rare gases, Table 1.2 lists rather inert molecules like CH_4 that physisorb on metals, and a number of other molecules, such as N_2O, that either chemisorb or physisorb on metals. Indeed, as indicated in our short discussion on chemisorption, once all "dangling" bonds have been used up at a surface for chemisorption, additional molecules will only be subject to dispersion forces of the van der Waals type, i.e., get physisorbed, unless bulklike condensation in the adsorbate takes place.

Before Langmuir (1916) stressed the monolayer character of chemisorption, it was generally believed that adsorption of a gas results in a transition layer which Eucken (1914) treated as a kind of miniature atmosphere through which particles must diffuse to react with the surface. At the 1932 Faraday Discussion "On the Adsorption of Gases and on Theories of the Adsorption of Gases", a clear distinction between physical adsorption due to weak van der Waals forces and chemical adsorption due to bonding forces was worked out [see articles by Lennard-Jones (1932) and Polanyi (1932)].

4

Table 1.2. Kinetic data for physisorbed gas-solid systems. Numbers in parentheses are estimates. References are given by first author only.

Gas	Solid	T [K]	θ_0 [10^{14}cm^{-2}]	θ/θ_0	E_d [kJ mole^{-1}]	ν [s^{-1}]	S	Reference
He	Ar	3.25	--	--	1.26±0.06	--	0.48	Lee(1974)
He	Kr	3.20	--	--	1.27±0.06	--	0.38	"
He	Xe	3.40	--	--	1.54±0.08	--	0.37	"
He	Nichrome	1.5-3.75	--	≤1	0.17-0.54	$(0.67-10)\times10^9$	--	Sinvani et al.(19872)
He		3.6	--	0 and 0.7	--	--	>0.66	Sinvani et al.(1983)
He	Constantan	3.5-3.8	--	≤1	0.49-0.36	$(2.5-51)\times10^9$	--	"
He	Sapphire	3.6	--	0 and 0.7	--	--	>0.66	"
Xe	C(0001)	85-102	6.35	0.002-0.08	20-23	--	--	Suzanne et al.(1973,1974)
Xe		74-80	5.7	0.3-0.9	25.1±1.7	--	0.65	Bienfait, Venables(1977)
Xe	Ni(100)	30-100	5.65	0.5	23.01	$(1-2)\times10^{12}$	--	Christmann, Demuth(1982)
Xe		30	5.65	≤1	18.83	(10^{12})	--	"
Xe		30-100	5.65	0.05-0.8	25.10-18.83	--	1-0.5	"
Xe	Cu(311)	77-130	5.85	≈1	19±2	--	--	Papp, Pritchard(1975)
Xe	Cu(211)	77	5.9	0.2-0.8	18±2	--	--	Roberts, Pritchard(1976)
Xe	Cu(100)	77-83	5.7	<0.8	--	--	0.9	Glachant, Bardi(1979)
Xe		77-80	5.7	<0.9	25.1±2.9	--	--	"
Xe	Ru(0001)	100	5.9	0-0.09	35.6-31.0	(10^{13})	--	Wandelt et al.(1981)
Xe		100	5.9	0.09-0.9	24.3-16.7	(10^{13})	1	"
Xe		107.4	5.9	0.1	(21.0)	6×10^{12}	(1)	"

Table 1.2 (continued)

Gas	Solid	T [K]	θ_0 [10^{14} cm^{-2}]	θ/θ_0	E_d [kJ mole^{-1}]	ν [s^{-1}]	S	Reference
Xe	Pd(100)	77	5.8	0-1	31.8-27.0	--	1	Palmberg(1971)
Xe	Pd(110)	100	5.8	0-0.6	--	--	1	Küppers et al.(1979)
		100	5.8	0.1	46.9	(10^{15})	--	"
		100	5.8	0.1	41.4	(10^{13})	--	"
		100	5.8	1	41.4	(10^{15})	--	"
		100	5.8	1	36.0	(10^{15})	--	"
Xe	Ag(111)	66-123	5.65	--	18±1	--	--	McElhiney et al.(1976)
		75-90	5.86	1	27±3	--	--	Cohen et al.(1976)
		60	6.14	0.92	28±1	--	--	Unguris et al.(1979)
		60	6.14	0.99	21±1	--	--	"
		60	6.14	0.96	21.7±0.5	--	--	"
		60	6.14	1	16.7±0.5	--	--	"
Xe	W(111)	104	(6)	0.1-1	38.9±5.6	10^{15}	0.5	Dresser et al.(1974)
		104	(6)	0.0025-0.1	42.8-38.9	10^{15}	0.5	"
		120	(6)	0-0.7	--	--	1	Yates, Erickson(1974)
		120	(6)	0.1-0.3	38.9±5.6	--	--	"
Xe	W(100)	60	5.5	0-1	--	--	1	Wang, Gomer(1979)
		65	4.2	0-1	--	--	1	"
		62	--	--	22.2	(10^{15})	--	Wang, Gomer(1980)
		20	--	>1	--	--	1	"

Gas	Solid	T [K]	θ_0 [10^{14}cm^{-2}]	θ/θ_0	E_d [kJ mole^{-1}]	ν [s^{-1}]	s	Reference
Xe	W(110)	62	5.5	0-1	--	--	1	Wang, Gomer(1979)
		62	5.5	1	25.9	(10^{15})	--	Wang, Gomer(1980)
		20	5.5	>1	--	--	1	"
		27	(6.6)	0-0.25	18.0	1×10^{12}	--	Opila, Gomer(1981)
		27	(6.6)	0-0.25	16.3	4×10^{10}	--	"
		27	(6.6)	1-1.25	13.8	6×10^{12}	--	"
		27	(6.6)	1-1.25	11.7	3×10^{10}	--	"
		27	(6.6)	2-2.25	10.5	1.5×10^{10}	--	"
Xe	Ir(111)	78-125	5.9	1	27.2±0.8	($k_B T/h$)	--	Nieuwenhuis(1974)
	and (100)	78-125	5.9	1	31.4±0.8	($k_B T/h$)	--	"
	Ir(110)	78-125	5.9	1	26.4±0.8	($k_B T/h$)	--	"
		78-125	5.9	<<1	29.3±0.8	($k_B T/h$)	--	"
	Ir(210)	78-125	5.9	1	28.0±0.8	($k_B T/h$)	--	"
		78-125	5.9	<<1	30.1±0.8	($k_B T/h$)	--	"
	Ir(321)	78-125	5.9	1	29.3±0.8	($k_B T/h$)	--	"
		78-125	5.9	<<1	32.6±0.8	($k_B T/h$)	--	"
	Ir(511)	78-125	5.9	1	27.2±0.8	($k_B T/h$)	--	"
		78-125	5.9	<<1	30.1±0.8	($k_B T/h$)	--	"
	Ir(531)-	78-125	5.9	1	28.0±0.8	($k_B T/h$)	--	"
	(731)	78-125	5.9	<<1	30.1±0.8	($k_B T/h$)	--	"

Table 1.2 (continued)

Gas	Solid	T [K]	θ_0 [10^{14} cm^{-2}]	θ/θ_0	E_d [kJ mole^{-1}]	ν [s^{-1}]	s	Reference
Xe	Au(100)	81–112	--	1	22±2	--	--	McElhiney et al.(1976)
Xe	ZnO	120–130	12.2	10^{-4}	24.1	--	--	Esser, Göpel(1980)
H₂	Graphite	90–140	--	<0.1	5.4	--	--	Constabaris et al.(1961)
D₂	Graphite	90–140	--	<0.1	5.6	--	--	"
H₂	Si	3	(10)	0–1	4.2–0.9	--	0.1–0.8	Govers, et al.(1980)
D₂	Si	3	(10)	0–1	4.7–1.1	--	0.26–0.9	"
H₂	ZnO	95–196	12.2	0.0001	3.5	--	--	Esser, Göpel(1980)
CH₄	Graphite	90–140	--	<0.1	12.7	--	--	Constabaris et al.(1961)
CD₄	Graphite	90–140	--	<0.1	12.6	--	--	"
CH₄	W(110)	110	--	--	29.1±4.2	3×10^{12}	--	Yates, Madey(1971)
CH₄	W(111)	125	--	--	33.9±4.2	3×10^{12}	--	Madey(1972)
CH₄	W(poly)	78	--	--	(33–50)	--	1	Shigeishi(1975)
CH₄	NaCl	83–90	--	0.25–1	14–16	--	--	Ross(1954)
NH₃	Al(poly)	128	(5.2)	0–1	37.7±8.4	--	0.13±0.08	Rogers, et al.(1980)
NH₃	Ru(0001)	100	3.0	0–1	30.5	(10^{13})	0.2	Danielson et al.(1978)
NH₃		100	3.0	0–1	44.4	(10^{13})	0.2	"
NH₃	Pt(111)	95–100	--	--	36±3	--	--	Gland, Kollin(1981)
N₂	K(film)	75	--	--	17.2±2.1	10^{11}–10^{12}	--	Mennicke et al.(1973)
N₂	Ni(110)	140	11.0	0–0.47	42	10^{13}	1	Grunze(1984)
N₂	Ni(110)	140	11.0	05–0.7	20	--	0.4±0.3	"

Gas	Solid	T (K)	θ_0 (10^{14}cm^{-2})	θ/θ_0	E_d (kJ/mole)	ν (s^{-1})	S	Reference
N_2	Fe(111)	85	---	---	24	10^{13}	0.7	Grunze(1984)
CO	ZnO	137-161	12.2	10^{-2}	26	---	---	Esser et al.(1980)
O_2	ZnO	127-241	12.2	10^{-2}	22.3	---	---	"
CH_3OH	Pd(100)	77	6	≤ 1	45.2	(10^{14})	---	Christmann, Demuth(1982)
		77	6	1-2	37.7	(10^{14})	---	"
		77	6	>2	30.1	---	---	"
CO_2	ZnO	209-235	(12.2)	10^{-2}	44.9	---	---	Esser, Göpel(1980)
SF_6	Ru(001)	80	(5.3)	1	21.1	(3×10^{12})	1	Fisher et al.(1977)

For most physisorbed gas-solid systems, the surface potential is rather constant along the surface, resulting in highly mobile adsorbates that, for submonolayer coverages, form a two-dimensional fluid as discovered by Volmer and co-workers (1926) [see also Volmer (1932)]. As monolayer coverage is approached, geometric restrictions might force adsorbing molecules into shallow adsorption sites forming a two-dimensional solid phase (or registered gas phase at high mobility), usually commensurate with the underlying solid. If the mutual interaction between adsorbed particles becomes dominant at low temperatures, incommensurate structures of tightly packed adsorbate particles might also appear; for recent discussions see Schick (1981) and Bienfait (1982). The phase diagrams of the helium isotopes adsorbing on graphite are reproduced in Fig. 1.3 as an example.

Our discussion so far has been concerned with the statics of the coupled gas-solid system, i.e., its equilibrium properties. If we want to know how fast and by what mechanism a gas particle adsorbs on, or desorbs from, a solid, we have to study its kinetics. Whereas the equilibrium properties of a large system are controlled by the minimum of its (free) energy, the kinetics involves questions of energy transfer. To establish the relevant time scales for the adsorption and desorption processes, let us follow a gas particle approaching the surface of a solid. If it rids itself of enough energy

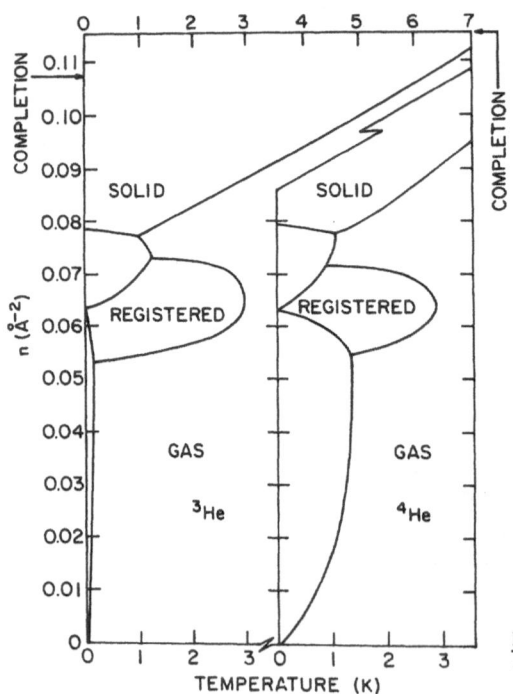

Fig. 1.3. Phase diagrams of ^{3}He and ^{4}He on graphite. (Schick 1981)

within the attractive region of the surface potential, it will get trapped. However, even if it descends all the way to the bottom of the surface potential well, it will eventually evaporate again; thus the concept of absolute trapping is meaningless. For time t_0 required for a particle to traverse the attractive potential well, the particle will remain close to the top of the well within a layer of width k_BT. In this time there is a fair chance that the particle will actually acquire enough energy from the heat bath of the solid to escape again. If this escape, which we may call inelastic scattering, has not happpened within a few round trips, the particle will begin its systematic descent to the bottom of the potential well. This adsorption process, characterized by an adsorption time t_a, is, of course, more likely at lower temperatures. After it has happened, the particle will try again and again to climb back out of the potential well, which it eventually will succeed in doing after a desorption time t_d. If t_a is much shorter than t_d, then adsorption and desorption are statistically independent, and the processes of sticking, energy accommodation (i.e., thermalization) and desorption can be well separated. This is most likely the case if the thermal energy k_BT is much less than the depth of the surface potential. However, for shallow adsorption states, such as for helium on most surfaces, the binding potential is so weak that this separation of time scales does not take place except at the lowest temperatures. In such situations, sophisticated statistical methods must be invoked for the description of the time evolution of a gas in front of a solid, the master equation approach being the most successful to date.

The energy necessary to desorb a particle from the adsorbate can either come from the solid or from some external source. As the latter, lasers or other sources of electromagnetic radiation have been used in photodesorption and photon stimulated desorption (e.g., Chuang 1983). Likewise, electron and ion beams are employed to cause electron and ion-stimulated desorption respectively (e.g., Tolk et al. 1983; Menzel 1982). Strong electric fields at field emission tips cause field desorption and field evaporation, sometimes used in conjunction with lasers or electrons to produce photon- and electron-stimulated field desorption (Block 1982).

If the solid itself acts as the reservoir from which the desorption energy is taken, we speak of thermal desorption. Lennard-Jones and Strachan (1935) and Lennard-Jones and Devonshire (1936a,b) argued that the thermal motion of the lattice should act as a time-dependent perturbation on the surface potential with which the adsorbate is bound to the solid, and can hence supply the desorption energy. It has been shown in recent years that this picture (in modern parlance called phonon-mediated desorption) is, indeed, appropriate and sufficient for physisorption, as we will see in detail in this book. Again, we should be aware that physisorption kinetics is much simpler than chemisorption kinetics where the additional questions concerning energy transfer between the electronic degrees of freedom of the solid and the adsorbate are still largely unanswered, and are currently under intense investigation (e.g., Schönhammer and Gunnarson 1983a,b).

At low coverage, physisorbed particles will desorb independently of each other. We can therefore assume that in the low coverage regime, physisorption kinetics proceeds via a reaction of first order. To quantify these ideas let us define the coverage,

$$\theta = \frac{N_a}{N_s} \quad , \tag{1.1}$$

as the number N_a of adsorbed gas particles per unit area of the surface of the solid, normalized to the surface concentration N_s of surface sites. Frequently, N_s is taken to be the number of atoms per unit area in the surface of the solid; sometimes N_s is the number of adsorbed particles per unit area at maximum monolayer coverage. Either convention, or any other, is acceptable as long as it is used consistently. If the coverage changes according to a first-order reaction, we can postulate a phenomenological rate equation

$$\frac{d\theta}{dt} = S \frac{P}{N_s(2\pi mk_BT)^{1/2}} - \frac{\theta}{t_d} \quad . \tag{1.2}$$

The first term on the right hand side of (1.2) is the rate of adsorption given by the flux $P/(2\pi mk_BT)^{1/2}$ of particles of mass m arriving from the gas phase, multiplied by the sticking probability S. The gas is assumed to be at a pressure P and a temperature T. The second term in (1.2) is the desorption rate, proportional to θ in a first order reaction, with t_d being the desorption time. Both S and t_d are usually functions of temperature T and coverage θ.

Experimental data on t_d are frequently given in the Frenkel-Arrhenius parametrization (sometimes also called the Wigner-Polanyi equation)

$$t_d = \nu^{-1} \exp(E_d/k_BT) = t_d^0 \exp(E_d/k_BT) \quad , \tag{1.3}$$

reflecting the fact that desorption is a thermally activated process in which an energy E_d is supplied to a desorbing particle with a Boltzmann probability $\exp(-E_d/k_BT)$. In a classical picture, ν can be interpreted as the attempt frequency with which the adsorbed particle tries to escape from the adsorption well. Thus E_d is roughly equal to the surface potential well depth and thus to the heat of adsorption. Whereas E_d is an equilibrium property of the gas-solid system, the dynamics of the energy transfer during the desorption process is contained in the prefactor ν. For physisorbed systems well below monolayer coverage, ν is typically of order $10^{12} - 10^{13}$ s^{-1}, dropping frequently below 10^{10} s^{-1} around and above monolayer coverage. Data on ν, E_d, and t_d for physisorbed gas-solid systems are collected in Table 1.2.

This book will be concerned with setting up microscopic models to calculate the time evolution of a gas-solid system and thus, in particular, the desorption time t_d. At this stage, we want to make brief contact with a widely used phenomenological approach based on transition state theory (or absolute rate theory). For this purpose, we view the desorption process as a chemical reaction where a particle A bound to a surface site Σ, forming a "molecule" (A-Σ), is removed into the gas phase via a transition state or an activated complex (A-Σ)*, leaving an empty site behind.

$$(A-\Sigma) \rightleftharpoons (A-\Sigma)^* \rightarrow A + \Sigma \tag{1.4}$$

This reaction corresponds to the desorption term in (1.2). If surface reactions such as dissociation and recombination, absent in physisorption, are taking place, (1.4) must be suitably modified. Eyring (1935,1938) assumed that the transition state is in statistical equilibrium with the reactants. The corresponding equilibrium constant is then given by

$$K^* = \frac{N^*}{N_a} = \frac{Q^*}{Q_{ad}} \quad . \tag{1.5}$$

Here N^* and N_a are the number of complexes and adsorbed particles per unit surface area, while Q^* and Q_{ad} are the molecular partition functions of the complex and the adsorbed particle, respectively. We next remove the zero point energies by writing

$$Q = q \exp(-E/k_B T) \tag{1.6}$$

and get for (1.5)

$$K^* = \frac{q^*}{q_{ad}} \exp(-E_d/k_B T) \quad , \tag{1.7}$$

where $E_d = E^* - E_{ad}$ is the activation energy of desorption.

The transition state is assumed to vibrate along the reaction path, per-pendicular to the surface, with frequency ν. With the transition state having the highest energy along the reaction coordinate, most vibrations will lead to desorption so that the rate of desorption becomes

$$r_d = \theta/t_d = \kappa\nu \frac{N^*}{N_s} \quad , \tag{1.8}$$

where κ is the transmission coefficient. Eliminating N^* with the help of (1.5) and (1.7) we get

$$t_d^{-1} = \kappa\nu \frac{q^*}{q_{ad}} \exp(-E_d/k_BT) \quad . \tag{1.9}$$

The partition function q^* of the transition state still contains a contribu-tion from the vibration along the reaction cordinate. Its frequency ν is typically so small that $h\nu \ll k_BT$. We therefore have

$$q^* = q^{**} \frac{1}{1-\exp(-h\nu/k_BT)} \approx q^{**} \frac{k_BT}{h\nu} \quad , \tag{1.10}$$

so that (1.9) reads

$$t_d^{-1} = \kappa \frac{k_BT}{h} \frac{q^{**}}{q_{ad}} \exp(-E_d/k_BT) \quad . \tag{1.11}$$

We note that the molecular partition functions q can have translational, rotational and vibrational contributions, see Table 1.3. In particular, if the transition state has the same degrees of freedom as the adsorbate, e.g., if both are mobile or localized, then $q^{**} \approx q_{ad}$. Thus, comparing (1.3) and (1.11), we find for the preexponential factor

$$\nu = \kappa \frac{k_BT}{h} \quad , \tag{1.12}$$

which, for $\kappa \approx 1$, is about $10^{11}T[K] \ s^{-1}$. Following similar arguments, we can calculate the rate of adsorption starting from the process

$$A + \Sigma * (A-\Sigma)^* \rightarrow (A-\Sigma) \quad . \tag{1.13}$$

The transmission coefficient κ will later on be identified with the sticking coefficient S. We should note that setting $\kappa=1$ reduces the expression (1.11) for the desorption time to one involving only static equilibrium properties. It will, therefore, be the task of a microscopic theory to improve upon this result by including dynamical effects connected with the process of energy transfer from the solid to the desorbing particle.

14

Table 1.3. Molecular partition functions. Here σ is the rotational symmetry number, I is the moment of inertia of a linear molecule and A,B,C are the principal moments of inertia of a nonlinear molecule.

Motion	Degrees of freedom	Partition function	Typical value
Translation	3	$q_{tr}^3 = \dfrac{(2\pi m k_B T)^{3/2} V}{h^3}$	$10^{30}-10^{33} V [m^3]$
Rotation (linear)	2	$q_r^2 = \dfrac{8\pi^2 I k_B T}{\sigma h^2}$	$10-10^2$
Rotation (nonlinear)	3	$q_r^3 = \dfrac{8\pi^2 (8\pi^3 ABC)^{1/2} (k_B T)^{3/2}}{\sigma h^3}$	10^2-10^3
Vibration	1	$q_v = [1-\exp(-h\nu/k_B T)]^{-1}$	$1-10$

We have said above that in the submonolayer regime desorption can be likened to a first-order reaction. This obviously presupposes that adsorbed particles desorb independently of each other. This is most likely not the case if the initial coverage is a fair fraction of a monolayer, or if clustering occurs in the adsorbate. In such situations, desorbing particles must not only break their surface bonds, but they must also overcome any interaction they might have with their neighbours. Rather than postulating some fractional-order desorption process, one usually keeps the first-order equation (1.2) and assumes that the sticking coefficient S and the desorption time t_d become functions of coverage θ and temperature T. In some systems, the adsorbate will segregate into clusters at some coverage. This might be a linear chain along a ledge between single crystal planes, or single and multilayer pools (Opila and Gomer 1981). In such situations, desorption may occur from the edges of such clusters, making desorption independent of coverage, i.e. apparently a zero-order process.

A particle adsorbed on the surface of a solid will desorb if it is supplied with enough energy to break its surface bond and, possibly, to overcome an activation barrier. In thermal desorption, this energy is drawn from the solid itself, either directly from the lattice degrees of freedom in phonon-mediated desorption or from the electronic degrees of freedom in chemisorbed systems. For laser- or particle-induced desorption, the energy is supplied by the laser or the particle beam. In the latter cases, it is obvious that the desorption rate will be a function of the beam intensities, i.e. of the rate at which the energy is supplied. It is also easily seen that the rate of thermal desorption does not only depend on the coupling of the adsorbate to the internal degrees of freedom of the solid (vibrational or electronic), but also on the capacity of the latter to supply the energy to the surface during the desorption time. A simple estimate can readily show that in typical systems no significant energy depletion, i.e. cooling, will occur in the surface region as a result of the desorption process (Gortel and Kreuzer 1983). To desorb N_a particles per unit area, each must be supplied with the desorption energy E_d; which will be drawn from a volume of surface A and depth L into the solid. If C_V is its specific heat per mole and V_m its molar volume, then this energy depletion results in a temperature drop at the surface

$$\Delta T \approx -\frac{N_a A E_d}{LA} \frac{V_m}{C_V} \ . \tag{1.14}$$

This kind of reasoning is, of course, only acceptable as long as the desorption process is slow on the time scale of establishing local equilibrium in the system, which in turn is determined by the phonon collision times. Also, one must demand that the relative temperature change ($\Delta T/T$) over a phonon mean free path is small. If these conditions are fulfilled, heat conduction will be subject to Fourier's law and L is given by

$$L = \sqrt{\chi t_d} \ , \tag{1.15}$$

where $\chi = \lambda V_m/C_V$ is the thermal diffusivity with λ being the coefficient of thermal conductivity, and t_d is the desorption time in which the initial adsorbate coverage drops to a fraction e^{-1}. The estimate (1.14) is uncertain (and presumably too high) to within a factor like e^{-1}.

Table 1.4 presents some estimates. The desorption time is parametrized according to Frenkel–Arrhenius (1.3). In none of the systems quoted does the temperature drop exceed a few percent. Neither are the conditions for local equilibrium violated. At the low temperatures, the mean free path for phonon collision is of the order of or somewhat smaller than L, whereas at the higher temperatures, it is 5 or 6 orders of magnitude smaller than L. Thus, the assumption in theories of adsorption and desorption kinetics that the solid acts as an (infinite) heat bath, is usually satisfied except possibly for very thin films. Whereas the arguments presented so far are general and apply to both chemisorption and physisorption, Gortel and Kreuzer (1983) have substantiated the above conclusions in greater detail within the cascade model of physisorption kinetics (Sect. 5.10).

Table 1.4. Parameters for typical gas-solid systems and estimates of cooling according to (1.14).(Gortel and Kreuzer 1983a)

System	T [K]	c_V [a] [J mol^{-1}K^{-1}]	χ [a] [m^{2}s^{-1}]	E_d [K]
He/graphite [b]	10	1.4×10^{-2}	4.1×10^{-4}	140
CO/Ru(001) [c]	500	25.0	3.6×10^{-5}	20000
Au/W(110) [d]	1450	31.0	5.2×10^{-5}	44000
O$_2$/W(110) [e]	1900	31.0	5.2×10^{-5}	46000

System	ν [s^{-1}]	t_d [s]	$L = \sqrt{\chi t_d}$ [Å]	$\Delta T/T$
He/graphite	5×10^{11}	2.0×10^{-6}	3.0×10^5	1×10^{-2}
CO/Ru(001)	1×10^{17}	2.3	9.0×10^7	1×10^{-7}
Au/W(110)	1×10^{13}	1.5	9.0×10^7	1×10^{-7}
O$_2$/W(110)	1×10^9	39.1	4.5×10^8	2×10^{-8}

[a] Data for c_V and χ from Keesom and Pearlman (1955) and Touloukian et al. (1970).
[b] Adsorption data from Gortel et al. (1980c).
[c] Adsorption data from Pfnür et al. (1978).
[d] Adsorption data from Bauer et al. (1975).
[e] Adsorption data from Kohrt and Gomer (1970).

1.2 Survey of Experimental Methods

Experimental methods developed for surface and adsorption studies have been reviewed in several books, e.g. Somorjai (1972), Ertl and Küppers (1974), Czanderna (1975), Roberts and McKee (1978), and Tompkins (1978).

Experimental studies of adsorption phenomena at solid surfaces are naturally grouped into those concerned with equilibrium properties and those investigating the kinetics. To get reproducible results, it is mandatory to prepare a clean, well-defined surface under ultra-high-vacuum conditions that is then brought into contact with a precisely known dose of a gas of known composition. To classify the equilibrium state of an adsorbate, three sets of information must be obtained, namely, on the chemical composition, on the geometry and on the thermodynamics of the adsorbate:

(i) the chemical composition of the solid surface and of the adsorbate must be identified. Various electronic fingerprinting techniques have emerged over the past decades, the most popular being AES (Auger

Electron Spectroscopy), UPS (Ultraviolet Photo Spectroscopy), XPS (X-ray Photo Spectroscopy), XAFS (X-ray Absorption Fine Structure Spectroscopy).

(ii) next, the geometrical arrangement of surface atoms and adsorbate particles must be determined, most frequently using LEED (Low Energy Electron Diffraction), field emission microscopy or Rutherford back-scattering. Infrared spectroscopy and EELS (Electron Energy Loss Spectroscopy) give information on molecular orientation and internal rearrangement.

(iii) to pin down the thermodynamics of adsorbates, one determines adsorption isotherms and various thermodynamic quantities like entropies, specific heats, and heats of adsorption. Good surveys of the thermodynamics of adsorption have been given, e.g., by Clark (1970) and Honig (1979).

The adsorption isotherm gives the amount of gas adsorbed as a function of pressure P at constant temperature. Because in almost all studies the gas can be treated as ideal, one frequently uses the chemical potential

$$\mu = k_B T \, \ln[h^3 P / k_B T (2\pi m k_B T)^{3/2}] \quad ,$$

(1.16)

in lieu of the pressure P. Isotherms for helium on graphite are reproduced in Fig. 1.4 as an example. They were determined by carefully measuring the dosage of admitted helium gas and the resulting gas pressure. Alternatively, one can look at the AES signal from the adsorbing species which increases (often nonlinearly) with the amount of gas adsorbed. Specific heats (Fig. 1.5), heats of adsorption (Fig. 1.6) and entropies are usually measured calorimetrically.

To study the kinetics of nonequilibrium phenomena connected with the adsorption and desorption processes, one has to take the gas-solid system out of equilibrium by changing external constraints. In an isothermal desorption experiment, one removes the gas phase above the surface, either by reducing the initial pressure $P = P_i$ to a final pressure $P_f \ll P_i$ by rapid pumping, or by chopping a molecular beam impinging onto the surface. Obviously, this removal of the gas must be much faster than the subsequent desorption process. This being achieved, one measures the time evolution of the desorbed particles as a pressure rise in a mass spectrometer or some other detector like a bolometer suitably placed above the surface. If the number of particles desorbed at any instant is measured, a desorption time can be determined. If, in addition, time of flight techniques are employed, their energy distribution can be determined as well, which may contain in-

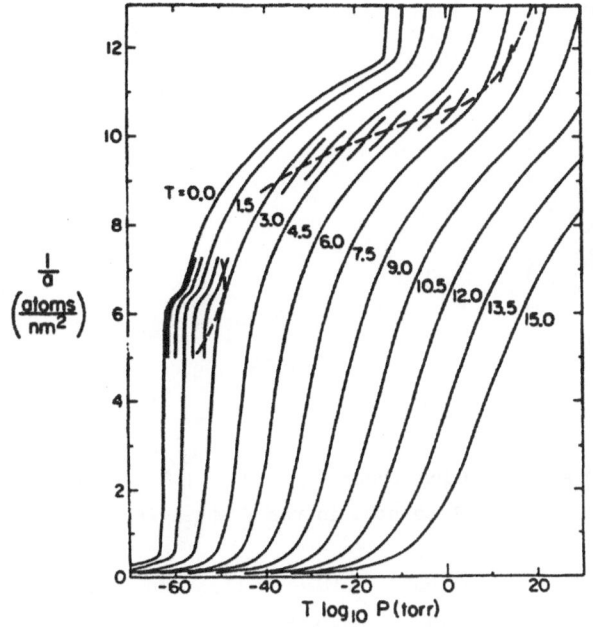

Fig. 1.4. Adsorption isotherms for helium on graphite. The dashed curve above 9 atoms/nm² is the melting transition; the dashed curve near 6 atoms/nm² is the lattice-gas ordering transition. The short solid curves in the regions of these transitions show isotherms at 0.5K intervals. (Elgin and Goodstein 1974)

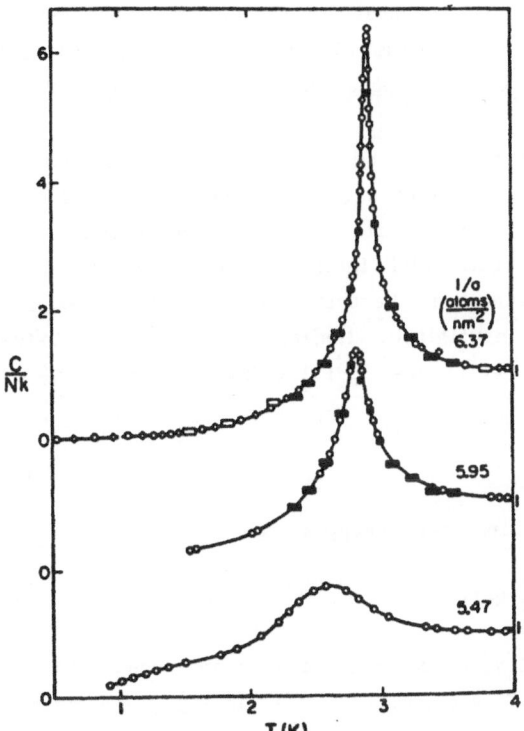

Fig. 1.5. Specific heat for helium on graphite in the vicinity of the lattice-gas ordering transition at various coverages. (Elgin and Goodstein 1974)

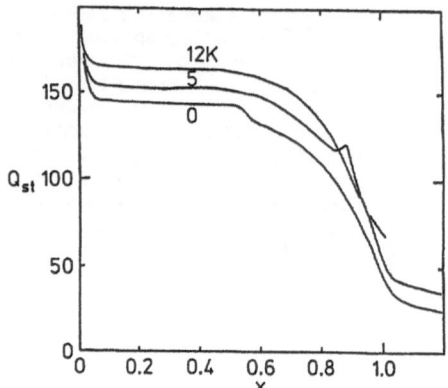

Fig. 1.6. Isosteric heat of adsorption for helium on graphite. (Elgin and Goodstein 1974).

formation on the dynamics of the desorption process. Table 1.2 contains desorption times for physisorbed systems presented in the Frenkel-Arrhenius parametrization (1.3).

Rapid removal of the gas phase is often not possible so that one, alternatively, creates a nonequilibrium situation by rapidly heating (flashing) the substrate from an initial temperature T_i to a final temperature T_f. Because physisorbed adsorbates do not necessarily thermalize to T_f before desorption starts, interesting nonequilibrium effects, like Taborek's (1982) critical cone in the angular and energy dependence of the desorption flux, can be observed.

The most frequently employed technique to study desorption kinetics is temperature-programmed desorption, in which the temperature of the solid is raised (usually linearly) as a function of time, the resulting pressure rise in the gas phase being recorded from which information on desorption time and desorption energies can be extracted. We briefly outline the method for a situation where readsorption is negligible. Menzel (1975, 1982) has twice reviewed the state of the art in recent years. Following Redhead (1962), we assume a linear temperature rise

$$T = T_0 + \beta t \quad , \tag{1.17}$$

which, inserted in (1.2) with the first term dropped, gives with (1.3)

$$- \frac{d\theta}{dT} = - \frac{d\theta}{dt} \frac{dt}{dT} = \frac{\nu}{\beta} \, \theta \, \exp(-E_d/k_B T) \quad . \tag{1.18}$$

The decrease with adsorbate density, of course, leads to an increase in the gas pressure above the surface. This signal is plotted in Fig. 1.7 as fitted to experimental data for the desorption of xenon desorbing from a W(111)

20

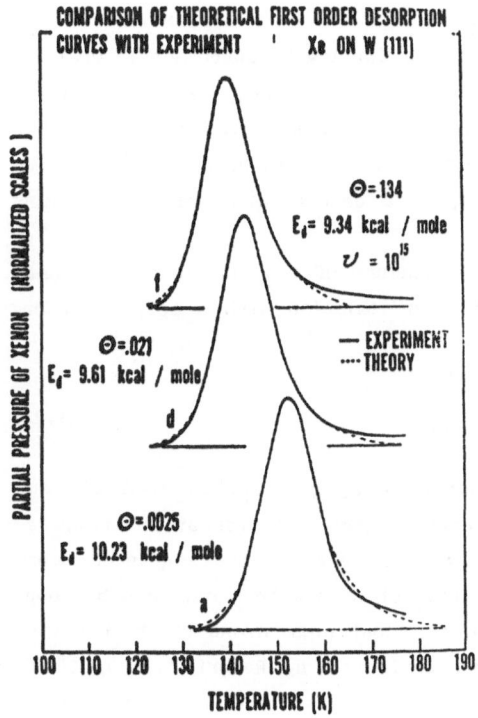

COMPARISON OF THEORETICAL FIRST ORDER DESORPTION CURVES WITH EXPERIMENT ' Xe ON W (111)

$\Theta = .134$
$E_d = 9.34$ kcal / mole
$\nu = 10^{15}$

$\Theta = .021$
$E_d = 9.61$ kcal / mole

$\Theta = .0025$
$E_d = 10.23$ kcal / mole

— EXPERIMENT
···· THEORY

PARTIAL PRESSURE OF XENON (NORMALIZED SCALES)

100 110 120 130 140 150 160 170 180 190
TEMPERATURE (K)

Fig. 1.7. Temperature programmed desorption of xenon from a W(111) surface at various coverages. Also plotted are theoretical first order fits from (1.18). (Dresser et al. 1974)

surface. The temperature T_p at the peak is given by

$$\frac{\beta}{\nu} = \frac{E_d}{k_B T^2} \exp(-E_d/k_B T)] \quad . \tag{1.19}$$

Using two different heating rates β_1 and β_2 - for practical purposes different by one or two orders of magnitude - one can, from the respective peak temperatures T_1 and T_2, determine E_d and ν via

$$E_d = k_B \frac{T_1 T_2}{T_1 - T_2} \ell n \left[\frac{\beta_1}{\beta_2} \frac{T_2^2}{T_1^2} \right] \quad . \tag{1.20}$$

The value of ν then follows from (1.19) with $T_p = T_1$ or T_2. Complications and uncertainties in the data analysis due to a coverage dependence of E_d and ν, and the occurrence of several adsorption sites, are discussed extensively by Menzel (1975, 1982) and King (1975).

As discussed in section 1.1, sticking and energy accommodation are two further processes that give information on the dynamics of the interaction gas-solid system. The sticking coefficient or sticking probability S, has been defined as the ratio of the rate of adsorption r_a on a surface to the rate of collision of particles from the gas phase at that surface, see the first term in (1.2). Thus, scattering a thermal beam of particles off a surface, one can determine S as the ratio of ingoing to outcoming intensities as long as desorption, i.e., the last term in (1.2) is negligible. Likewise, one can introduce the adsorbing gas into the evacuated chamber and follow the pressure drop from which the increase in energy can be determined, leading, again via the first term of (1.2), to a determination of the sticking coefficient S.

The accommodation coefficient α is a measure of the efficiency of energy and momentum transfer of gas molecules on collision with a solid surface. It is defined as

$$\alpha = \lim_{T_s \to T_g} \alpha(T_s, T_g) = \lim_{T_s \to T_g} \frac{E' - E(T_g)}{E(T_s) - E(T_g)} \quad . \tag{1.21}$$

Here, $E(T_g)$ is the average energy of the incoming gas particle beam at temperature T_g, $E(T_s)$ is the average energy of gas particles at the temperature T_s of the solid, and E' is the average energy in the scattered beam. Experimental methods for the determination of α have been reviewed by Goodman (1975) and Saxena and Joshi (1980); the most common one is the hot wire method invented by Knudsen (1911) and perfected by Roberts (1935, 1939). For a system comprising a metal filament of known surface area, and an inert gas at a pressure P, the energy loss $E' - E(T_g)$ is equated to the electrical energy supplied to the filament in order to maintain it at a constant temperature. Extensive data on the accommodation coefficient α have been collected by Saxena and Joshi (1981).

Before closing this section, we should mention that Toennies and his collaborators (1982) have, in recent years, succeeded in measuring the microscopic transition probabilities for phonon-mediated inelastic scattering of rare gases off alkali-halide surfaces. These experiments will be described extensively in Sect. 2.1.

2. Gas-Solid Interaction

2.1 The Static Surface Potential $V_S(r)$

We begin our review of the gas-solid interaction by looking at the static surface potential $V_S(r)$. This is the interaction that a particle experiences when it adiabatically approaches the surface of a solid at zero temperature. Far from the surface, the atomic structure of the solid is smeared out and $V_S(r)$ becomes a function of only the distance z that the particle is away from the surface. If the gas particle is charged, or carries a permanent dipole moment or some significant higher multipole, then classical electromagnetic theory tells us how such an object in front of, e.g., a metal, interacts with its image. To see how a neutral gas particle in its ground state approaching a solid will interact with it, we realize that over time scales $\tau_0 \approx h/\Delta E$ it can actually be excited into a state ΔE above the ground state. Rapidly fluctuating dipole and higher multipole moments are the result; they induce similar ones in the solid, leading to an overall attractive force whose potential energy is given in the dipole approximation by

$$V_{vdW}(z) = - C_3 z^{-3} \quad . \tag{2.1}$$

This is the van der Waals interaction. The most encompassing theory has been developed by Dzyaloshinskii et al. (1961) building on earlier work by London (1930), Lennard-Jones (1932), Bardeen (1940), Margenau and Pollard (1941), Prosen and Sachs (1942), and Casimir and Polder (1948). Relevant to our present understanding of the van der Waals forces are also the papers by Mavroyannis (1963) and McLachlan (1964). Recent reviews have been compiled by Langbein (1974), Steele (1974), Mahanty and Ninham (1974), Takaishi (1975), and Schmeits and Lucas (1983). A nonlocal theory of the van der Waals dispersion force, including dipole and quadrupole fluctuations, has been proposed by Apell and Holmberg (1984) and by Jiang et al. (1984 a,b).

The van der Waals C_3 can be written as

$$C_3 = \frac{\hbar}{4\pi} \int_0^\infty \frac{\varepsilon(i\omega)-1}{\varepsilon(i\omega)+1} \, \alpha(i\omega) \, d\omega \quad , \tag{2.2}$$

23

where α is the polarizability of the adsorbing atom evaluated at imaginary frequency and

$$\varepsilon(i\omega) = 1 + \frac{2}{\pi} \int_0^\infty \frac{x\varepsilon''(x)}{x^2+\omega^2} \, dx \quad , \tag{2.3}$$

is the dielectric function of the solid, again at imaginary frequency, which can be evaluated from the above Kramers – Kronig relation involving the imaginary part of the dielectric constant. Values for C_3 have recently been collected and analysed by Bruch (1983), by Vidali et al.(1983) and by Nath et al.(1985) and are given in Table 2.1. Bruch (1983) has also reviewed the van der Waals part of the substrate-mediated dispersion energy between two particles close to a surface.

The van der Waals potential (2.1) is a general result valid for a particle at a large distance z from the (flat) surface of a solid. As the particle approaches to within a few angstroms of the solid, details of the atomic structure become important and microscopic models must be constructed to account for the short-range repulsion. Two main approaches have been followed, namely: (i) density functional theory for atoms adsorbing on metals and (ii) summation of two-body forces for the adsorption of atoms on molecular or ionic solids. We begin with the latter model; it was first proposed by Lennard-Jones (1928, 1932). One writes the surface potential as

Table 2.1. Van der Waals constants C_3 [meV$Å^3$] for an atom in front of a flat surface.

	H	H_2	He	Ne	Ar	Kr	Xe
Cu [a]	468	664	222	450	1510	2140	3050
Ag [a]	493	711	246	500	1640	2300	3240
Au [a]	529	769	270	553	1780	2490	3490
Al [b]	496	670	219	443	1548	2201	3274
Pd [b]	452	613	211	439	1460	2064	3056
Graphite [a]	397	550	184	346	1208	1730	2460
LiF [a]	192	274	92	187	625	883	1260
NaF [a]	154	218	73	148	497	703	1000
Ge [c]	449	615	201	421	1429	2047	3529
InSb [c]	399	541	173	360	1244	1790	3071
ZnO [c]	322	450	155	332	1072	1519	2648

[a] Vidali et al. (1983)   [b] Bruch (1983)   [c] Nath et al. (1985)

24

$$V_S(\mathbf{r}) = \sum_i V_2(\mathbf{r},\mathbf{r}_i) \tag{2.4}$$

where $V_2(\mathbf{r},\mathbf{r}_i)$ is the two-body interaction between a gas particle at position $\mathbf{r}$ and a constituent particle of the solid at lattice site $\mathbf{r}_i$. One usually chooses for V_2 a central Lennard-Jones potential

$$V_2(\rho=|\mathbf{r}-\mathbf{r}_i|) = 4\varepsilon\left[(\frac{\sigma}{\rho})^{12} - (\frac{\sigma}{\rho})^6\right] \quad . \tag{2.5}$$

Carlos and Cole (1980b) also considered an isotropic 6-8-12 potential

$$V_2(\rho) = \frac{\varepsilon}{2s+3}\left[(4s+3)\,(\frac{\rho_m}{\rho})^{12} - 6s(\frac{\rho_m}{\rho})^8 - 6(\frac{\rho_m}{\rho})^6\right] \quad , \tag{2.6}$$

the ρ^{-8} term representing a dipole-quadrupole dispersion force. They also investigated an exponential – six pair potential

$$V_2(\rho) = \frac{Fa_S\alpha^2}{4\pi}\,e^{-\alpha\rho} - \frac{3a_S dC_3}{\pi\rho^6} \quad , \tag{2.7}$$

which, however, necessitates a short range cutoff to avoid the negatively divergent attraction. Their Yukawa-six potential replaces the exponential in (2.7) by a Yukawa repulsion $\rho^{-1}\exp(-\alpha\rho)$. For solids such as graphite where strong directional hybridization of the bonding orbitals occurs, Carlos and Cole (1979,1980b) make the point that V_2 should be nonspherical and aniso-tropic. They choose

$$V_2(\mathbf{r}-\mathbf{r}_i) = A\,\frac{\exp(-\alpha|\mathbf{r}-\mathbf{r}_i|)}{|\mathbf{r}-\mathbf{r}_i|}\,(1+\gamma_R\cos^2\theta) - \frac{3a_S dC_3}{\pi|\mathbf{r}-\mathbf{r}_i|^6}\left[1+\gamma_A(1-\frac{3}{2}\cos^2\theta)\right] \tag{2.8}$$

for the He-C interaction above the basal plane of graphite for which $\gamma_A = 0.4$ and $\gamma_R = 0.29$. Here θ is the polar angle of $\mathbf{r}$.

Once the two-body interaction V_2 is chosen and the lattice structure of the solid is determined, it is a simple matter to evaluate the sums in (2.4) and to calculate $V_S(\mathbf{r})$ after making proper use of the translational symme-tries along the surface, as we will review below. For now we present a rough estimate for (2.4) and replace the lattice sums by an integration over a continuum of mass density ρ_0 giving a volume M_S/ρ_0 to each solid atom of mass M_S so that we get (London 1930)

$$V_S(\mathbf{r}) = \frac{\rho_0}{M_S}\int d^3\mathbf{r}'V_2(\mathbf{r},\mathbf{r}') \quad . \tag{2.9}$$

For a Lennard-Jones potential (2.5) one finds (Hill 1952)

$$V_S(\mathbf{r}) = V_S(z) = 4\pi\epsilon\sigma^3 \frac{\rho_0}{M_S} \left[\frac{1}{15} \left(\frac{\sigma}{z}\right)^9 - \frac{1}{2} \left(\frac{\sigma}{z}\right)^3 \right] \ . \tag{2.10}$$

This 9-3 potential is useful, though not too accurate if the parameters ϵ and σ are taken from two-body scattering data. If, however, it is known that $V_S(\mathbf{r})$ shows little lateral variation and if the adsorbate bound states are known experimentally then they can be fitted to the bound states of (2.10) to fix its depth ($-V_0$) and the position z_m of its minimum phenomenologically as

$$V_S(z) = \frac{1}{2} V_0 \left[\left(\frac{z_m}{z}\right)^9 - 3\left(\frac{z_m}{z}\right)^3 \right] \ . \tag{2.11}$$

We will elaborate this point further at the end of this section.

For a proper evaluation of the lattice sums in (2.4), one introduces a two-dimensional vector along the surface [we follow Steele (1973a,b, 1974); for an account of the early work see Young and Crowell (1962)],

$$\mathbf{R_1} = l_1\mathbf{a}_1 + l_2\mathbf{a}_2 \ , \tag{2.12}$$

where l_1 and l_2 are integers and $\mathbf{a}_1$ and $\mathbf{a}_2$ are lattice vectors in two dimensions spanning a surface unit cell. Translational symmetry along the surface then implies

$$V_S(z,\mathbf{R} + \mathbf{R_1}) = V_S(z,\mathbf{R}) \ , \tag{2.13}$$

where we adopt the notation $\mathbf{r}=(z,\mathbf{R})$. As a Fourier series, (2.13) implies

$$V_S(\mathbf{r}) = V_0(z) + \sum_{\mathbf{m}\neq0} V_\mathbf{m}(z) \exp(i\mathbf{K_m}\cdot\mathbf{R}) \ , \quad \text{where} \tag{2.14}$$

$$V_\mathbf{m}(z) = \sum_{\alpha=0}^{\infty} w_\mathbf{m}(z_\alpha) \ , \quad \text{and} \tag{2.15}$$

$$\mathbf{K_m} = 2\pi(m_1\mathbf{b}_1 + m_2\mathbf{b}_2) \ , \tag{2.16}$$

where m_1 and m_2 are integers and the reciprocal lattice vectors $\mathbf{b}_j$ are such that

$$\mathbf{a}_i\cdot\mathbf{b}_j = \delta_{ij} \ ; \ i,j = 1,2 \ , \tag{2.17}$$

The sum over α in (2.15) enumerates the various lattice planes which are a distance $z_\alpha = z+\alpha d_L$ below the gas particle, where d_L is the distance between lattice planes parallel to the surface. Also

$$w_{\mathbf{m}}(z_\alpha) = \frac{1}{a_s} \int_{a_s} \exp(-i\mathbf{K_m} \cdot \mathbf{R}) \, V_s(z_\alpha, \mathbf{R}) d^2\mathbf{R}$$

$$= \frac{1}{a_s} \sum_{k=1}^{q} \exp(-i\mathbf{K_m} \cdot \mathbf{m}_k) \int_A \exp(-i\mathbf{K_m} \cdot \mathbf{R}) \, V_2(z_\alpha, \mathbf{R}) d^2\mathbf{R} \quad , \qquad (2.18)$$

where the first integral goes over the area a_s of a unit cell and the second one over the total area A of the surface. In (2.18) $\mathbf{m}_k$ is a vector to the k-th of q atoms in the unit cell. For a two-body potential V_2 of Lennard-Jones type (2.5) these integrals can be done explicitly and yield

$$V_s(\mathbf{r}) = \frac{2\pi}{a_s} \varepsilon \sum_{\alpha=0}^{\infty} \{ q(\frac{2}{5} \frac{\sigma^{12}}{z_\alpha^{10}} - \frac{\sigma^6}{z_\alpha^4}) + \sum_{\mathbf{K_m} \neq 0} \sum_{k=1}^{q} \exp[i\mathbf{K_m} \cdot (\mathbf{m}_k + \mathbf{R})]$$

$$* \left[\frac{\sigma^{12}}{60} (\frac{K_m}{2z_\alpha})^5 \, K_5(K_m z_\alpha) - 4\sigma^6 (\frac{K_m}{2z_\alpha})^2 \, K_2(K_m z_\alpha) \right] \} \quad . \qquad (2.19)$$

Here $K_2(x)$ and $K_5(x)$ are modified Bessel functions of the second kind and $K_m = |\mathbf{K_m}|$. We note that changing sums to integrals in (2.19) results again in the 9-3 potential (2.10). Equation (2.19) is easily generalized to solids containing several types of atoms since the only change required is the insertion of differing values of $\varepsilon(k)$ and $\sigma(k)$ for the k-th atom in the unit surface cell. Another generalization of (2.19) can be made when the atoms in an underlying layer are at different positions from those in the first layer; if this is the case, one defines q_α to be the total number of atoms in the projection of the unit cell in the α-th layer, and $\mathbf{m}_k(\alpha)$ to be the position vectors of these atoms.

Fig. 2.1 shows the surface potential for a helium atom interacting with the basal plane of graphite above the positions indicated in Fig. 2.2. To show the variation of $V_s(z,\mathbf{R})$ along the surface more clearly, we have plotted it in Fig. 2.3 as a function of z for $\mathbf{R}$ along the long diagonal across the unit cell going from adsorption site to surface atom to saddle point to surface atom to adsorption site. To show the corrugation at the bottom of the surface potential, we calculate the minimum of $V_s(z_{min},\mathbf{R})$ for fixed $\mathbf{R}$ and plot it in Fig 2.4 as a function of $\mathbf{R}$ over the section of the surface indicated in Fig. 2.2. Note that the corrugation amounts to less than 8% of the absolute potential minimum. A graph of the position z_{min} of the bottom of the surface potential over the R-plane looks very much the same.

Fig. 2.5 shows $V_s(z_{min},\mathbf{R})$ as a function of $\mathbf{R}$ for an atom on a fcc(111) surface. Compared to Fig. 2.4 we see, as anticipated in Fig. 2.2, that maxima and minima are reversed. However, it should be noted that overall

27

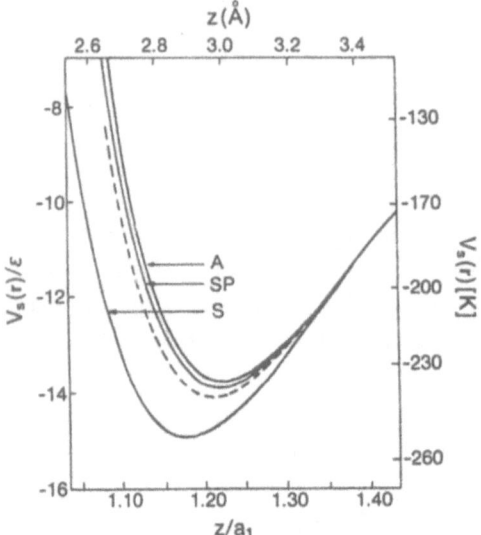

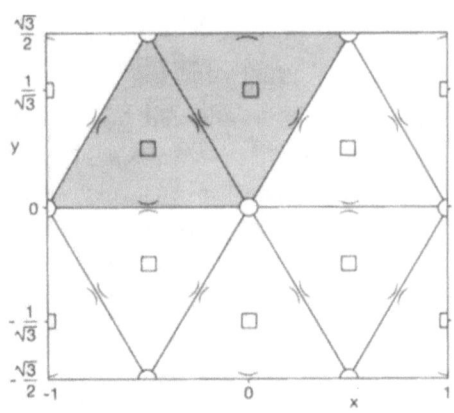

Fig.2.1. The surface potential $V_S(\mathbf{r})$ for helium above three points of the basal plane of graphite, from (2.17). The dashed line is $V_0(z)$. Parameters of the He–C 6-12 potential: $\varepsilon/k_B = 16.9$ K, $\sigma = 2.98$ Å, lattice spacing $d_L = 3.39$ Å, area of the unit cell $a_S = 5.24$ Å². Number of atoms in the surface unit cell of graphite $q=2$, its side $a_1=2.46$ Å. S: above a surface atom; A: above an adsorption site; SP: above a saddle point. (Steele 1973b)

Fig.2.2. Structure of a (111)fcc surface and that of the basal plane of graphite. The shaded area is a unit cell. Circles are surface atom positions for (111)fcc and adsorption sites for graphite. Likewise, squares denote surface atoms for graphite and adsorption sites for (111)fcc. Saddle points of the surface potential are also indicated.

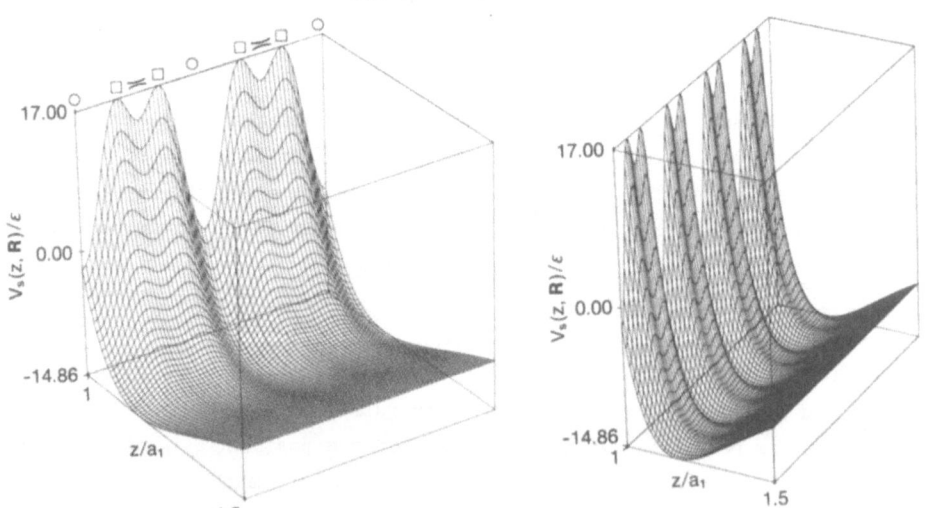

Fig.2.3. He-graphite surface potential $V_S(z,\mathbf{R})$ as a function of z and for $\mathbf{R}$ along the long diagonal of the unit cell from adsorption site (circle) to atom (square) to saddle point to atom, etc.

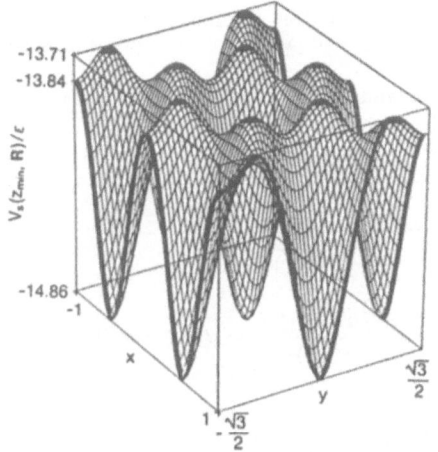

Fig.2.4. The bottom of the He-graphite surface potential $V_s(z_{min}, R)$ over the surface area of Fig.2.2. The mountain tops (of equal heights) are the positions of the surface atoms.

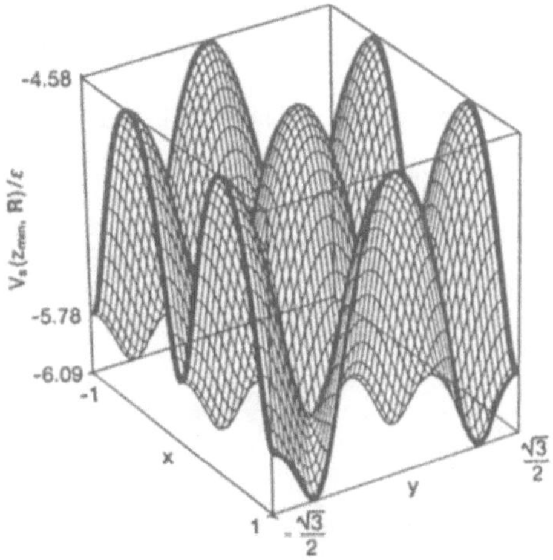

Fig.2.5. The bottom of the surface potential $V_s(z_{min}, R)$ over a fcc(111) surface with $\sigma/a_1 = 1$.

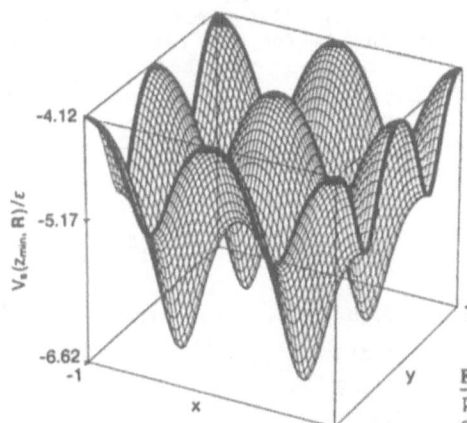

Fig.2.6. The bottom of the surface potential $V_s(z_{min}, R)$ over a fcc(100) surface with $\sigma/a_1 = 1$.

corrugation is much larger, namely about 25%, although the saddle points are now much lower. For the open fcc(100) surface, the corrugation becomes larger yet, as seen in Fig. 2.6. The surface atoms are again at the mountain peaks as indicated.

The results so far reported are for a 6-12 pair potential (2.5). Carlos and Cole (1979,1980) have performed a similar analysis for the pair potentials (2.6-8) for helium interacting with the basal plane of graphite. The functions $V_m(z)$ entering (2.14) can be found in Table 2.2. Table 2.3 gives the relevant parameters for the two-body He-C interactions (2.5-8) and also

Table 2.2. Fourier components $V_m(z)$ in (2.13) for various pair potential forms, $\beta_m = \exp(-iK_m \cdot m_1) + \exp(-iK_m \cdot m_2)$. Here m_1 and m_2 are the positions of carbon atoms in the unit cell of graphite and $\zeta(n,x) = \sum_{j=0}^{\infty} (j+x)^{-n}$ is a Riemann zeta function. (Carlos and Cole 1980b)

$U(\rho)$	$V_0(z)$	$V_m(z)$
$\rho^{-(2n+2)}$	$\dfrac{4\pi}{2na_s}\,\dfrac{\zeta(2n,z/d)}{d^{2n}}$	$\dfrac{2\pi\beta_m}{n!a_s}\left(\dfrac{K_m}{2z}\right)^n K_n(K_m z)$
$e^{-\alpha\rho}$	$\dfrac{4\pi(1+\alpha z)}{a_s\alpha^2}\,e^{-\alpha z}$	$2\pi\beta_m\,\dfrac{1+z(\alpha^2+K_m^2)^{1/2}}{a_s(\alpha^2+K_m^2)^{3/2}}\,\exp[-z(\alpha^2+K_m^2)^{1/2}]$
$\rho^{-1}e^{-\alpha\rho}$	$\dfrac{4\pi}{a_s\alpha}\,e^{-\alpha z}$	$\dfrac{2\pi\beta_m}{a_s(\alpha^2+K_m^2)^{1/2}}\,\exp[-z(\alpha^2+K_m^2)^{1/2}]$
$\dfrac{\cos^2\theta}{\rho^6}$	$\dfrac{2\pi}{3a_s d^4}\,\zeta(4,z/d)$	$K_m^3\pi\beta_m\,\dfrac{K_3(K_m z)}{48a_s z}$

Table 2.3. Parameters for the two-body interactions $V_2(\rho)$ given in (2.5-8): they vanish at ρ_0 and have a minimum V_2^{min} at ρ_m. Also given are the zeros z_0, minima V_{min}, and van der Waals constant C_3 for the resulting $V_0(z)$. (Carlos and Cole 1980b)

	6-12	6-8-12	Exp-6
ρ_0 [Å]	2.74	3.32	2.34
ρ_m [Å]	3.07	3.68	2.74
$-V_2^{min}$ [meV]	1.40	1.12	1.40
z_0 [Å]	2.34	2.88	1.78
z_{min} [Å]	2.74	3.31	2.28
$-V_{min}$ [meV]	16.38	16.09	15.78
C_3 [mev Å³]	140	127	86

gives zeros, minima and van der Waals constant C_3 for the resulting laterally averaged surface potentials $V_0(z)$. The parameters were adjusted to optimize the energy eigenvalues for ^{3}He and ^{4}He in $V_0(z)$ with those determined from the bound state resonances in the scattering cross sections. Table 2.4 shows the results. The helium-graphite interaction has been reviewed by Cole et al. (1981).

Recently, Vidali et al.(1983) noted that for a great number of different physisorption gas-solid systems the surface potentials averaged along the surface can be reduced to a universal shape as a function of z,

Table 2.4. Bound state energies [meV] of ^{3}He and ^{4}He interacting with a graphite surface. Experimental values from scattering data (Derry et al. 1979), compared with those derived by summing pair potentials of the form indicated. (Carlos and Cole 1980b)

n	^{4}He			
	Exp.	6-12	6-8-12	Exp-6
0	-12.06	-12.12	-12.08	-12.10
1	-6.36	-6.18	-6.27	-6.40
2	-2.85	-2.78	-2.80	-2.81
3	-1.01	-1.06	-1.01	-0.96
4	-0.17	-0.32	-0.27	-0.23
5	--	-0.07	-0.06	-0.05

n	^{3}He			
	Exp.	6-12	6-8-12	Exp-6
0	-11.62	-11.53	-11.52	-11.58
1	-5.38	-5.19	-5.27	-5.39
2	-1.78	-1.95	-1.94	-1.92
3	--	-0.57	-0.52	-0.46
4	--	-0.10	-0.09	-0.05

$$V_S(z) = V_0(z) = V_0\, g(z^*) \quad ,$$

$$z^* = (z - z_m)/\ell \quad , \tag{2.20}$$

in terms of the minimum position z_m and the well depth V_0. To make (2.20) compatible with (2.1) one chooses

$$\ell = (C_3/V_0)^{1/3} \quad ,$$

$$g(z^*) \propto -(z^*)^{-3} \quad . \tag{2.21}$$

Thus the reduced potentials $g(z^*)$ are assured of having both a common minimum value (-1) at $z^* = 0$ and the same form for large z^*. Since C_3 can be calculated rather accurately, only two parameters, say V_0 and z_m, are needed to characterize a specific system. How well the hypothesis of a universal shape for physisorption potentials works can be gathered from Fig. 2.7 where a specific analytic form for $g(z^*)$ is also plotted, namely

$$g(z^*) = \frac{3}{u-3}\, e^{-uz^*/a} - \frac{1}{(z^*+a)^3} \quad , \qquad \text{with}$$

$$a = (1-3/u)^{1/3} \quad . \tag{2.22}$$

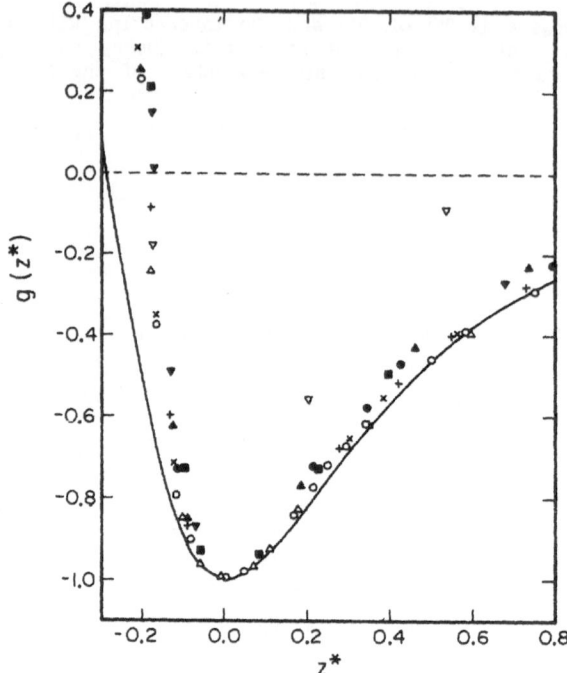

Fig.2.7. Reduced potential g(z) as a function of z. Full squares: He on Cu; full triangles: He on Ag; inverted triangles: He on Au; rare gases on gra‸ phite: He (full circles), Ne (plusses), Ar (crosses), Kr (open circles) and Xe (open triangles). The curve is from (2.22) with u=5.25. (Vidali et al. 1983)

It emerges from this analysis that noble gases adsorbed on alkali halides experience a softer repulsion than on noble metals and noble-gas solids. This difference is traced back to an additional induced dipole attraction of the atom due to the ions' electric field.

Returning to the three-dimensional surface potential $V_S(\mathbf{r})$, we observe that its periodicity along the surface induces a two-dimensional band structure for the motion of the adparticle along the surface (Carlos and Cole 1978, 1980a). Remembering the two-dimensional Fourier series for the surface potential $V_S(\mathbf{r})$ in (2.14), we expand the adparticle wave function for a given two-dimensional wave vector $\mathbf{K}$ in the surface plane in plane waves

$$\psi(\mathbf{K};\mathbf{r}) = \sum_{n,\mathbf{K_m}} \alpha(\mathbf{K};n,\mathbf{K_m}) \, \phi_n(z) \, e^{i(\mathbf{K}+\mathbf{K_m})\cdot\mathbf{R}} \quad . \tag{2.23}$$

With $\mathbf{K}$ restricted to the two-dimensional Brillouin zone spanned by $\mathbf{b_1}$ and $\mathbf{b_2}$ in (2.16) we note that $\psi(\mathbf{K};\mathbf{r})$ satisfies the Bloch theorem in two dimensions

$$\psi(\mathbf{K};z,\mathbf{R}+\mathbf{R_1}) = \psi(\mathbf{K};z,\mathbf{R}) \, e^{i\mathbf{K}\cdot\mathbf{R_1}} \quad . \tag{2.24}$$

With $\phi_n(z)$ being a solution in the potential $V_o(z)$ with energy E_n, we insert (2.23) into the Schrödinger equation to get

$$\sum_{n'K_{m'}} \left[\left[E_n + \frac{\hbar^2}{2m} (K+K_m)^2 - E(K) \right] \delta_{m,m'} \, \delta_{n,n'} + \langle n|V_{m-m'}|n'\rangle \right] \alpha(K';n',K_{m'}) = 0$$

(2.25)

where

$$\langle n|V_{m-m'}|n'\rangle = \frac{1}{a_s} \int_0^\infty \phi_n^*(z) \, V_{m-m'}(z) \, \phi_{n'}(z) dz \quad , $$

(2.26)

a_s being the area of the surface unit cell. Carlos and Cole (1978, 1980a) calculate the band structure for helium on the basal plane of graphite, taking E_n and the lowest Fourier transforms from scattering experiments, with higher ones calculated from model potentials. A typical example is given in Fig. 2.8 where the broken lines indicate the free particle dynamics

$$E^{(o)}(K) = \frac{\hbar^2}{2m} (K+K_m)^2 + E_n \quad .$$

(2.27)

Obviously the free-particle model is rather adequate except for the appearance of a small band gap in the density of states as evidenced in Fig. 2.9 where the free two-dimensional particle density of states is also given.

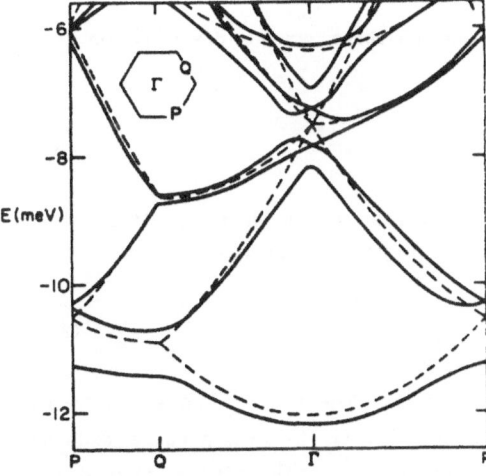

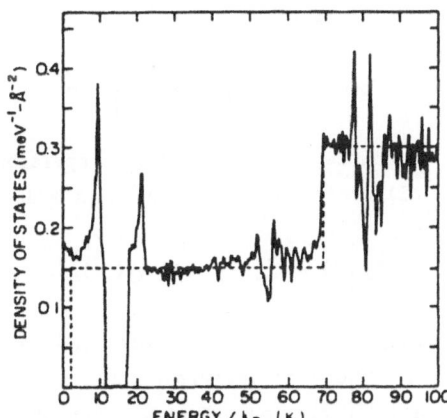

Fig.2.8. Band structure of ⁴He on graphite (full curve) for the two-dimensional wave vector **K** along symmetry lines in the two-dimensional Brillouin zone. Dashed curve is the free-particle result which neglects lateral variations in the surface potential. (Carlos and Cole 1980a)

Fig.2.9. Density of states for ⁴He on graphite (full curve). The ground state represents the zero of the abscissa. Dashed curve is the two-dimensional free-particle model (2.28). (Carlos and Cole 1980a)

$$\rho_2(E) = A^{-1} \sum_n \sum_K \delta(E - E_n - \hbar^2 K^2 / 2m)$$

$$= \frac{m}{2\pi\hbar^2} \sum_n \theta(E - E_n) \quad . \tag{2.28}$$

The model contained in (2.4), which obtains the surface potential by summing pairwise potentials, largely loses its meaning for physisorption on metals where the principal interaction is with the conduction electrons. For this situation, extensive calculations have been carried out within the framework of functional density theory (Kleiman and Landman 1973a,b, 1976; Ying et al. 1975; Lang and Williams 1978; Lundqvist et al. 1979). One treats the conduction electrons of the metal as a free gas dispersed in the background of a uniform positive charge (jellium) filling a half space. A nucleus of charge (Ze) is placed at a distance z above its surface. Within the local density approximation, the one-electron wave functions are then subject to a Schrödinger equation

$$\left[-\frac{\hbar^2}{2m_e} \nabla^2 + V_{eff}[n_e(r), r] \right] \psi_i(r) = E_i \, \psi_i(r) \quad , \tag{2.29}$$

where the mean field potential is given by

$$V_{eff}[n_e(r), r] = V_+(r) + V_H(r) + V_{xc}(r) \quad . \tag{2.30}$$

Here V_+ is the potential energy of the positive charge distribution. The Hartree term

$$V_H(r) = -2e^2 \int \frac{n_e(r')}{|r - r'|} \, d^3r' \tag{2.31}$$

accounts for the Coulomb repulsion between an electron at position r and all others, of which there is a density $n_e(r')$ at r', where

$$n_e(r') = \sum_i \psi_i^*(r') \, \psi_i(r') \quad . \tag{2.32}$$

The sum runs over all occupied energy levels. The factor two accounts for the two spin polarizations. The last term in (2.30) incorporates exchange and correlation effects. Both of these being many-body effects, they can be treated only approximately in the single-particle approximation of mean field theory.

In Fig. 2.10, we present an energy diagram for an electron of the coupled atom-metal system, roughly to scale for helium in front of an aluminum surface. On the left we have the conduction band with the zero of energy chosen so that an electron very far outside the metal has zero kinetic

34

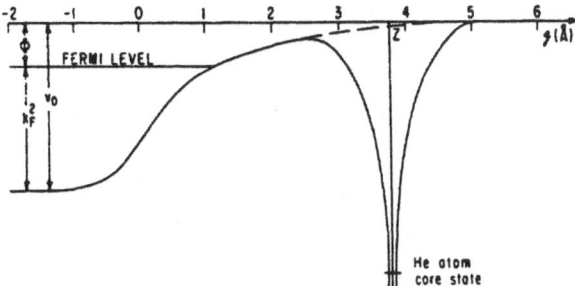

Fig.2.10. Schematic illustration of the nonlocal potential of a helium atom near the surface of a metal. As drawn, the metallic potential is representative of aluminum. The position of the atomic potential is at the calculated equilibrium position of a He atom on an aluminum surface. The origin is chosen at the edge of the positive background. (Zaremba and Kohn 1977)

energy. States up to the Fermi energy E_F are occupied; Φ is the work function. On the right is the potential well for electrons around the helium nucleus; note that the 1s levels are much lower than the bottom of the conduction band. Within the local density approximation (2.29-32), the asymptotic form of the surface potential $V_s(r)$ is exponential rather than of the van der Waals power law (2.1). Zaremba and Kohn (1977) have, therefore assumed a form

$$- \frac{C_3}{(z-z_{im})^3} \; , \tag{2.33}$$

for large z, with the image position z_{im} chosen about 1 bohr outside the positive jellium background. The short range part is then treated in the Hartree-Fock approximation. Their results for helium adsorbing on various metals are reproduced in Figs. 2.11, 12. We take note that the properties of the metal enter such a free electron gas theory via the work function Φ

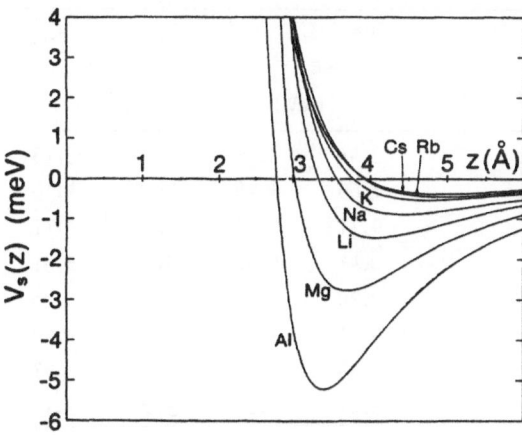

Fig.2.11. He-metal potentials for various simple metals. The solid lines are the results using the Hartree-Fock potential. The origin is chosen at the edge of the jellium background. (Zaremba and Kohn 1977)

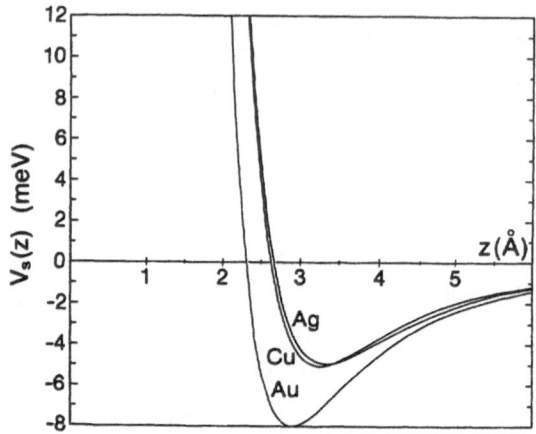

Fig.2.12. Physisorption potentials for He on the noble metals. (Zaremba and Kohn 1977)

and a parameter r_s given via the electron density by

$$\frac{1}{n_e} = \frac{4\pi}{3} (r_s r_B)^3 \quad , \tag{2.34}$$

where $r_B = 0.529$ Å is the Bohr radius. Both are listed in Table 2.5 together with the van der Waals constants, the He binding energy $(-E_0)$ and the position of the potential minimum. Lang and Nørskov (1983) have recently developed a much simplified effective-medium theory for such systems.

So far we have reviewed theories that calculate the static gas-surface interaction. We now briefly review gas-surface scattering experiments that probe it directly. The vast literature has recently been reviewed, e.g., by

Table 2.5. Characteristic properties of He adsorbed on various metals: work function Φ, van der Waals constant C_3, binding energy $-E_0$, and equilibrium position $\langle z \rangle_0$ of the adsorbed atom. (Zaremba and Kohn 1977)

Metal	r_s	Φ [eV]	C_3 [meV Å³]	³He $-E_0$ [meV]	⁴He $-E_0$ [meV]	³He $\langle z \rangle_0$ [Å]	⁴He $\langle z \rangle_0$ [Å]
Al	2.07	4.19	202	3.44	3.65	3.79	3.73
Mg	2.65	3.66	153	1.65	1.78	4.27	4.18
Li	3.28	3.1	117	0.77	0.85	4.91	4.77
Na	3.99	2.7	92	0.41	0.46	5.51	5.31
K	4.96	2.39	70	0.20	0.24	6.23	5.95
Cu	2.67	4.65	225	3.34	3.55	3.66	2.08
Ag	3.02	4.0	249	3.34	3.53	3.73	3.67
Au	3.01	5.22	274	5.61	5.91	3.23	3.17

Goodman and Wachman (1976), Hoinkes (1980), and Benedek and Valbusa (1982). The experiment consists of a nozzle beam of light particles, typically H, H_2, He, and Ar scattered from surfaces with negligible sticking probability. The kinetic energy of the beam particles is typically 20 meV corresponding to a temperature T=100 K and a de Broglie wavelength of 1 Å suitable for probing the corrugation of the surface and typical physisorption surface wells. The former, being a periodic structure, as expressed, e.g., in (2.13-14), gives rise to diffraction as illustrated in Fig. 2.13 where other collision processes are also indicated. By solving the Schrödinger equation for a laterally periodic scattering wave function (2.23), the diffraction intensity can be calculated. A laborious fit to experimental data then yields surface potential parameters. To illustrate how this comes about, we recall (Estermann and Stern 1930; Lennard-Jones and Devonshire 1937b) that the scattering of mono-energetic molecular beams from well-characterized crystal surfaces often shows sharp features that can be ascribed to resonant scattering, both in the elastic and inelastic intensities. In resonant scattering, an imping-ing molecule skims along the surface for times of the order of 10^{-12} s and distances of the order of 10 Å before rescattering outwards (path C in Fig. 2.14). The probability of inelastic collisions being enhanced by the long interaction time leads to a decrease in the total elastic scattering intensity, hence the term "selective adsorption". Fig. 2.15 shows energy and momentum conservation for elastic and one-phonon inelastic scattering. For the former we get

$$E_{in} = \hbar^2 (K_{in} + K_m)^2/2m + E_n \tag{2.35}$$

where E_{in} is the kinetic energy of the incident beam; K_{in} its momentum com-

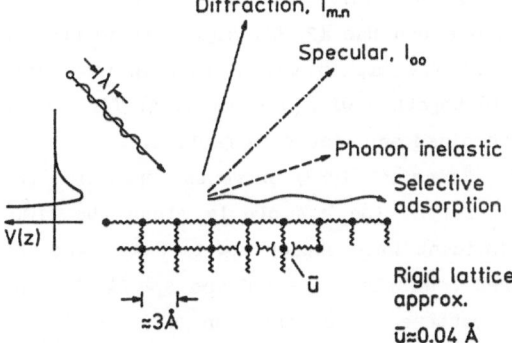

Fig.2.13. Diagram showing the different collision processes which can occur in the nonreactive scattering of a light atom with a de Broglie wavelength comparable to the lattice dimensions. Since the lattice vibrational ampli-tudes are small, phonon inelastic scattering is expected to be improbable relative to elastic diffractive scattering. (Toennies 1982)

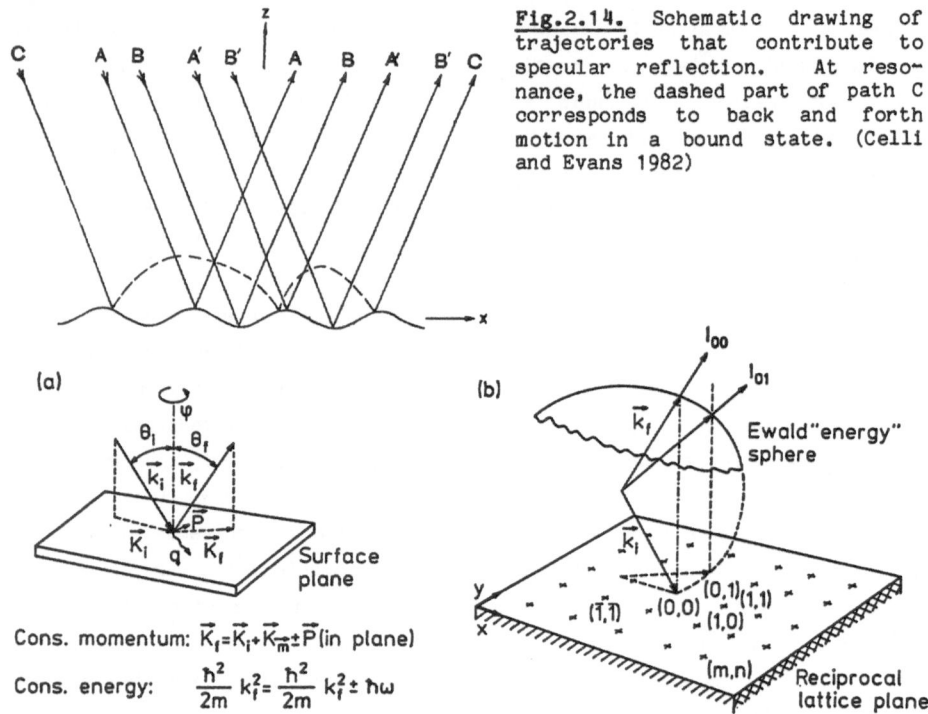

Fig.2.14. Schematic drawing of trajectories that contribute to specular reflection. At resonance, the dashed part of path C corresponds to back and forth motion in a bound state. (Celli and Evans 1982)

(a)

(b)

Surface plane

Cons. momentum: $\vec{K}_f = \vec{K}_i + \vec{K}_{\vec{m}} \pm \vec{P}$ (in plane)

Cons. energy: $\dfrac{\hbar^2}{2m}\, k_i^2 = \dfrac{\hbar^2}{2m}\, k_f^2 \pm \hbar\omega$

I_{00}

I_{01}

Ewald "energy" sphere

$(\bar{1},1)$ $(0,0)$ $(0,1)$ $(1,1)$
$(1,0)$

(m,n) Reciprocal lattice plane

Fig.2.15. In **a** the usual definition of angles and particle momenta and the conservation equations used in single phonon inelastic scattering are shown. In **b** the surface Ewald diagram is shown for the case of elastic scattering. Note that elastic events are limited to specific discrete directions indicated, for example, I_{00} is a specular peak and I_{01} a possible diffraction peak. (Toennies 1982)

ponent parallel to the surface; K_m a reciprocal lattice vector (2.16); and E_n a bound state energy in the surface potential. With K_{in} being related to E_{in} via the angle of incidence θ_i, one can use (2.35), e.g., for scattering along a direction K_m to find E_n from the dip in the scattering intensity. An example is furnished in Fig. 2.16 together with a theoretical fit. Some examples of binding energies so determined are listed in Table 2.6.

Turning to inelastic scattering, Toennies (1982) suggests replotting the Ewald diagram as a side view, Fig. 2.17. The example is along the (100) direction for a LiF cyrstal. The incident wave vector k_{in} is 6 Å^{-1}, and the angle between incident and scattered beams is θ_{SD} = 90° and θ_i= 64.2°. The detector thus sees all particles scattered in a direction θ_f = 25.8°. The point 1 indicates one of many possible inelastic events, in which a phonon has been annihilated and the atom has gained energy. The phonon momentum transfer P to the atom is in the forward direction. The point 6 indicates

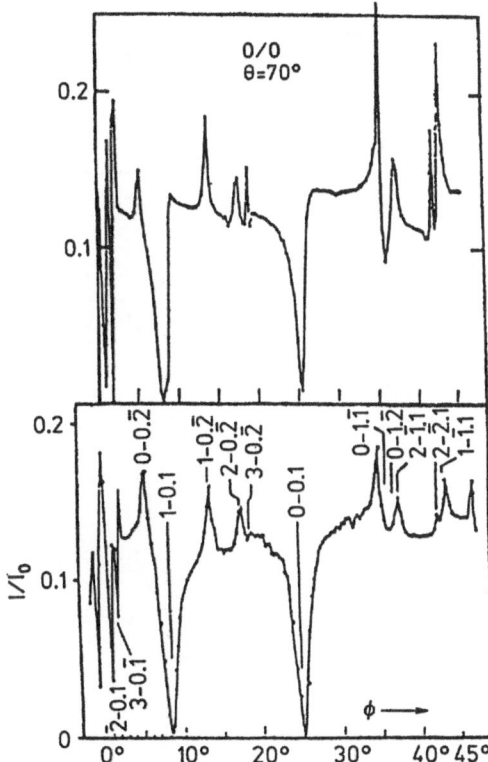

Fig.2.16. Theory (above) and experiment (below) for the azimuthal dependence of the specular intensity for He-LiF. The incident wave vector was 5.76 Å^{-1}, and the incident polar angle was θ_i=70°. The label 1-0,1, for example, indicates a resonant transition to the first excited state through $\mathbf{K_m}$ = (2π/a)(0,1). The theoretical model is a hard corrugated wall with an attractive well in front of it. Inelastic effects are not included, except for an overall Debye-Waller scaling of the diffracted intensity. (Garcia, et al. 1979)

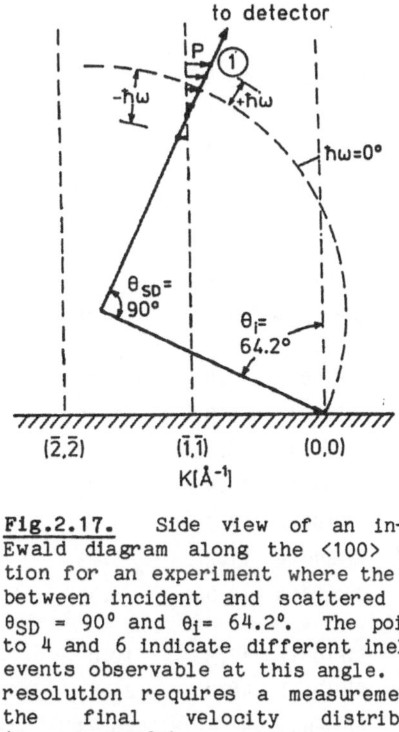

Fig.2.17. Side view of an in-plane Ewald diagram along the <100> direction for an experiment where the angle between incident and scattered beams θ_{SD} = 90° and θ_i= 64.2°. The points 1 to 4 and 6 indicate different inelastic events observable at this angle. Their resolution requires a measurement of the final velocity distribution. (Toennies 1982)

another possibility, in which a phonon has been created and the atom has lost energy. The momentum transfer P is in the backward direction. Note that P is always measured to the nearest reciprocal lattice vector. Thus, in the example of Fig. 2.17, we would speak of phonons associated with the (1,1) reciprocal lattice vector. For given scattering angles θ_i and θ_f, there is a set of values (ω,P) seen by the detector, of which those events will be registered that match the dispersion relations of the surface pho-

Table 2.6. Binding energies [meV] in the surface potential $V_0(z)$ as determined from scattering data. From Hoinkes (1980), where references to the original literature can be found.

Gas	Surface	$-E_0$	$-E_1$	$-E_2$	$-E_3$	$-E_4$	$-E_5$
H	LiF(001)	12.3	3.9	0.5			
	NaF(001)	11.8	3.0	0.4			
	Graphite	31.6	15.3				
	NaCl	30.3	20.0	13.8	9.0	6.2	
D	LiF(001)	13.4	6.6	2.4	0.5		
	LiF(001)	14.0	6.7	2.3	0.5		
	NaF(001)	13.3	5.8	1.6	0.3		
	Graphite	35.4	21.35	12.0	5.9		
^{3}He	LiF(001)	5.59	2.00				
	NaF(001)	4.5	1.38				
	Graphite	11.62	5.38	1.78			
^{4}He	LiF(001)	5.8	2.2	0.6	0.1		
	LiF(001)	5.9	2.46	0.78	0.21		
	NaF(001)	4.92	1.87	0.54			
	NaCl(001)	10.4	3.7				
	NiO(001)	7.9	4.0	1.6			
	Graphite	11.77	6.13	2.68	0.83		
	Graphite	12.06	6.38	2.85	1.01	0.17	
H_2	LiF(001)	17.3	10.0	4.3			
	NiO(001)	48.0	24.5	14.5	7.5	3.7	
	Graphite	41.61	26.43	15.33	7.96	3.61	1.46

nons. To separate these different contributions, we must measure one additional quantity such as the energy of the scattered particles. This is most conveniently done by measuring the time of flight distribution. It is easy to derive from the conservation equations, the following equation relating the total momentum transfer $\Delta K = K_m \pm P$ to the experimental quantities $\hbar\omega$ and θ_f for in-plane scattering:

$$\frac{\hbar\omega}{E_n} = -1 + (1 + \frac{\Delta K}{K})^2 \frac{\sin^2\theta_i}{\sin^2\theta_f} \ . \tag{2.36}$$

For a given bound state energy, one thus gets a scan curve (ω, P) of possible

40

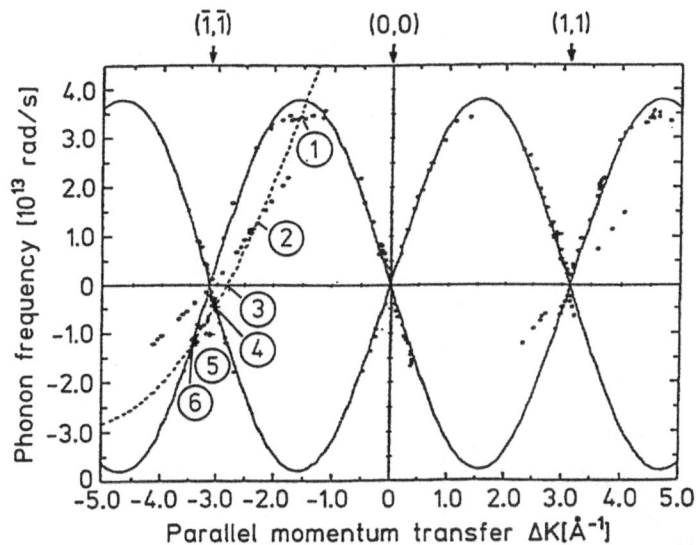

$(\bar{1},\bar{1})$ $(0,0)$ $(1,1)$

Fig.2.18. The scan curve for $\theta_i = 64.2°$ and the assignment of the different points of the time of flight spectrum are shown in an extended-zone plot for the <100> direcon of LiF. Also shown are the results of all measured time of flight spectra, which can be compared with the sine approximation for the Rayleigh phonon dispersion curves. (Toennies 1982)

phonon energies and momenta (in the surface plane) which are nothing but phonon dispersion curves, see Fig. 2.18.

We have seen at the beginning of this section that the static surface potential can be calculated from first principles, e.g., by summing two-body potentials of the Lennard-Jones type, (2.4,5). With the parameters of the latter known from two-body scattering experiments, $V_S(r)$ is essentially parameter-free. It turns out, however, that it does not give a satisfactory fit to the bound state energies or to the surface corrugation function needed to reproduce the data of gas-surface scattering. The main reason is the fact that the interaction between, e.g., a helium atom and a free, isolated carbon atom, is different from that of a helium and a carbon embedded in the graphite lattice where substantial hybridization into σ and π orbitals takes place, as mentioned above equation (2.8). It has therefore become customary to assume a form for the surface potential and the corrugation function and fit a limited number of parameters to the gas-surface scattering data. The most popular choice for the surface potential is a Morse potential

$$V_S(z) = V_0 \left[e^{-2\gamma(z-z_0)} - 2e^{-\gamma(z-z_0)} \right] \; , \tag{2.37}$$

for which all wave functions and the bound state energies can be determined

41

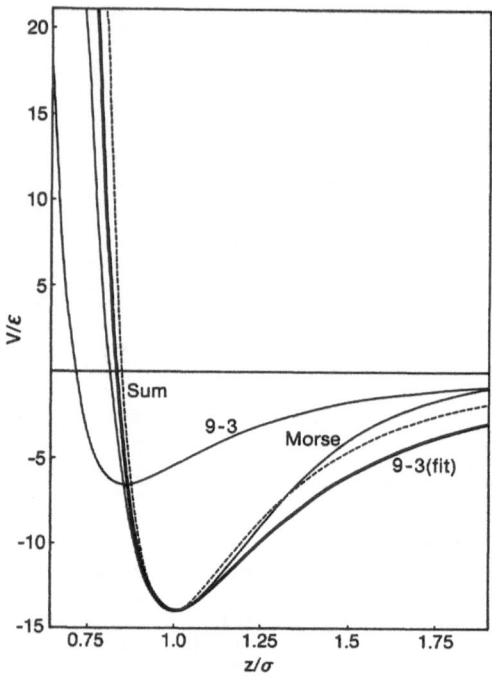

Fig.2.19. Surface potentials for a physisorbed particle: Sum from (2.19); 9-3 from (2.10); 9-3(fit) from (2.11) with parameters fitted to Sum; likewise for Morse from (2.37).

analytically. The latter are given by

$$E_n = - \frac{\hbar^2\gamma^2}{2m} (\sigma_0 - n - 1/2)^2 \quad , \quad \text{with}$$

$$\sigma_0^2 = 2mV_0/(\hbar\gamma)^2 \quad ,$$

$$n = 0,1,2,\ldots,N_{max} \; ; \quad \sigma_0 - 3/2 < N_{max} < \sigma_0 - 1/2 \quad . \tag{2.38}$$

In Fig. 2.19 we compare a Morse potential with a 9-3 potential introduced in (2.10) and (2.11). For the latter, Cole and Tsong (1977) have calculated the bound state energies using the WKB approximation; they find

$$E_n \approx -\frac{V_0}{\sigma_1^6} (\sigma_1 - n - 1/2)^6 \quad , \quad \text{with}$$

$$\sigma_1 = \frac{2.79}{\pi} (2mV_0 z_m^2/\hbar^2)^{1/2} \quad ,$$

$$n = 0,1,2,\ldots < \sigma_1 - 1/2 \quad , \tag{2.39}$$

where the depth V_0 and the position z_m of the potential minimum were introduced in (2.11).

2.2 Mean-Field Surface Potential at Finite Coverage

In this section we review the mean-field approach to multilayer physisorption as developed by Summerside et al.(1982) and Sommer and Kreuzer (1982a-d). We will construct an effective surface potential in which the interaction between adsorbed particles is taken into account in a self-consistent mean field. It will be used in Chaps. 4 and 5 to study multilayer physisorption kinetics. We start from the many-body Hamiltonian

$$H = T + V_S + V_2$$

$$= \sum_{i=1}^{N} \frac{p_i^2}{2m} + \sum_{i=1}^{N} V_S(r_i) + \sum_{i<j=1}^{N} V_2(r_i-r_j) \quad , \tag{2.40}$$

where T is the kinetic energy of the N gas particles of mass m, V_S is the bare surface potential, and V_2 contains the two-body interactions between gas particles. As coverage builds up, e.g., upon lowering the temperature or raising the gas pressure, the latter ensures saturation at monolayer coverage and causes such quantities as the heat of adsorption Q , the desorption time t_d, and specific heats to become coverage dependent. Several different mechanisms can be identified.

For strongly localized adsorption as occurs in many chemisorption systems, only one particle can be adsorbed per adsorption site due to the finite size of the adsorbing particle and due to bond saturation restricting chemisorption typically to monolayer coverage. As the monolayer fills up, the long-range interaction between adsorbed particles can lead to a variety of ordered structures reflected, in particular, in characteristic changes in Q and t_d. For recent experiments, see Pfnür et al.(1978), Pfnür and Menzel (1983), and Pfnür et al. (1983); a microscopic model has been analyzed by Zhdanov (1981a) and by Leuthäuser (1981).

Next we look at the adsorption of a gas onto a solid surface without pronounced adsorption sites. For physisorption systems, this implies that the surface potential is more or less uniform along the surface, i.e., a function of the distance z above the surface only. At high temperatures, gas particles in the adsorbate will then be highly mobile. With monolayer densities typically of the order of liquid densities, collisions of adparticles will probe predominantly the short-range repulsion between them, leading to a decrease in the heat of adsorption as coverage builds up. A dramatic decrease typically occurs at monolayer completion because the adparticles in the second layer are bound far more weakly to the adparticles in

43

the first layer than the latter are to the solid. For multilayer adsorbates, Q will eventually approach the heat of vaporization of the corresponding liquid. The mean field theory demonstrates these features explicitly. At lower temperatures, small lateral variations in the surface potential due to the lattice structure of the solid can lead to a two-dimensional
commensurate crystallization in the adsorbate. In addition, the attractive
part of the two-body interaction between adparticles can lead to a crystalline phase, typically as an adsorbate superstructure, accompanied most often
by a rise in the heat of adsorption.

Equilibrium theories accounting for these effects have been developed
(e.g. the review by Schick 1981) in which it is argued that the surface gives
rise to an external potential $V_S(\mathbf{r})$ which provides the basis for establishing
a quasi-two-dimensional adsorbed structure. Once reduced to a two-dimensional problem, a virial expansion can be used to include the effects of
adparticle-adparticle interactions at low coverage (fluid regime). At near
monolayer coverages, strictly two-dimensional lattice models of, for example, the Ising or Potts type can be employed successfully to study phase
transitions in films adsorbed on surfaces (Schick 1981). The ground-state
properties of completed monolayers, as well as second- and higher-order
layers have been studied by describing the quasi-two-dimensional nature of
an adsorbed layer as having a spatial (Gaussian, for example) distribution
normal to the surface (Novaco 1973). In particular, such a distribution
makes it possible to account for the anomalously high density in an adsorbed
helium monolayer on graphite.

In two-dimensional theories of adsorbed films, all coupling associated
with energy and particle exchange between the adsorbate and the gas phase
is suppressed. This may be quite appropriate for such specific systems as
helium adsorbing on grafoil, where the distance between opposing surfaces in
grafoil is much shorter than the mean free path of helium away from the
surface so that the notion of a gas phase becomes irrelevant. For the study
of adsorption and desorption kinetics at open surfaces, however, the explicit
coupling of the adsorbate to the gas phase must form an integral part of
the theory. Thus, for example, in the study of adsorption kinetics it is
important to know what changing environment additional particles arriving
from the gas phase will experience as the coverage on the surface builds up.
In a single-particle picture, this necessitates the construction of an effective coverage-dependent surface potential given by

$$V_S(\mathbf{r},\theta) = V_S(\mathbf{r}) + V_{MF}(\mathbf{r},\theta) \quad , \tag{2.41}$$

where $V_S(r)$ is the interaction of a single gas particle with the solid, referred to from now on as the bare surface potential. $V_{MF}(r,\theta)$ is the potential arising from the mean field experienced by a gas particle in the presence of all other gas particles already in the surface region at a given coverage θ. Once this mean-field potential is determined, it is fairly straightforward to extend the quantum-statistical theory of phonon-mediated physisorption kinetics to nonnegligible coverage. In a quantum-statistical theory, the mean-field potential $V_{MF}(r,\theta)$ may be determined by employing temperature-dependent Hartree-Fock theory. Before this can be done, two problems have to be considered.

First, the two-body interaction $V_2(r)$ between the physisorbing (neutral) particles has a strong short-range repulsive singularity so that $V_2(r) \to \infty$ as $|r| \to 0$ (for example, as r^{-12} in the Lennard-Jones potential), leading to an infinite Hartree-Fock energy, a difficulty which must be avoided by softening the core of the two-body potential. The second and more subtle problem concerns the fact that the particle density in the adsorbate, say around unit coverage, is of the order of that in liquids, so that the two-body correlations, totally neglected in a straightforward Hartree-Fock theory, become very important. Both of these difficulties may be avoided simultaneously in a systematic way by extending the Hartree-Fock theory to the Brueckner-Hartree-Fock (BHF) theory (Brueckner et al.1958) in which the short-range singularity in V_2 is removed by partially including two-body correlations in the construction of a K matrix which takes the place of V_2 in the Hartree-Fock equations. Brueckner's theory has been devised for fermionic particles, such as nucleons and ^{3}He, but arguments can be advanced that the resulting effective two-body interaction can also be used for bosons (Brueckner and Frohberg 1965). Brueckner's **K** matrix is defined by

$$V_2\psi = K\phi \quad , \tag{2.42}$$

where $\phi(r_1,r_2)$ is a free two-particle state and $\psi(r_1,r_2)$ is the fully correlated wave function for two gas particles at positions r_1 and r_2, interacting via $V_2(r_1,r_2)$ in the background of the (N-2) other gas particles. The K matrix satisfies the Brueckner integral equation

$$K = V_2 - V_2 \frac{\hat{Q}}{e} K \quad , \tag{2.43}$$

with $\hat{Q}$ being the Pauli exclusion operator and

$$e = H^{SCF} - \omega \quad , \tag{2.44}$$

where H^{SCF} is the Hamiltonian of the self-consistent mean field and ω is called the starting energy. Replacing V_2 in the total Hamiltonian (2.40) by the K matrix, a variational calculation leads to the spin-averaged temperature-dependent Brueckner-Hartree-Fock equations (Summerside et al. 1982)

$$\left[-\frac{\hbar^2}{2m} \frac{d^2}{dr_1^2} + V_S(r_1) - E_1 \right] \psi_1(r_1) + N_g \sum_{1'} n_{1'} \int d^3r_2 d^3r_3 d^3r_4 \psi^*_{1'}(r_2)$$

$$*\langle r_1, r_2 | K | r_3, r_4 \rangle [(2s + 1)\psi_1(r_3)\psi_{1'}(r_4) \pm \psi_{1'}(r_3)\psi_1(r_4)] = 0 \quad , \tag{2.45}$$

where E_1 and ψ_1 are the single-particle energies and wave functions, respectively, and s is the spin. N_g is the total number of gas particles. In equilibrium, the occupation functions are given by Bose-Einstein or Fermi-Dirac statistics.

In principle, one now seeks a self-consistent solution to the nonlinear equations (2.45). However, this requires a knowledge of the K matrix which, in turn, depends on the single-particle energies E_1 through the energy denominator in (2.43). It seems impracticable to deal with this double self-consistency problem numerically, and certain approximations must be invoked. To decouple the Brueckner self-consistency in (2.43) from the Hartree-Fock self-consistency in (2.45), one invokes a local density approximation whereby (2.43) is solved in a fictitious infinite system at the (average) adsorbate density as determined by the solution of (2.45). Moreover, the nonlocal K matrix is approximated by a local effective potential

$$\langle r_1, r_2 | K | r_3, r_4 \rangle = V_{eff}(r_1 - r_4) \, \delta(r_1 - r_3) \, \delta(r_1 - r_2) \quad , \tag{2.46}$$

where V_{eff} is obtained from the bare two-body interaction V_2 by averaging with the Brueckner correlation function. Equations (2.45) then read

$$\left[-\frac{\hbar^2}{2m} \frac{d^2}{dr^2} + V_S(r) - E_1 \right] \psi_1(r) + N_g \sum_{1'} n_{1'} \int d^3r' \psi_{1'}^*(r) V_{eff}(r - r')$$

$$*[(2s+1)\psi_1(r)\psi_{1'}(r') \pm \psi_{1'}(r)\psi_1(r')] = 0 \quad , \tag{2.47}$$

These are the standard Hartree-Fock equations for particles obeying Fermi-Dirac (−) or Bose-Einstein (+) statistics and interacting via $V_{eff}(r)$.

Further simplifications can now be introduced by considering the case of mobile adsorption. This reduces (2.47) to a one-dimensional problem by considering gas-solid systems with a surface potential

$$V_S(r) = V_S(z) \quad , \tag{2.48}$$

depending on the distance z from the surface only. As long as the adsorbate

46

remains fluid, i.e., does not crystallize into a two-dimensional structure, we can assume that

$$\psi_\iota(\mathbf{r}) = \frac{1}{\sqrt{A}} \phi_\iota(z) e^{i\mathbf{Q}\cdot\mathbf{R}} \quad , \tag{2.49}$$

where A is the surface area, $\mathbf{Q} = (q_x, q_y)$ is a two-dimensional wave vector, $\mathbf{r} = (\mathbf{R}, z)$, and $\iota = (q_x, q_y, \iota)$ with ι enumerating the bound states and the continuum. Inserting (2.49) into (2.47) and integrating out the lateral degrees of freedom one eventually gets after some further approximations (Summerside et al. 1982),

$$\left[-\frac{\hbar^2}{2m} \frac{d^2}{dz^2} + V_S(z) - \varepsilon_\iota \right] \phi_\iota(z) + \sum_{\iota'} \bar{n}_{\iota'} \int dz' \bar{V}(z-z') \phi_{\iota'}{}^*(z')$$

$$*[(2s+1)\phi_\iota(z)\phi_{\iota'}(z') \pm \phi_\iota(z')\phi_{\iota'}(z)] = 0 \quad , \tag{2.50}$$

where

$$\bar{n}_\iota = \frac{\sigma_g^2}{A} N_g \sum_{\mathbf{Q}} n_{\mathbf{Q}\iota} = \frac{1}{2\pi} \sigma_g^2 \int Q dQ \frac{1}{\exp[\beta[\varepsilon_\iota + \hbar^2 Q^2/2m - \mu]] \mp 1}$$

$$= \mp \frac{1}{2\pi} \frac{m k_B T \sigma_g^2}{\hbar^2} \ln(1 \mp e^{-\beta(\varepsilon_\iota - \mu)}) \quad , \tag{2.51}$$

and

$$\bar{V}(z) = \sigma_g^{-2} \int d^2R \, V_{eff}(z, \mathbf{R}) \quad . \tag{2.52}$$

The powers of the range σ_g of the two-body interaction $V_{eff}(\mathbf{r})$ have been introduced so as to render the occupation functions $\bar{n}_\iota$ dimensionless and to keep units of energy for the effective one-dimensional potential $\bar{V}(z)$.

We can now identify the mean-field potential $V_{MF}(z, \theta)$. From (2.50) we see that it receives one contribution from the Hartree term, namely

$$V_{MF}{}^{(H)}(z, \theta) = \sum_\iota \bar{n}_\iota \int dz' \bar{V}(z-z') \phi_\iota{}^* (z') \phi_\iota(z') \quad , \tag{2.53}$$

where the coverage is obtained by summing over the bound states only:

$$\theta = \sum_i \theta_i = \sum_i \bar{n}/\bar{n}_i{}^{max} \quad . \tag{2.54}$$

The contribution of the exchange or Fock term in (2.50) to the mean-field potential is less straightforward due to its nonlocality and state dependence. We therefore prefer to include as part of the mean-field potential $V_{MF}(z, \theta)$ its statistical average (Slater 1951)

$$\overline{V}_{MF}^{(F)} = \sum_{l,l'} \bar{n}_l \bar{n}_{l'} \int dz' \overline{V}(z-z') \phi_{l'}^{*}(z') \phi_l(z') \phi_l^{*}(z) \phi_{l'}(z)$$

$$* \frac{1}{\sum_{l''} \bar{n}_{l''} |\phi_{l''}(z)|^2} \quad , \tag{2.55}$$

so that finally

$$V_S(z,\theta) = V_S(z) + V_{MF}^{(H)}(z,\theta) + \overline{V}_{MF}^{(F)}(z,\theta) \quad . \tag{2.56}$$

Numerical examples of this coverage-dependent surface potential will be discussed below.

Before one can proceed with a numerical solution of the Hartree-Fock equations (2.50), one must specify the bare surface potential $V_S(z)$ and the effective two-body interaction $\overline{V}(z)$ for specific gas-solid systems. For the former, Summerside et al. (1982) and Sommer and Kreuzer (1982a-d) use the ζ-potential (Table 2.2),

$$V_S(z) = 2\pi\varepsilon_S \sigma_S^{6} c_S a_S^{-1} d_L^{-4} \left[\frac{2}{5} (\sigma_S/d_L)^6 \zeta(10, z/d_L) - \zeta(4, z/d_L) \right] \quad , \tag{2.57}$$

where

$$\zeta(n,x) = \sum_{j=0}^{\infty} (j+x)^{-n} \quad , \tag{2.58}$$

is a Riemann ζ function, ε_S and σ_S are the well depth and range of the two-body adatom-surface atom interaction (2.5), d_L is the distance between crystal planes parallel to the surface and $n_S = c_S/\sigma_S$ is the average lateral density of the surface plane whose two-dimensional unit cell of area a_S contains c_S atoms.

A numerical example of the effective potential V_{eff}, obtained by averaging a Lennard-Jones potential (2.5)

$$V_2(r) = 4\varepsilon_g \left[(\sigma_g/r)^{12} - (\sigma_g/r)^6 \right] \quad , \tag{2.59}$$

with the Brueckner correlation function, is given in Fig. 2.20a. Important to note is the fact that the repulsive core in V_2 has been reduced substantially. Because ³He atoms cannot approach each other much closer than σ_g, the short-range repulsion for $r < \sigma_g$ should not contribute too much to the ground-state energy as evaluated by a Brueckner-Hartree-Fock theory.

Carrying out the lateral integration of $V_{eff}(r)$ and dividing by σ_g^2, we derive from the result depicted in Fig.2.20a, the effective one-dimensional interaction $\overline{V}(z)$ as shown in Fig.2.20b. Because $\overline{V}(z)$ represents the effec-

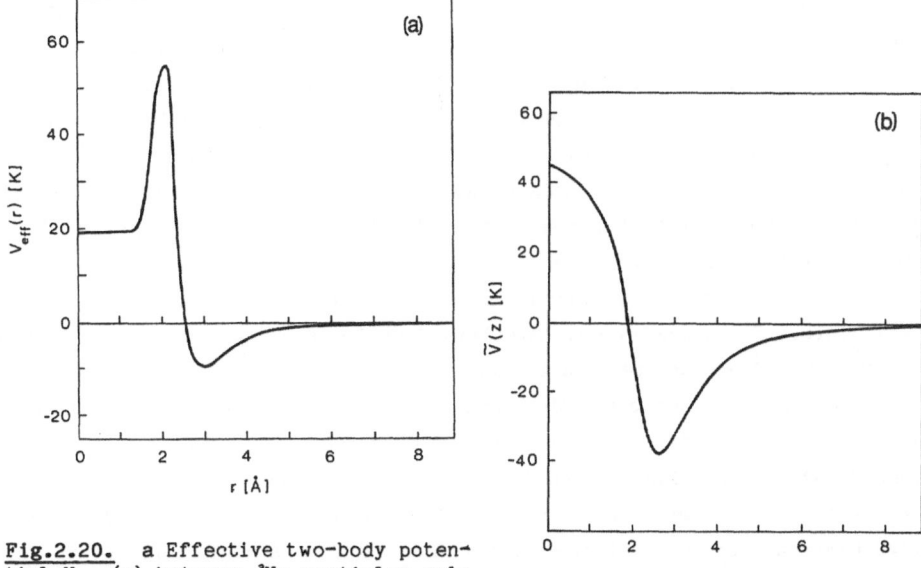

Fig.2.20. a Effective two-body potential $V_{eff}(r)$ between ^{3}He particles calculated in the effective mass approx-
imation. Parameters taken from Østgaard (1968): $\epsilon_g/k_B=10.22$ K, $\sigma_g=2.490$ Å.
b One-dimensional effective two-body potential obtained by integrating
$V_{eff}(r)$ above over x and y. (Summerside et al. 1982)

tive interaction between two ^{3}He layers of liquid density a distance z apart,
it must develop a repulsive barrier for $z < \sigma_g$ so as to prevent these layers
from penetrating each other. The height $\tilde{V}(0)$ is found to vary more or less
linearly with the lateral density to account for the fact that at higher
densities, i.e., at shorter mean separation between the particles, more of
the repulsive core contributes to the total energy of the system. In a completely self-consistent Brueckner-Hartree-Fock theory, $\tilde{V}$ would have to be
determined at each density as calculated from the Hartree-Fock equations
(2.50). Since $\tilde{V}$ changes little as a function of density, apart from the
linear dependence of $\tilde{V}(0)$, Summerside et al. (1982) parametrized $\tilde{V}(z)$, keeping $\tilde{V}(0)$ as an input parameter. They chose

$$\tilde{V}(z)=2\pi\epsilon_g \frac{z^{10}}{z^{10}+A\sigma_g^{10}\exp[-(z/z_1)^\alpha]} \left[\frac{2}{5}(\frac{\sigma_g}{z})^{10}-(\frac{\sigma_g}{z})^4\right] \quad . \tag{2.60}$$

The parameters z_1 and α are fairly well determined: z_1 is always close to
the zero of the unscreened potential, obtained by lateral integration of
(2.50), and α is of the order of 10 to 15 so that the attractive well of
(2.60) is not affected. In the limit $z \to 0$, (2.60) yields $\tilde{V}=4\pi\epsilon_g/5A$ so that A
alone determines the finite barrier height and is thus the only adjustable
parameter.

To obtain the self-consistent solution to the Hartree-Fock equation (2.50) numerically, one encloses the system in a finite one-dimensional box of length L and discretizes the distance from the surface wall into a mesh, thus casting the integro-differential Hartree-Fock equations into a set of finite-difference matrix equations. The self-consistent solutions are found by an iterative procedure initiated at high temperatures, where all $\bar{n}_l$ and thus the Hartree and Fock terms, are negligibly small. The data so obtained will include the temperature and pressure dependence of eigenvalues ε_l, wave functions ϕ_l, adlayer separations, coverage, and coverage-dependent surface potentials. The calculations have been done for several gas-solid systems: ^{3}He-graphite, ^{4}He-graphite, and Ar-Ag.

We now present the discussion by Summerside et al. (1982) of the formation of multilayered mobile ^{3}He physisorbed on graphite in a Gedanken experiment in which, starting at high temperatures, one lowers the temperature at constant gas pressure, fixed in the following example at P = 1.33 Pa. At high temperatures the Hartree-Fock terms in (2.50) are negligible due to the smallness of the bound state occupation factors $\bar{n}_i$. Thus isolated gas particles will find themselves in the bare surface potential $V_S(z)$ which develops five bound states. Lowering the termperature to about 10 K we see in Fig. 2.21 that the coverage θ rises to about 0.1 of a monolayer, all adparticles occupying the lowest bound state with energy $\varepsilon_0 \approx \mu$ and $\theta \approx 0.2$. The energy ε_1 of the first excited state has by now moved up considerably so that its occupation $\bar{n}_1$ remains negligible. With about half the monolayer volume occupied by gas particles trapped into ε_0, the two-body repulsion starts to become important. This causes ε_0 to rise substantially as we further lower T to about 6 K at which stage the first monolayer is complete. Most adsorbates, including the ^{3}He-C system, at low temperature crystallize into a two-dimensional solid before a monolayer is completed. Keeping the ansatz (2.49) with a momentum cutoff q_c, we infer that a summation over the lattice sites in this first monolayer can again be replaced by an integration over a uniform plane of ^{3}He. By now the wave functions of the higher bound states have been expelled from the immediate vicinity of the surface as evidenced by the average position

$$\langle z_1 \rangle = \int z \left| \phi_1(z) \right|^2 dz \quad , \tag{2.61}$$

depicted in the lowest panel of Fig. 2.21. Additional particles approaching from the gas phase now see a new effective surface potential $V_S(z, \theta=1)$ considerably modified from the bare surface potential depicted in the upper

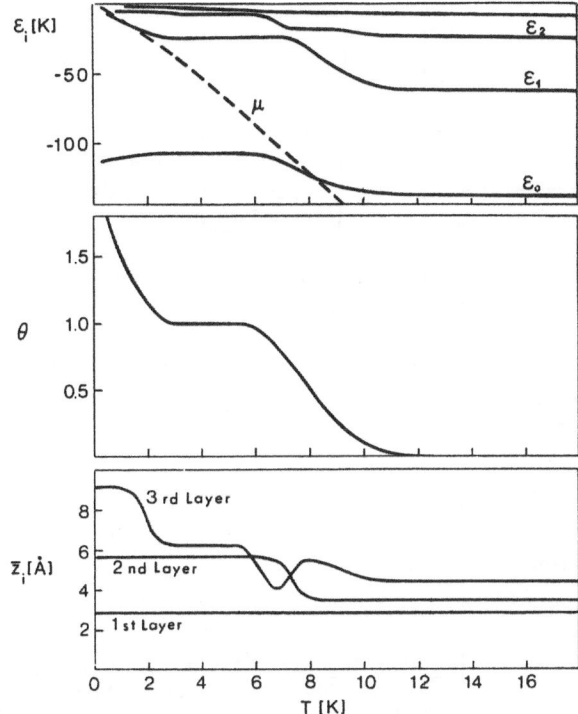

Fig.2.21. Single-particle energies ε_i from (2.50), coverage θ from (2.54), and mean adlayer positions $\langle z_i \rangle$ from (2.61) for a model of ^{3}He on graphite for a weakly repulsive two-body interaction (2.60) with $z_1=2.2$ Å, $A=0.3$, $\alpha=15$, so that $\bar{V}(0)=-2\bar{V}(z_{min})$. Pressure $P=1.33$ Pa. (Summerside et al.1982).

panel of Fig. 2.22 by the mean field $\bar{V}_{MF}(z,\theta)$ due to the particles already adsorbed in the monolayer, as shown in the central panel of Fig. 2.22. A definite repulsive barrier appears at $z\approx4.8$ Å to exclude particles from the region of the filled monolayer. To illustrate this point further we contrast, in the upper and central panels of Fig. 2.22, the lowest three (squared) wave functions at $\theta=0$ and 1, respectively. Whereas $|\phi_0(z)|^2$ remains relatively unchanged, $|\phi_1(z)|^2$ changes dramatically, its inner peak having diminished to negligible size and its second one shifted out. Indeed, it appears very much like the new ground-state wave function at the position where the second adlayer will eventually form. Lowering the temperature below 3 K, we see a similar development to that just described with the ε_1 level taking over the role of the lowest bound state energy of interest and the ε_2 level representing the first excited state as far as the formation of the second adlayer is concerned. By the time the latter is nearing comple-

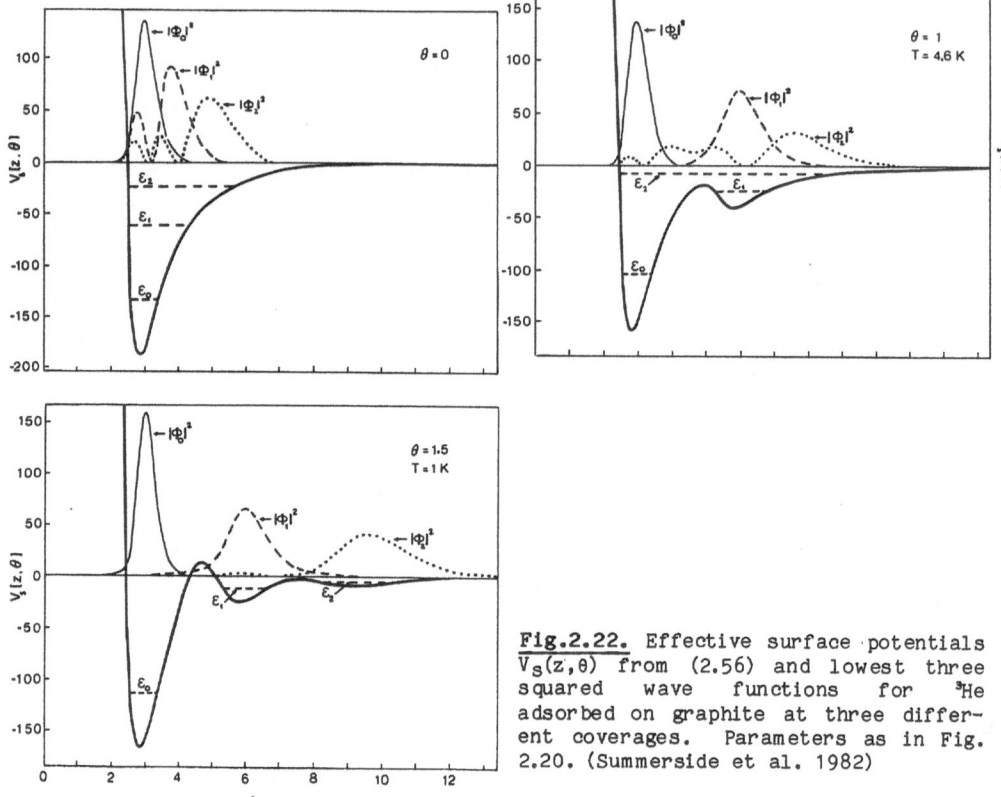

Fig.2.22. Effective surface potentials $\overline{V}_S(z,\theta)$ from (2.56) and lowest three squared wave functions for ^{3}He adsorbed on graphite at three different coverages. Parameters as in Fig. 2.20. (Summerside et al. 1982)

tion at T≤1K, $\langle z_3 \rangle$ has moved out to about 10.2 Å, as seen in the lowest panel of Fig. 2.21, $|\phi_1(z)|^2$ has narrowed considerably, and the effective surface potential develops three minima for the three adlayers separated by two repulsive barriers, as seen in the lowest panel of Fig. 2.22.

To obtain a complete picture of the development of physisorbed multilayers, Summerside et al. (1982) and Sommer and Kreuzer (1982a-d) have prepared a series of three-dimensional perspective views of the squared wave functions and of the effective coverage-dependent surface potential $V_S(z,\theta)$ as a function of z and θ. In Fig. 2.23 we present a view of $V_S(z,\theta)$. It is impressive to see how abruptly the repulsive barrier forms in front of the first monolayer as θ reaches unity with saturation of the first adsorbed layer; the position of the lowest minimum in $V_S(z,\theta)$, however, stays fixed at $z_{min} \approx 2.75$ Å. Figure 2.24 showing $|\phi_0(z)|^2$ illustrates through its lack of striking features that nothing much happens to the position and shape of the wave function for particles in the first adlayer, the reason being that

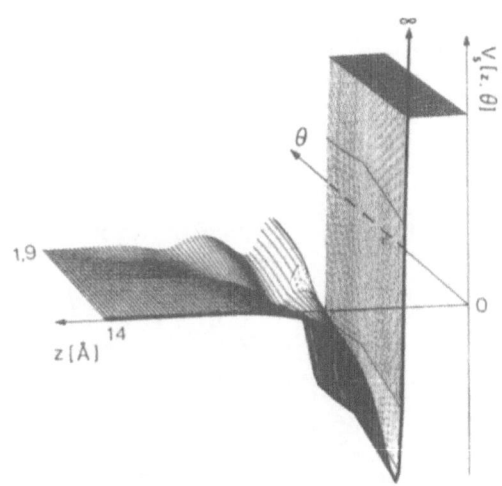

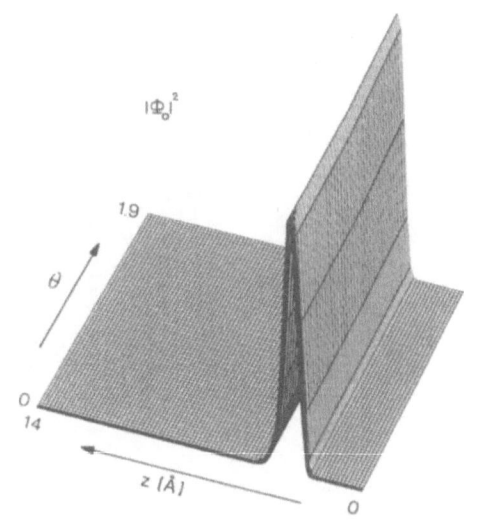

Fig.2.23. Perspective view of the effective surface potential $V_S(z,\theta)$ for ^{3}He on graphite with parameters as in Fig. 2.20 from (2.56) plotted over the (z,θ) plane. For $z \to 0$, $V_S(z,\theta) \to \infty$. The plateau for small z is used to indicate the distance from the wall. (Summerside et al. 1982)

Fig.2.24. Perspective view of $\left|\phi_0(z)\right|^2$ over the (z,θ) plane for ^{3}He on graphite with parameters as in Fig. 2.20. (Summerside et al. 1982)

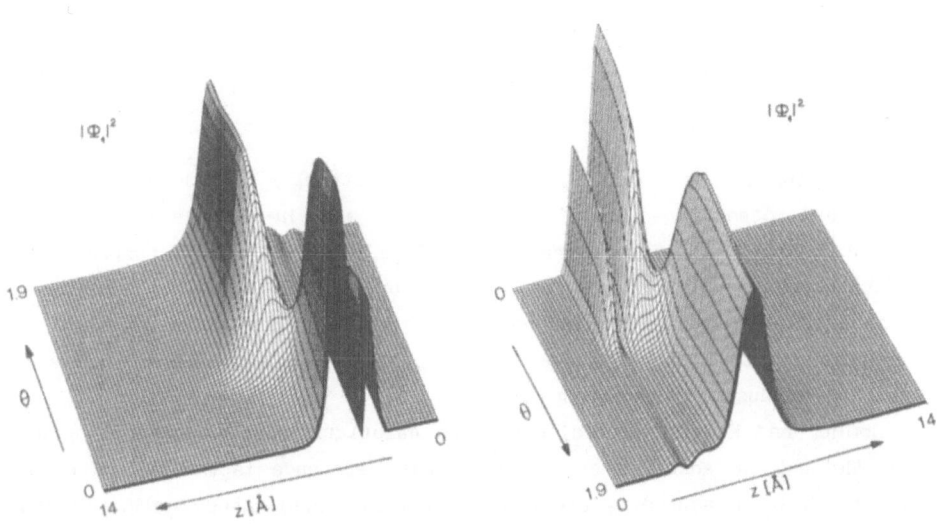

Fig.2.25. Perspective views of $\left|\phi_1(z)\right|^2$ over the (z,θ) plane for ^{3}He on graphite with parameters as in Fig. 2.12. (Summerside et al. 1982)

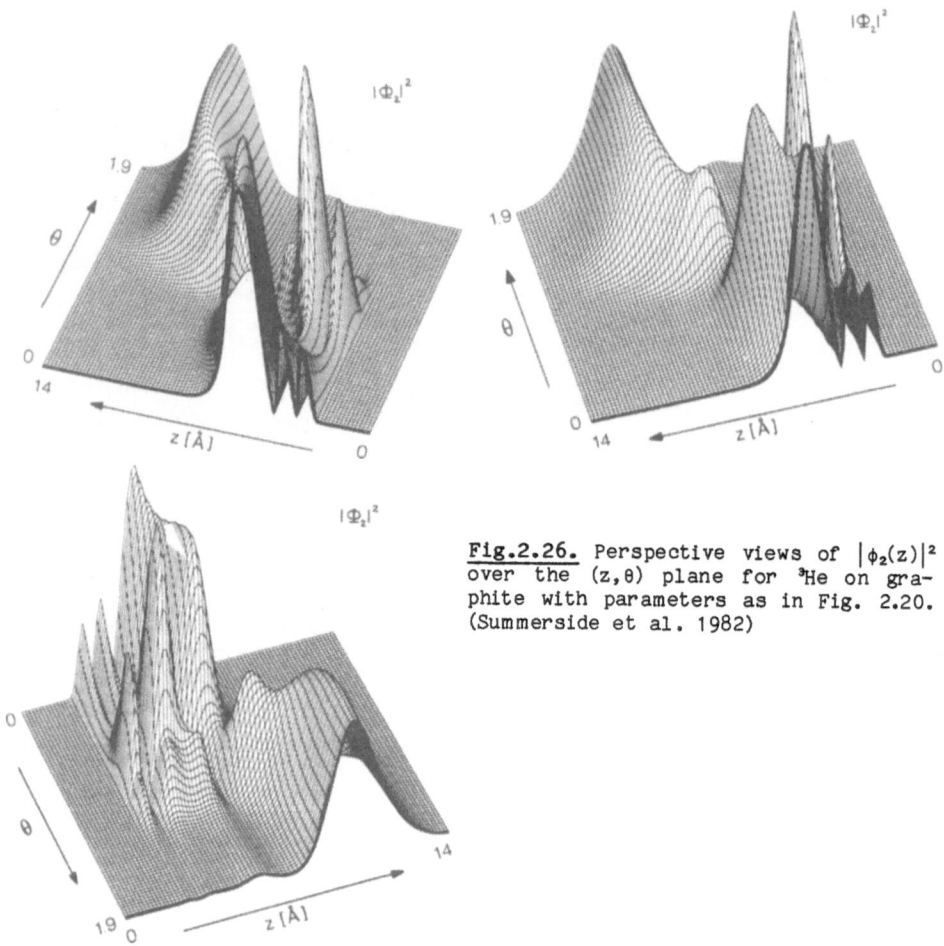

Fig.2.26. Perspective views of $|\phi_2(z)|^2$ over the (z,θ) plane for ^{3}He on graphite with parameters as in Fig. 2.20. (Summerside et al. 1982)

the bare surface potential $V_s(z)$ is so much stronger and deeper than the mean-field potential arising from the rather weak He-He interaction. Figures 2.25 and 26 showing several views of $|\phi_1(z)|^2$ and $|\phi_2(z)|^2$, respectively, demonstrate very clearly how, as coverage builds up, the higher wave functions move out to form the second and, eventually, the third adlayer. We now also see that the decrease in $<z_3>$ at $T \approx 7$ K in the lowest panel of Fig. 2.21 is due to a spreading out of $|\phi_2(z)|^2$ as $|\phi_1(z)|^2$ moves out.

Sommer and Kreuzer (1982a) have also calculated thermodynamic functions for ^{3}He and ^{4}He adsorbed on graphite that reproduce isotherms and excess specific heats measured by Elgin and Goodstein (1974) rather well. We will show in Chaps. 4 and 5 how this mean-field theory can be used to study physisorption kinetics from multilayer adsorbates.

2.3 The Dynamic Atom-Solid Interaction

At the microscopic level, the interaction between an atom and a solid is simply the sum total of all electronic and ionic Coulomb interactions. Trying to describe it as an effective surface potential, in which an atom of mass m moves, one assumes, of course, that the structures of the atom and the solid are well specified. In particular, in the construction of the static atom-solid interaction, as outlined above, one assumes that the solid has a rigid lattice structure thus suppressing, in particular, all of its vibrational degrees of freedom. It is, however, the latter that participate in the energy transfer between the solid and the atom in physisorption kinetics. Well below the Debye temperature T_D, thermal lattice vibrations have an amplitude

$$u_{thermal} \approx \frac{\hbar}{2} \sqrt{\frac{3}{M_s k_B T_D}} \quad , \tag{2.62}$$

where M_s is the mass of a unit cell of the solid. The amplitude (2.62) is typically a fraction of an angstrom and thus very much smaller than lattice spacings and smaller than the typical distance of closest approach of atoms to a surface. On the other hand, electrons follow the thermal vibrations of the ionic cores in solids adiabatically. Lennard-Jones and Devonshire (1936a,b) therefore argued that all the thermal vibrations do is to modulate the distance between the gas particle and the solid surface. For a structureless flat surface they replaced the static surface potential by

$$V_S(z) \rightarrow V_S(z-u_z(t)) \quad , \tag{2.63}$$

and then

$$V_S(z-u_z(t)) \approx V_S(z) - u_z(t) \frac{dV_S(z)}{dz} \quad , \tag{2.64}$$

producing the derivative coupling of the adsorbate to the lattice vibrations of the solid. In this model, it is assumed that the surface of the solid moves as a whole, with the displacements in the z-direction being important for the coupling to the adsorbate. It obviously ignores the fact that lattice vibrations of finite wavelength will introduce density modulations into the solid and laterally along its surface. This can be accounted for by starting from (2.4) and treating $\mathbf{r}_i$ as the instantaneous position of an atom at lattice site i, i.e.,

$$\mathbf{r}_i(t) = \mathbf{r}_i + \mathbf{u}_i(t) \quad , \quad \cdot \tag{2.65}$$

where $u_i(t)$ is the value of the displacement at r_i, i.e., $u_i(t) = u(r_i;t)$. Instead of (2.64), we then get for a central two-body interaction $V_2(r,r_i) = V_2(|r-r_i|)$,

$$V_S(r,t) \approx V_S(r) - \sum_i u(r_i;t) \cdot \frac{\partial}{\partial r} V_2(|r-r_i|) \quad . \tag{2.66}$$

In the long wavelength limit, $u(r_i;t)$ varies slowly over the range of the two body potential. One can therefore expand $u(r_i;t)$ about the point in the surface directly below the gas particle, i.e., around $(R, z = 0)$, so that one gets

$$V_S(r,t) = V_S(r) - \sum_i \frac{\partial}{\partial r} V_2(|r-r_i|) \cdot [\ u(R,0;t)$$

$$+ (r_i-R) \cdot \frac{\partial}{\partial r} u(r;t)|_{(R,0)} + \ldots] \quad . \tag{2.67}$$

The first two terms then give the generalization of (2.64),

$$V_S(r,t) \approx V_S(r) - u(R,0;t) \cdot \frac{\partial V_S(r)}{\partial r} \quad . \tag{2.68}$$

To calculate the higher corrections, Cole and Toigo (1982) work in the continuum limit (2.9) treating the density ρ as spatially varying

$$\rho(r') \approx \rho_0 - \rho_0 \nabla \cdot u(r',t) \tag{2.69}$$

within the solid, i.e., for $z < 0$, and adding a surface term

$$\delta \rho(r) \approx \rho_0 u_z \delta(z) \quad , \tag{2.70}$$

to account for the discontinuity of the density at the surface. Again expanding the displacement vector $u(r',t)$ around a point directly below the gas particle one gets, in the continuum limit for a Lennard-Jones two-body potential (2.5),

$$V_S(r,t) = V_S(z) - u_z \frac{dV_S}{dz} - V_S(z) \nabla \cdot u$$

$$+ V_0 z_m \beta(z) \frac{\partial}{\partial z} (\nabla \cdot u) + V_0 z_m \gamma(z) \frac{\partial^2 u_z}{\partial R^2} + \ldots \quad , \tag{2.71}$$

where $V_S(z)$ is the 9-3 potential (2.11) and

$$\beta(z) = \frac{1}{16} (\frac{z_m}{z})^8 - \frac{3}{4} (\frac{z_m}{z})^2 \quad , \tag{2.72}$$

$$\gamma(z) = \frac{9}{32} (\frac{z_m}{z})^8 - \frac{9}{8} (\frac{z_m}{z})^2 \quad . \tag{2.73}$$

For computational reasons, it is advantageous to parametrize the static surface potential for a flat, structureless surface by a Morse potential (2.37)
for which the Schrödinger equation can be solved analytically. Within the
continuum approximation (2.9), it originates from a two-body interaction

$$V_2(\rho) = \frac{M_S}{\pi \rho_0} \gamma^2 V_0 \frac{1}{\rho} [e^{-2\gamma(\rho-z_0)} - 2e^{-\gamma(\rho-z_0)}] \quad , \tag{2.74}$$

where $\rho = |r-r'|$. In this case, the dynamic interaction is again given by
(2.71), with z_m replaced by γ^{-1}, and the functions $\beta(z)$ and $\gamma(z)$ are given
now by

$$\beta(z) = \frac{1}{2} e^{-2\gamma(z-z_0)} - 2e^{-\gamma(z-z_0)} \qquad and \tag{2.75}$$

$$\gamma(z) = \frac{1}{4} (2\gamma z+1) e^{-2\gamma(z-z_0)} - (\gamma z+1)e^{-\gamma(z-z_0)} \quad . \tag{2.76}$$

Figures 2.27,28 show the functions $V_S(z)$, dV_S/dz, $\beta(z)$ and $\gamma(z)$ for the Lennard-Jones and for the Morse case, respectively.

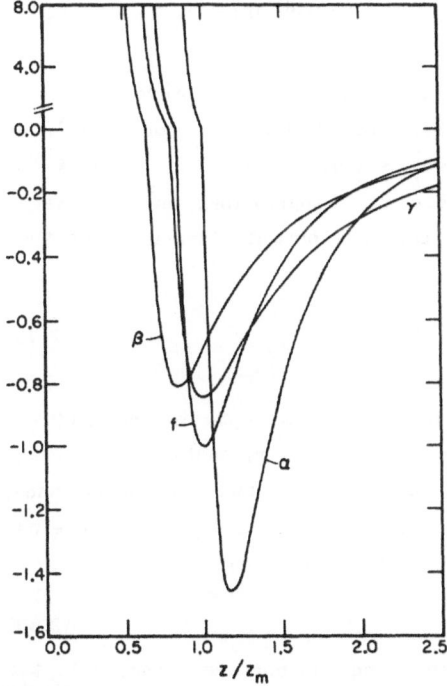

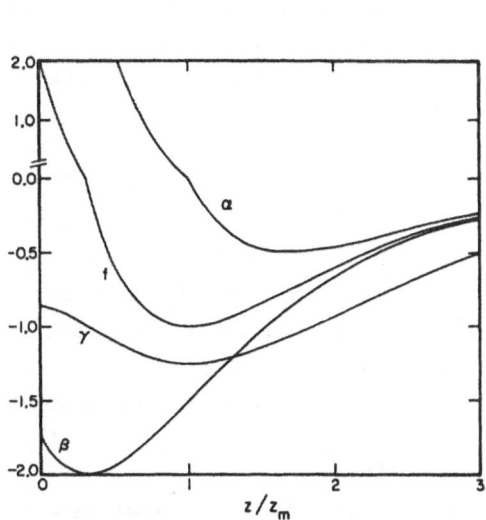

Fig.2.27. The functions α =
$-(z_m/V_0)dV_S/dz$, β from (2.72), γ from
(2.73) and $f=V_S(z)/V_0$ in (2.71) for a
Lennard-Jones pair potential. (Cole and
Toigo 1982)

Fig.2.28. Same as Fig. 2.27 but for
the Morse potential. (Cole and Toigo
1982)

57

We now turn to a derivation of the dynamic atom-solid interaction based on the discrete lattice sum (2.4). We assume that the two-body interaction V_2 is central and displace all lattice vectors by $u_i(t)$, which in turn we use as an expansion parameter to get

$$V_S(r,t) = V_S(r) - \sum_i u_i(t) \cdot \nabla V_2(|r-r_i|)$$

$$+ \sum_i \sum_{\beta,\beta'=1}^{3} u_{i\beta}(t)\, u_{i\beta'}(t)\, \frac{\partial^2}{\partial r_{i\beta'}\, \partial r_{i\beta}} V_2(|r-r_i|) + \dots \quad . \quad (2.77)$$

We note in passing that the higher-order terms in the above expansion are essential for calculations going beyond the one-phonon approximation (Brenig and Schönhammer 1979; Gortel et al. 1980b). For now, we again concentrate on the first order term. To use the lateral symmetry of the solid surface, we expand the displacement vectors in a set of normal vibrational modes $u^{(J)}(r)$,

$$u_i(t) = u(r_i,t)$$

$$= \sqrt{\frac{\hbar}{2\rho_0}} \sum_J \frac{1}{\sqrt{\omega_J}} \left[b_J(t)\, u^{(J)}(r_i) + b_J{}^\dagger(t)\, u^{(J)*}(r_i) \right] \quad . \quad (2.78)$$

The coefficients b_J and $b_J{}^\dagger$ will later on be identified as phonon annihilation and creation operators, respectively. To identify the modes $u^{(J)}$, we first recall that in an infinite harmonic solid, the label J contains the wave vector p and an index σ to denote the longitudinal and transverse acoustic branches and possible optical ones. The normal vibrations are then given by

$$u^{(J=(p,\sigma))}(r) = \frac{1}{\sqrt{AL_S}}\, e_{p\sigma}\, e^{ip\cdot r} \quad , \quad (2.79)$$

where A is the surface area of the solid and L_S its depth in the z-direction. The polarization $e_{p\sigma}$ is a unit vector that for longitudinal waves ($\sigma=L$) is in the direction of propagation p, and for the two transverse ones, ($\sigma=T_1,T_2$) orthogonal to it and mutually orthogonal. For a semi-infinite elastic continuum, the normal modes can be classified by (Ezawa 1971)

$$J=(P,c,\sigma) \quad , \quad (2.80)$$

where P is a two-dimensional wave vector along the surface plane, c is the apparent sound velocity along the surface, and σ labels five different acoustic modes. The dispersion relation is given by

$$\omega_J = c|P| \quad (2.81)$$

58

and the displacements are given by

$$u^{(J)}(R,z) = \frac{1}{\sqrt{A}} e^{iP \cdot R} g(P,c,\sigma;z) \quad . \tag{2.82}$$

The vector functions g are given in Sect.2.4. [(2.126-135)]. Inserting (2.78,82) into the second term of (2.77) we get

$$\delta V_S(r,t) = V_S(r,t) - V_S(r)$$

$$= -\sqrt{\frac{\hbar}{2\rho_o A}} \sum_{J=(P,c,\sigma)} \frac{1}{\sqrt{\omega_J}} [f(P;R,z)b_J(t) + c.c.] \quad , \tag{2.83}$$

where c.c. stands for complex conjugate and

$$f(P;R,z) = \sum_1 \sum_{\alpha=0}^{\infty} e^{iP \cdot R_1} \nabla V_2(R-R_1,z+\alpha d_L) \cdot g(P,c,\sigma;\alpha d_L) \quad . \tag{2.84}$$

In order to avoid unnecessary complications, we have again assumed that there is only one atom per surface unit cell, its position being R_1, and that all atoms in the underlying planes parallel to the surface occupy the same lateral positions R_1, their distance from the surface being αd_L.

We note that f is an eigenfunction of the translation operator along the surface

$$f(P;R-R_1,z) = \exp(-iP \cdot R_1) f(P;R,z) \quad , \tag{2.85}$$

so that it can be written in a Bloch-like form

$$f(P;R,z) = e^{iP \cdot R} \phi(P;R,z) \quad , \qquad \text{where} \tag{2.86}$$

$$\phi(P;R-R_1,z) = \phi(P; R,z) \tag{2.87}$$

is periodic. It can thus be expanded as

$$\phi(P;R,z) = \sum_m e^{iK_m \cdot R} a_m(P;z) \quad , \tag{2.88}$$

where K_m are the translation vectors (2.16) of the two-dimensional reciprocal lattice, satisfying

$$e^{iK_m \cdot R_1} = 1 \quad . \tag{2.89}$$

With the help of (2.86) and (2.88), we then get from (2.83)

$$\delta V_S(r,t) = -\sqrt{\frac{\hbar}{2\rho_o A}} \sum_{J=(P,c,\sigma)} \frac{1}{\sqrt{\omega_J}}$$

$$* \left[b_J(t) \sum_m e^{i(P+K_m) \cdot R} a_m(P;z) + c.c. \right] \quad . \tag{2.90}$$

Then $a_m(P;z)$ can be obtained by inverting (2.88) and using the definitions (2.86) and (2.84). The sum over 1 can be carried out explicitly resulting in

$$a_m(P;z) = \frac{1}{a_s} \int_A d^2R' \; e^{-i(P+K_m)\cdot R'} \sum_{\alpha=0}^{\infty} \nabla V_2(R',z+\alpha d_L) \cdot g(P,c,\sigma;\alpha d_L) \quad . \tag{2.91}$$

Equations (2.90,91) are exact for the linear term in the displacement within the harmonic approximation of the solid.

It is instructive to see what approximations are necessary to recover the Lennard-Jones result (2.64) from (2.90,91). We first observe that if the deformation field in the solid varies slowly over the range of the two-body potential $V_2(R,z)$, we can set $\alpha = 0$ in $g(P,c,\sigma;\alpha d_L)$ in the expression (2.91) for $a_m(P,z)$. Simultaneously, P in the exponent in the same expression can be set equal to zero. Equations (2.90,91) then give

$$\delta V_S(r,t) \approx -\sqrt{\frac{\hbar}{2\rho_0 A}} \sum_J \frac{1}{\sqrt{\omega_J}} \; [b_J(t) \; g(P,c,\sigma;0)$$

$$\cdot \sum_m e^{i(K_m+P)\cdot R} \; d_m(z) + c.c.] \quad , \quad \text{where} \tag{2.92}$$

$$d_m = \frac{1}{a_s} \int_A d^2R' \; e^{-iK_m\cdot R'} \sum_{\alpha=0}^{\infty} \nabla V_2(R',z+\alpha d_L) \quad . \tag{2.93}$$

Comparing (2.93) with (2.15,18), we immediately see that the z-component of $d_m(z)$ is

$$d_m{}^z(z) = \frac{dV_m(z)}{dz} \quad . \tag{2.94}$$

Integrating the x- or y-component of (2.93) by parts and observing that the two-body potential $V_2(R,z)$ vanishes as R becomes large, we observe for the components of $d_m(z)$ parallel to the surface denoted by $D_m(z)$

$$D_m(z) = iK_m V_m(z) \quad . \tag{2.95}$$

Next we observe that for relatively structureless, flat surfaces the Fourier components $V_m(z)$ of the surface potential become rapidly smaller with increasing m. We therefore keep only the term with $K_m = 0$ in (2.92) and obtain

$$\delta V_S(r,t) \approx -\sqrt{\frac{\hbar}{2\rho_0 A}} \sum_J \frac{1}{\sqrt{\omega_J}} \frac{dV_0(z)}{dz} \left[b_J(t) \; g_z(P,c,\sigma;0) \; e^{iP\cdot R} + c.c. \right] \quad . \tag{2.96}$$

Recalling the definitions (2.78) and (2.82), we finally get

$$\delta V_S(\mathbf{r},t) \approx -u_z(\mathbf{R},z=0,t) \; \frac{dV_0(z)}{dz} \tag{2.97}$$

which is the Lennard-Jones result (2.64) or the second term in the Cole and Toigo expansion (2.71).

In Sect.2.1., we introduced the static surface potential as the interaction of a gas particle with the solid at zero temperature. Lattice vibrations at nonzero temperatures then yield the dynamic, i.e., time-dependent interaction derived above. Efrima et al. (1983) have recently argued that in the construction of a dynamical model for the gas-solid interaction, one should start from the thermally averaged potential energy of the coupled gas-solid system as the zero-order approximation, rather than from the static interaction derived in Sect.2.1. The former is used for elastic scattering calculations (Cabrera et al. 1970). We recall that the time-dependent surface potential (2.63) can, alternatively, be written as the potential energy $V(\mathbf{r},Q)$ of the interacting many-body system, depending on the position $\mathbf{r}$ of the gas particle, and on a set of $3N_S$ coordinates for the solid assumed to contain N_S constituent particles. So far we have identified Q with the positions $\mathbf{r}_i(t)$ of the solid constituents at lattice sites i; their time dependence was further assumed to be known from classical mechanics and given by (2.78). To calculate the thermal average of (2.4) with $\mathbf{r}_i(t)$ given by (2.65), we write

$$V_S(\mathbf{r},t) = \frac{AL_S}{(2\pi)^3} \sum_i \int v_p \, e^{i\mathbf{p}\cdot(\mathbf{r}-\mathbf{r}_i(0)-u_i(t))} \, d^3p \quad , \quad \text{where} \tag{2.98}$$

$$v_p = \frac{1}{AL_S} \int V_2(\mathbf{r}) \, e^{-i\mathbf{p}\cdot\mathbf{r}} \, d^3r \quad . \tag{2.99}$$

Denoting thermal averages by pointed brackets we get

$$\langle V_S \rangle = \frac{AL_S}{(2\pi)^3} \sum_i \int v_p \, e^{i\mathbf{p}\cdot(\mathbf{r}-\mathbf{r}_i(0))} \, \langle e^{[i\mathbf{p}\cdot u_i(t)]} \rangle \, d^3p \quad . \tag{2.100}$$

For a harmonic solid we have (e.g., Messiah 1961; Callaway 1974)

$$\langle e^{[i\mathbf{p}\cdot u_i(t)]} \rangle = e^{-W_i(\mathbf{p})} \quad , \tag{2.101}$$

where the Debye-Waller factor

$$W_i(\mathbf{p}) = \tfrac{1}{2} \langle (\mathbf{p}\cdot u_i)^2 \rangle$$
$$= \tfrac{1}{2} \, (p_x^2 \langle u_{ix}^2 \rangle + p_y^2 \langle u_{iy}^2 \rangle + p_z^2 \langle u_{iz}^2 \rangle) \quad , \tag{2.102}$$

has been introduced. It actually depends only on the distance αd_L of the equilibrium position of the i-th atom from the surface. The subscript i in W_i can therefore be replaced by α.

Splitting the integration over p in (2.98) into p_z and P, and using the closure relation

$$\sum_l e^{iP \cdot R_l} = \frac{N_s^{(2)}(2\pi)^2}{A} \sum_m \delta(P - K_m) \quad , \tag{2.103}$$

with $N_s^{(2)}$ being the number of atoms in the surface layer, we can write the thermally averaged surface potential as a two-dimensional Fourier series, in analogy to (2.14,15), to obtain

$$\langle V_s(r) \rangle = \sum_{\alpha=0}^{\infty} \langle w_0(z_\alpha) \rangle + \sum_{m \neq 0} e^{iK_m \cdot R} \langle \sum_{\alpha=0}^{\infty} w_m(z_\alpha) \rangle \quad , \tag{2.104}$$

where

$$\langle w_m(z_\alpha) \rangle = \frac{N_s^{(2)} L_s}{2\pi} \int_{-\infty}^{\infty} dp_z \, \exp(ip_z z_\alpha) \, v_{K_m,p_z} \exp[-W_\alpha(K_m,p_z)] \quad . \tag{2.105}$$

The sum over α extends over the crystal planes parallel to the surface and $z_\alpha = z + \alpha d_L$. Setting $W_\alpha = 0$ in (2.105), one obtains $w_m(z_\alpha)$ appropriate for T=0. Indeed, using (2.99) and performing the integration over p_z, one immediately recovers the second line of (2.18) with q=1 and $m_k=0$. Assuming that $W_\alpha(K_m,p_z)$ and thus $\langle u_i^2 \rangle$ do not depend on α, and that the displacements are isotropic within each plane, i.e.,

$$\langle u_{ix}^2 \rangle = \langle u_{iy}^2 \rangle = \langle u_x^2 \rangle \quad \text{and} \tag{2.106}$$

$$\langle u_{iz}^2 \rangle = \langle u_z^2 \rangle \quad , \tag{2.107}$$

one gets from (2.105)

$$\langle w_m(z_\alpha) \rangle = \frac{1}{\sqrt{2\pi \langle u_z^2 \rangle}} \exp(-\tfrac{1}{2} K_m^2 \langle u_x^2 \rangle)$$

$$* \int_{-\infty}^{\infty} dz' w_m(z_\alpha') \exp\left[-\frac{(z'-z)^2}{2\langle u_z^2 \rangle}\right] dz' \quad . \tag{2.108}$$

To see the effect of thermal averaging, let us evaluate the sum over all lattice layers in (2.108) for $K_m = 0$, i.e., for the laterally averaged surface potential. For simplicity, we take for $\Sigma w_0(z_\alpha')$, a Morse potential (2.37) and get

$$\langle V_s(z) \rangle = \sum_\alpha \langle w_0(z_\alpha) \rangle = \bar{V}_0 \left[e^{-2\gamma(z-\bar{z})} - 2 e^{-\gamma(z-\bar{z})} \right] \quad , \tag{2.109}$$

i.e., again a Morse potential, but with a reduced depth

$$\tilde{V}_0 = V_0 \, e^{-\gamma^2 \langle u_z^2 \rangle} \quad , \tag{2.110}$$

and shifted to a new minimum

$$\tilde{z} = z_0 + \frac{3}{2} \, \gamma \langle u_z^2 \rangle \quad . \tag{2.111}$$

With the range γ^{-1} typically being less than one angstrom and z_0 a few angstroms, we find with (2.62) that $\tilde{V}_0$ and $\tilde{z}$ differ from V_0 and z_0 by only a few percent. This is also true for the higher Fourier components $\langle w_m(z_\alpha) \rangle$.

Let us then return to the argument of Efrima et al. (1983) that a dynamical model for phonon induced desorption should start from the thermally averaged surface potential (2.104) rather than from the static potential (2.4). Using a correlation function approach to calculate phonon-mediated transition rates, they observe that the relevant time correlation functions, calculated with δV_s defined in the first line of (2.83), tend to a constant for large times, implying divergent time integrals, whereas those calculated for

$$\delta V_s = V_s - \langle V_s \rangle \quad , \tag{2.112}$$

tend to zero. This observation seems to have been included in earlier work by Bendow and Ying (1973a and b), Böheim and Brenig (1981), and Brako and Newns (1981).

So far we have considered the interaction of a gas particle with a molecular solid, in which case the surface potential is the sum total (2.4) of two-body interactions between the gas particle and each one of the constituent particles of the solid. If the solid is an ionic crystal, such as an alkali halide, such sums are still meaningful as atom-ion potentials are rather well known. If, however, the solid is a metal, more care has to be exercised to determine the dynamic gas-solid coupling. A starting point is the observation that the conduction electrons follow the thermal motion of the ion cores around their respective lattice sites adiabatically. Stutzki and Brenig (1981), Grimmelmann et al. (1981), and others before them, therefore again write down a sum like (2.4), taking for the constituents of the solid just the metal ions, with the two-body interactions being some effective Morse or 6-12 potentials. De et al. (1980), have gone beyond this simple approach by deriving, under certain rather vague approximations, an expression for the coupling between a point charge on the adsorbing particle, and the fluctuating part of the metal substrate. This approach is obviously more suited to a chemisorbed system, where a charge transfer indeed takes place. Considering physisorbed systems, charge transfer can be neglected as a first approximation, so that the above mechanism should be reworked for the dynamic coupling of an induced dipole, and the fluctuating part of the metal substrate.

2.4 Phonon Dynamics of a Solid

As we have seen in the previous section, the dynamical interaction of a gas particle with a solid surface is largely controlled by the vibrational degrees of freedom of the solid. We therefore review here the theory of phonon dynamics, starting with an elastic continuum approach, and then report results from lattice dynamics in the second part of this section.

Letting $u(x)$ be the displacement of a mass element at point x, we define a strain tensor

$$\varepsilon_{ij} = \frac{1}{2} \left[\frac{\partial u_i}{\partial x_j} + \frac{\partial u_j}{\partial x_i} \right] \quad . \tag{2.113}$$

We collect the forces in the elastic medium in a stress tensor whose element $\sigma_{ij}(x)$ contains the j-th component of the force per unit surface area, perpendicular to the i-th direction, on a volume element at x. We assume a linear stress-strain relationship

$$\sigma_{ij} = C_{ijk\ell} \, \varepsilon_{k\ell} \quad , \tag{2.114}$$

where Einstein's summation convention over double indices is implied. The symmetries $\varepsilon_{ij} = \varepsilon_{ji}$ and $\sigma_{ij} = \sigma_{ij}$ imply that

$$C_{ikj\ell} = C_{ikj\ell} = C_{ik\ell j} = C_{j\ell ik} \quad , \tag{2.115}$$

so that there are at most 21 independent elastic constants in an anisotropic elastic medium. With (2.114), Newton's equations of motion are closed and read

$$\rho_0 \frac{\partial^2 u_i}{\partial t^2} = \sum_j \frac{\partial \sigma_{ij}}{\partial x_j} = C_{ijk\ell} \frac{\partial^2 u_k}{\partial x_\ell \partial x_j} \quad . \tag{2.116}$$

For an isotropic continuum, there are only two independent constants which one conventionally chooses as the Lame constants

$$\mu = C_{1122} = C_{1133} = C_{2233}$$

$$\lambda = C_{2323} = C_{1313} = C_{1212} \quad \text{and}$$

$$\lambda + 2\mu = C_{1111} = C_{2222} = C_{3333} \quad , \tag{2.117}$$

so that the equations of motion read

$$\rho_0 \frac{\partial^2 u_i}{\partial t^2} = \mu \nabla^2 u_i + (\mu + \lambda) \frac{\partial^2 u_k}{\partial x_i \partial x_k} \quad , \tag{2.118}$$

where the Laplacian is defined as

$$\nabla^2 = \frac{\partial^2}{\partial x_1^2} + \frac{\partial^2}{\partial x_2^2} + \frac{\partial^2}{\partial x_3^2} \quad . \tag{2.119}$$

We can also introduce the longitudinal and transverse sound velocities

$$c_\ell^2 = \frac{\lambda + 2\mu}{\rho_0} \quad , \tag{2.120}$$

$$c_t^2 = \frac{\mu}{\rho_0} \quad , \tag{2.121}$$

and rewrite (2.118) as a vector equation

$$\frac{\partial^2 u}{\partial t^2} = c_t^2 \; \text{grad div} \; u - c_\ell^2 \; \text{curl curl} \; u \quad . \tag{2.122}$$

To solve it, we have to specify boundary conditions which, for a free surface, reflect the fact that there are no forces acting on it. With n_j being
the j-th component of a unit vector normal to the surface, they read

$$\sigma_{ij} \; n_j = 0 \quad , \tag{2.123}$$

implying for an isotropic medium

$$\left[(c_\ell^2 - 2c_t^2) \frac{\partial u_k}{\partial x_k} \delta_{ij} + c_t^2 (\frac{\partial u_i}{\partial x_j} + \frac{\partial u_j}{\partial x_i}) \right] n_j = 0 \quad . \tag{2.124}$$

To solve (2.122), subject to (2.124), we make an ansatz [compare (2.78)]

$$u(r,t) = \sum_J \sqrt{\frac{\hbar}{2 \rho_0 \omega_J}} \; [b_J u^{(J)}(R,z) \; e^{-i\omega_J t} + b_J^\dagger \; u^{(J)*}(R,z) \; e^{i\omega_J t}] \quad . \tag{2.125}$$

The index J labels the different eigenmodes with eigenvalues ω_J and eigenvectors (2.82)

$$u^{(J)}(R,z) = \frac{1}{\sqrt{A}} \; e^{iP \cdot R} \; g(P,c,\sigma;z)$$

$$= \frac{1}{\sqrt{A}} \; e^{iP \cdot R} \; \hat{0}(\frac{P}{P}) \cdot f(P,c,\sigma;z) \quad , \tag{2.126}$$

where $P = |P|$ is the magnitude of a two-dimensional wave vector parallel to
the exposed surface area A of the solid. We adopt the convention in which
$z < 0$ inside the solid. We also introduced the rotation operator around the
z-axis

$$\bar{\sigma}(\tfrac{P}{P}) \cdot \begin{bmatrix} P \\ 0 \\ 0 \end{bmatrix} = \begin{bmatrix} p_x \\ p_y \\ 0 \end{bmatrix} \quad , \tag{2.127}$$

where $P = (p_x, p_y)$ so that the f-functions can be calculated for a wave propagating in the direction of P.

To ensure that the above $u^{(J)}(R,t)$ form a complete and orthonormal set of eigenmodes, Ezawa (1971) has identified J as consisting of the two-dimensional wave vector P, the apparent velocity c with which the elastic wave propagates along the surface, and a discrete index σ that labels five different modes. We now list the f-functions for a wave propagating in the x-direction. For the shear horizontal mode ($\sigma=H$) we have

$$f(P,c,\sigma=H;z) = \theta(c-c_t)\sqrt{\frac{2cP}{\pi c_t{}^2 \beta}}\begin{bmatrix} 0 \\ \cos(P\beta z) \\ 0 \end{bmatrix} \quad , \quad \text{with} \tag{2.128}$$

$$\beta = \sqrt{(c/c_t)^2 - 1} \quad . \tag{2.129}$$

Its apparent velocity along the surface is restricted to $c_t<c<\infty$. It executes shearing vibrations parallel to the surface and perpendicular to the direction of propagation. Its amplitude oscillates as a function of distance z into the solid, as a standing wave.

There are two mixed P-SV modes ($\sigma = \pm$) given by

$$f(P,c,\sigma = \pm;z) = \theta(c-c_\ell)\sqrt{\frac{P}{4\pi c}}$$

$$* \begin{bmatrix} \mp \dfrac{1}{\sqrt{\alpha}}(e^{iP\alpha z} - \xi_\pm e^{-iP\alpha z}) + i\sqrt{\beta}(e^{iP\beta z} + \xi_\pm e^{-P\beta z}) \\ 0 \\ \pm\sqrt{\alpha}(e^{iP\alpha z} + \xi_\pm e^{-iP\alpha z}) + \dfrac{1}{\sqrt{\beta}}(e^{iP\beta z} - \xi_\pm e^{-iP\beta z}) \end{bmatrix} \quad , \tag{2.130}$$

with

$$\alpha = \sqrt{(c/c_\ell)^2 - 1} \quad , \quad \text{and}$$

$$\xi_\pm = \frac{(\beta^2 - 1 \pm 2i\sqrt{\alpha\beta})^2}{(\beta^2-1)^2 + 4\alpha\beta} \quad . \tag{2.131}$$

They have a component in the direction of propagation (x). In bulk they would be longitudinal or pressure (P) waves. In addition, they have a z-component transverse to the direction of propagation (shear S) with vertical (V) polarisation. They are propagating into the solid with two different wave vectors αP and βP; the mixing upon reflection at the surface is controlled by $\xi_\pm$.

The generalized Rayleigh mode ($\alpha = GR$)

$$\mathbf{f}(P,c,\sigma = GR;z) = \theta(c_\ell - c)\ \theta\ (c-c_t\ )\ i\ \sqrt{\frac{P}{2\pi c\beta}}$$

$$*\begin{bmatrix} B\ e^{P\gamma z} + \beta(e^{iP\beta z} + Ae^{-P\beta z}) \\ 0 \\ i\gamma B\ e^{P\gamma z} + e^{iP\beta z} - Ae^{-iP\beta z} \end{bmatrix}\ ,\quad \text{with} \qquad (2.132)$$

$$\gamma = \sqrt{1-(c/c_\ell)^2}\ ,$$

$$A = \frac{(\beta^2-1)^2 - 4i\gamma\beta}{(\beta^2-1)^2+4i\gamma\beta}\ ,$$

$$B = \frac{4\beta(\beta^2-1)}{(\beta^2-1)^2+4i\gamma\beta}\ , \qquad (2.133)$$

differs from the $\sigma = \pm$ modes in that part of its amplitude decays exponentially into the solid, with a length scale $1/\gamma P$ proportional to the wavelength along the surface $(2\pi/P)$. Note that the apparent velocity c along the surface is bounded from above and below, $c_t < c < c_\ell$.

Lastly there is the Rayleigh mode ($\sigma=R$)

$$\mathbf{f}(P,c,\sigma=R;z) = i\delta(c-c_R)\sqrt{\frac{2\gamma_R\eta_R^2}{(\gamma_R-\eta_R)(\gamma_R-\eta_R+2\gamma_R\eta_R^2)}}$$

$$*\begin{bmatrix} e^{P\gamma_R z} - \dfrac{2\gamma_R\eta_R}{1+\eta_R^2}\ e^{P\eta_R z} \\ 0 \\ i\gamma_R\ e^{P\gamma_R z} - \dfrac{2i\gamma_R}{1+\eta_R^2}\ e^{P\eta_R z} \end{bmatrix}\ ,\quad \text{with} \qquad (2.134)$$

$$\gamma_R = \sqrt{1-(\frac{c_R}{c_\ell})^2}\ ,$$

$$\eta_R = \sqrt{1-(\frac{c_R}{c_t})^2}\ , \qquad (2.135)$$

where c_R is the speed of the Rayleigh wave along the surface. It is a solution of the equation

$$4\gamma_R\eta_R - (1+\eta_R^2)^2 = 0\ . \qquad (2.136)$$

The Rayleigh mode is completely confined to the surface region. With the wave propagating along the surface, a given mass element executes excursions along an elliptical path.

To take care of the fact that the solid is infinite in the lateral direction, we impose two-dimensional Born-von Karman boundary conditions on a finite surface A. As a result, P can assume only discrete values. Introducing the two-dimensional lattice periodicity explicitly with R_1 being lattice translation vectors, one must restrict P to the two-dimensional first Brillouin zone. To construct a Debye model for surface waves, we recall that in the bulk model one equates the volume of the three-dimensional Brillouin zone to that of a sphere of radius p_D such that

$$\frac{4\pi}{3} p_D^3 = \frac{8\pi^3}{V_C} \quad , \tag{2.137}$$

where V_C is the volume of a crystal unit cell. In analogy, in the surface Debye model we equate the area of the two dimensional Brillouin zone to that of a disc of radius P_D so that

$$\pi P_D^2 = \frac{(2\pi)^2}{a_s} \quad , \tag{2.138}$$

where a_s is the area of a surface unit cell. There are five different two-dimensional Bravais nets, namely oblique, primitive rectangular, centered rectangular, square and hexagonal (see, e.g., D.P. Woodruff 1981). They are depicted in Fig. 2.29, together with their corresponding two-dimensional Brillouin zones. For example, the (100) surface of a fcc crystal with bulk lattice constant a, has a square surface unit cell of area $a_s = (a/\sqrt{2})^2$. For a fcc (111) surface, the surface lattice is hexagonal with the surface unit cell being a parallelogram with equal sides of length $a\sqrt{2}$ and angles 60° and 120° so that $a_s = \sqrt{3}\, a^2$. Consequently, in a surface Debye model, we

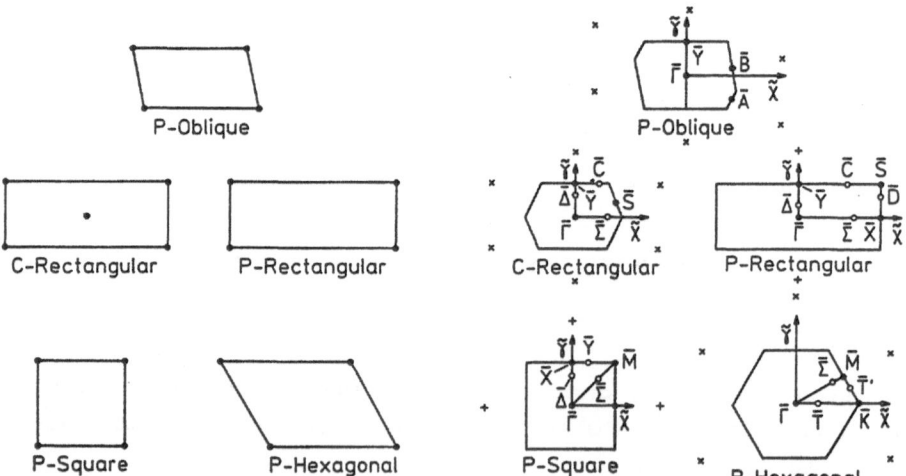

Fig.2.29. The five two-dimensional Bravais nets and their respective Brillouin zones. The labeling of the symmetry points is adopted from Koster (1957).

must restrict the two-dimensional wave vector to discs

$$P < P_D = 2\sqrt{2\pi/a} \qquad (2.139)$$

for a square surface unit cell of the (100) surface of a fcc crystal and to

$$P < P_D = \frac{2\sqrt{\pi}}{3^{1/4}a} \qquad (2.140)$$

for a hexagonal surface lattice. Whereas all phonons have the same cutoff in their two-dimensional wave vectors, the different modes have different frequency cutoffs. Because an elastic continuum has no dispersion so that

$$\omega_J = \omega_\sigma(c,P) = c|P| \quad , \qquad (2.141)$$

we get

$$P = \frac{\omega}{c} \le \frac{\omega}{c_{min}(\sigma)} \le P_D = \frac{\omega_D(\sigma)}{c_{min}(\sigma)} \quad , \qquad (2.142)$$

where $c_{min}(\sigma)$ is the smallest apparent velocity of the mode σ along the surface, i.e., $c_{min}(H) = c_{min}(GR) = c_t$, $c_{min}(\mp) = c_\ell$ and $c_{min}(R) = c_R$. The Debye cutoff frequencies $\omega_D(\sigma)$, as listed in Table 2.7 for different $\ell_s = c_t/c_\ell$, are thus different for different modes and also different on different surfaces and also different from bulk values. The surface Debye model is, for geometrical reasons, more adequate for the high-symmetry surfaces, i.e., square and hexagonal. How well it represents the phonon dynamics, of course, depends on the details of the lattice interactions, as we will see below.

To return to the dispersion relation (2.141), we again note that the five modes have different ranges of apparent velocities c along the surface as indicated in Fig.2.30. Thus the summation over J for the surface Debye model amounts to

$$\sum_J = \sum_\sigma \int_0^\infty dc \sum_P = \frac{A}{(2\pi)^2} \sum_\sigma \int_0^\infty dc \int_0^{2\pi} d\phi_P \int_0^{\omega_D(\sigma)/c} P dP \quad , \qquad (2.143)$$

where ϕ_P is the angle between P and the x-axis. To visualize the various modes, it is advantageous to look at the local density of states. We recall that in an infinite bulk medium the density of states is defined as the number of eigenstates per unit volume and unit frequency,

$$d(\omega) = \frac{1}{AL_s} \sum_J \delta(\omega - \omega_J) \quad . \qquad (2.144)$$

For a bulk Debye model this gives

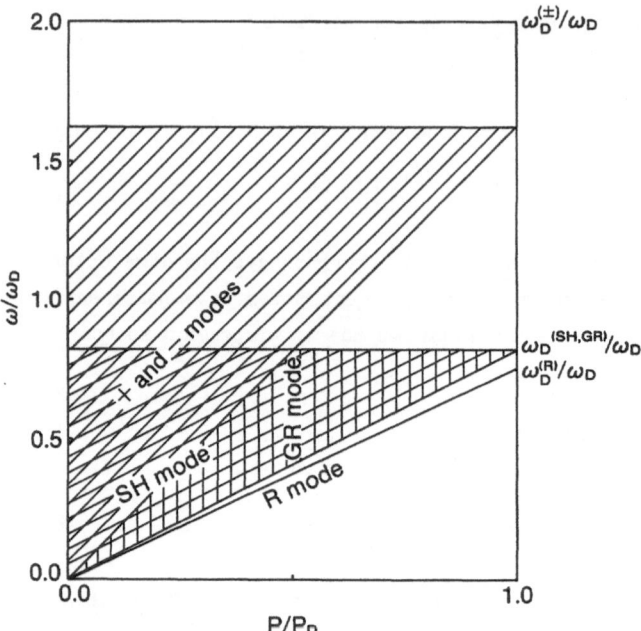

Fig.2.30. The dispersion relations for the five surface modes in the surface Debye model.

$$d(\omega) = \frac{1}{(2\pi)^3} \sum_\sigma \int_{|p|<p_D} d^3p \; \delta(\omega - c_\sigma p)$$

$$= 2d_t(\omega) + d_\ell(\omega) \;, \quad \text{where} \tag{2.145}$$

$$d_t(\omega) = \frac{1}{2\pi^2 c_t^3} \; \omega^2 \theta(\omega_D(t) - \omega)$$

$$d_\ell(\omega) = \frac{1}{2\pi^2 c_\ell^3} \; \omega^2 \theta(\omega_D(\ell) - \omega) \tag{2.146}$$

are the densities of states for the transverse and longitudinal modes, respectively.

The maximum frequencies for the longitudinal and transverse modes are given in the Debye model by

$$\omega_D(\ell,t) = \left(\frac{6\pi^2 \rho_0}{M_s}\right)^{1/3} c_{\ell,t} \;, \tag{2.147}$$

where ρ_0 and M_s are the mass density of the solid and the mass of its unit cell, respectively. Let us also recall that the average Debye frequency is usually defined as

70

$$\frac{1}{\omega_D^3} = \frac{1}{3}\left[\frac{1}{(\omega_D^{(\ell)})^3} + \frac{2}{(\omega_D^{(t)})^3}\right] \quad . \tag{2.148}$$

To introduce a local density of states in an inhomogeneous medium, like a solid with a surface, we recall that $d(\omega)$ is used to calculate the average of a quantity like the energy E as

$$\langle E \rangle = \int \hbar\omega\, d(\omega)\left[n^{(ph)}(\omega) + \tfrac{1}{2}\right]d\omega \quad , \tag{2.149}$$

where $n^{(ph)}(\omega)$ is the thermal occupation function, i.e., the Bose - Einstein distribution

$$n^{(ph)}(\omega) = \frac{1}{e^{\beta\hbar\omega} - 1} \tag{2.150}$$

for phonons. In an inhomogeneous medium one would define the local density of E by

$$\langle E \rangle = \int e(r)d^3r \quad . \tag{2.151}$$

For a harmonic solid, the total energy is twice its kinetic energy so that

$$e(r) = \frac{1}{\rho_0}\langle \pi^2(r,t)\rangle \quad , \quad \text{where} \tag{2.152}$$

$$\pi(r,t) = -i\sqrt{\frac{\hbar\rho_0}{2}}\sum_J \sqrt{\omega_J}\,[b_J u^{(J)}(r)e^{-i\omega_J t} - b_J^\dagger u^{(J)*}(r)e^{i\omega_J t}] \tag{2.153}$$

is the momentum conjugate to $u(r,t)$. We can then write

$$
\begin{aligned}
e(r) &= \sum_J \hbar\omega_J |u^{(J)}(r)|^2\left[n^{(ph)}(\omega_J) + \tfrac{1}{2}\right]\\
&= \sum_\sigma \int d\omega\, d_\sigma(\omega,z)\hbar\omega\left[n^{(ph)}(\omega) + \tfrac{1}{2}\right] \quad ,
\end{aligned} \tag{2.154}
$$

so that the local density of states is given by

$$d_\sigma(\omega,z) = d_\sigma^x(\omega,z) + d_\sigma^y(\omega,z) + d_\sigma^z(\omega,z) \quad , \quad \text{where} \tag{2.155}$$

$$d_\sigma^i(\omega,z) = \frac{1}{2\pi}\int_0^\infty \frac{dc}{c^2}\left|f_i(\frac{\omega}{c},c,\sigma;z)\right|^2 \omega\,\theta(\omega_D^{(\sigma)} - \omega) \quad . \tag{2.156}$$

is the density of states for the polarization in the i-th direction of the σ mode propagating in the x-direction. The integrals over c must be done numerically, except for the Rayleigh mode and for the SH mode. For the latter we get

71

$$d_{SH}{}^y(\omega,z) = \frac{1}{2\pi^2 c_t^3} \; \omega^2 \left[1 + \frac{\sin(2\omega z/c_t)}{2\omega z/c_t}\right] \; \theta(\omega_D{}^{(SH)} - \omega) \quad , \qquad (2.157)$$

All partial densities of states, for a given polarization and branch, can be written as

$$d_\sigma{}^i(\omega,z) = S_\sigma{}^i(\omega z) \; \frac{\omega^2}{2\pi^2 c_t^3} \; \theta(\omega_D{}^{(\sigma)} - \omega) \quad , \qquad (2.158)$$

stressing the fact that, even at z = 0, all of them are proportional to ω^2, including the Rayleigh mode, i.e., have the same behavior as for the modes of an infinite solid.

We begin the discussion of the functions $S_\sigma{}^i(\omega z)$ with the σ=SH mode, which is independent of ℓ_s; it is plotted in Fig.2.31. For a wave propagating in the x-direction along the surface, it is nonzero only for the i=y component. The oscillations represent, for fixed ω, standing waves that decay into the solid towards the bulk value 1. Figures 2.32,33 give the functions $S_\sigma{}^i(\omega z)$ for ℓ_s= 0.5 for σ=± and σ=R, respectively. The localization of the

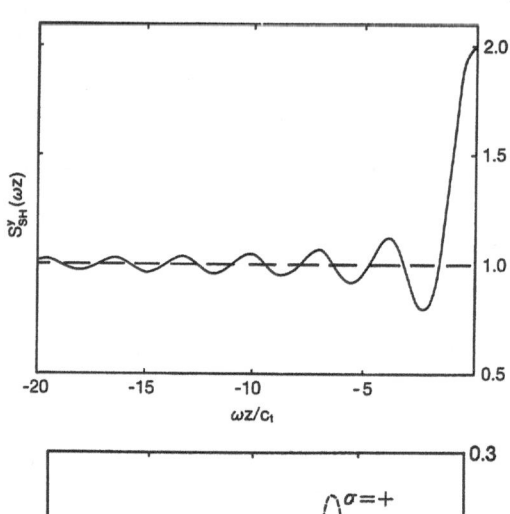

Fig.2.31. The function $S_\sigma{}^i(\omega z)$, defined in the partial densities of states (2.158) for σ=SH.

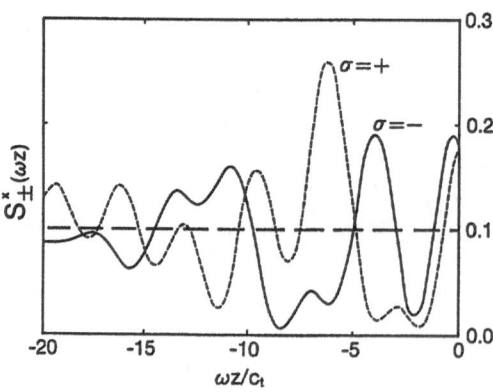

Fig.2.32. The functions $S_\sigma{}^i(\omega z)$, defined in the partial densities of states (2.158) for σ=±.

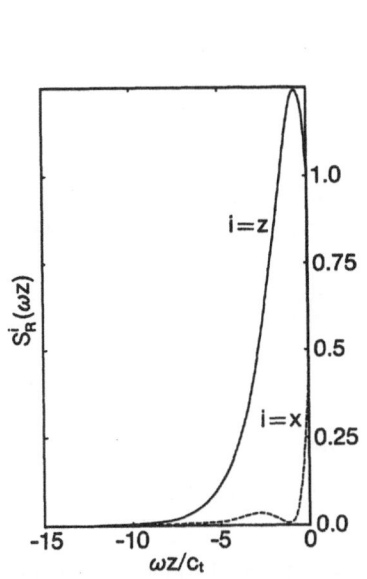

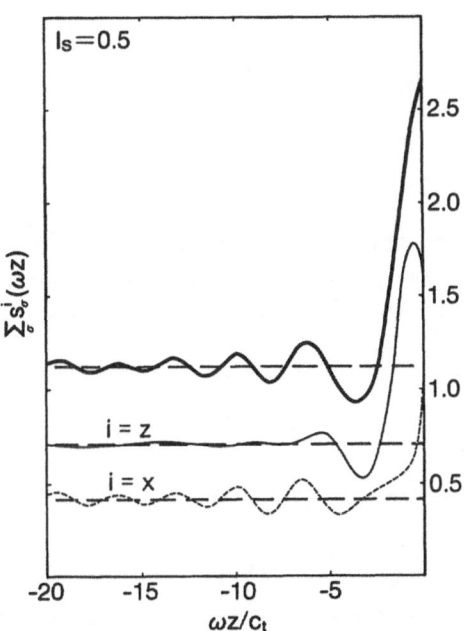

Fig.2.33. The $S_\sigma^i(\omega z)$, defined in the partial densities of states (2.158) for the Rayleigh mode $\sigma=R$.

Fig.2.34. The sums of the functions $S_\sigma^i(\omega z)$, defined in the partial densities of states (2.158).

Rayleigh wave in the surface region is clearly displayed. Note that the spatial decay constant is proportional to ω^{-1} so that the long-wavelength Rayleigh mode can penetrate deeply into the solid. The generalized Rayleigh modes decay to their asymptotic value over the same depth. Note that their z-polarization is depleted in the surface region. The ± modes exhibit two oscillation periods and oscillate far into the solid. Only their x-polarization, displayed in Fig. 2.32, is significant; their z-polarization is smaller by a factor of 10. In Fig. 2.34, we finally give the total contributions for y- and z-polarizations and their sum. Some of the features displayed by the individual modes actually cancel in their sum. For $\ell_s = 0.95$, the two periods of oscillation in the ± modes are well separated. The sum total looks much like that for $\ell_s = 0.5$, except that the contribution from the Rayleigh mode is about a factor of 30 larger, and decays much more rapidly.

Deep inside the solid, the weight functions S_σ^i can be evaluated analytically to give

$$S_{SH}^y(-\infty) = 1$$

$$S_\pm^x(-\infty) = \frac{1}{6}\left(2\ell_s^3 + 1 - (1-\ell_s^2)^{3/2}\right)$$

73

$$S_{\pm}{}^{z}(-\infty) \;=\; \tfrac{1}{6}\,(\ell_s{}^3 + 2 - (\ell_s{}^2 + 2)\,(1 - \ell_s{}^2)^{1/2})$$

$$S_{GR}{}^{x}(-\infty) \;=\; \tfrac{1}{3}\,(1 - \ell_s{}^2)^{3/2}$$

$$S_{GR}{}^{z}(-\infty) \;=\; \tfrac{1}{3}\,(\ell_s{}^2 + 2)\,(1 - \ell_s^{2})^{1/2}$$

$$S_R{}^{x}(-\infty) \;=\; S_R{}^{z}(-\infty) \;=\; 0 \;. \tag{2.159}$$

Note that in the common ω domain, the contribution to the total density of states for GR modes deep inside the solid cancels part of the contributions from the (two) $\pm$ modes, so that for $\omega < \omega_D(GR)$,

$$d^{x}(\omega, z=-\infty) = d_{+}{}^{x}(\omega, z=-\infty) + d_{-}{}^{x}(\omega, z=-\infty) + d_{GR}{}^{x}(\omega, z=-\infty)$$

$$= \frac{\omega^2}{2\pi^2 c_t{}^3}\,\frac{2\ell_s{}^3 + 1}{3}$$

$$d^{z}(\omega, z=-\infty) = d_{+}{}^{z}(\omega, z=-\infty) + d_{-}{}^{z}(\omega, z=-\infty) + d_{GR}{}^{z}(\omega, z=-\infty)$$

$$= \frac{\omega^2}{2\pi^2 c_t{}^3}\,\frac{\ell_s{}^3 + 2}{3} \;. \tag{2.160}$$

Let us for comparison, look at waves travelling in the x direction in an infinite solid. The longitudinal waves have a density of states

$$d_{\ell}{}^{x}(\omega) = \frac{1}{2\pi^2 c_t{}^3}\,\omega^2\,\ell_s{}^3 \;, \tag{2.161}$$

whereas, for each of two transverse waves, we have

$$d_t{}^{z}(\omega) = d_t{}^{y}(\omega) = \frac{1}{2\pi^2 c_t{}^3}\,\omega^2 \;. \tag{2.162}$$

We note that d^{x} and d^{z}, deep inside the solid in the surface model, are mixed waves consisting of a 2/3 longitudinal and 1/3 transverse contribution, and 1/3 longitudinal and 2/3 transverse respectively. The density of states of the second transverse mode is taken up by the SH mode.

Adding the density of states of all modes and integrating over ω, we get the total number of vibrational states per unit volumes,

$$N_{surface}(z=-\infty) = N_{bulk}\left[\frac{4}{3\sqrt{\pi}}\,\frac{1}{3}\,\frac{1 + 2\ell_s{}^3 - \sqrt{1 - \ell_s{}^2}\,(1 - \ell_s{}^3)}{\ell_s{}^3} \right] \;. \tag{2.163}$$

Let us note that the overall geometrical factor $(4/3\sqrt{\pi})$, can be traced back to replacing a cubic Brillouin zone of side $2\pi/a$ by a spherical one in the bulk model, and the square of side $2\pi/a$ by a disc in the surface model. This factor can be different for different crystals and surfaces, depending on the particular symmetries. The factor in square brackets changes from

74

about 2 to 0.75 for $0.1 < \ell_S < 1$, being about 1 for $\ell_S = 0.5$. This difference in normalization is really not too important because Debye models are intended to be used in the long wavelength regime, i.e., for $\omega \ll \omega_D(\sigma)$.

Let us finally discuss the densities of states for different modes for $z = 0$, i.e., at the surface of the solid. We then get

$$d_\sigma^1(\omega,0) = S_\sigma^1(0) \frac{\omega^2}{2\pi^2 c_t^3} \theta(\omega_D(\sigma) - \omega) \quad , \tag{2.164}$$

where $S_\sigma^1(0)$ are the following integrals:

$$S_{SH}^y(0) = 2 \ s_\parallel^{(SH)} = 2$$

$$S_\pm^x(0) = \frac{1}{2} \int_0^{\ell_S^2} dx \ \frac{\sqrt{1-x}}{(1-2x)^2 + 4x\sqrt{(1-x)(\ell_S^2-x)}} = s_\parallel^{(B)}$$

$$S_\pm^z(0) = \frac{1}{2} \int_0^{\ell_S^2} dx \ \frac{\sqrt{\ell_S^2-x}}{(1-2x)^2 + 4x\sqrt{(1-x)(\ell_S^2-x)}} = \frac{1}{2} \ s_\perp^{(B)}$$

$$S_{GR}^x(0) = \int_{\ell_S^2}^1 dx \ \frac{(1-2x)^2\sqrt{1-x}}{(1-2x)^4+16x^2(x-\ell_S^2)(1-x)} = 2 \ s_\parallel^{(GR)}$$

$$S_{GR}^z(0) = 4 \int_{\ell_S^2}^1 dx \ \frac{x(x-\ell_S^2)\sqrt{1-x}}{(1-2x)^4+16x^2(x-\ell_S^2)(1-x)} = s_\perp^{(GR)}$$

$$S_R^x(0) = \frac{\pi}{2} \ \frac{(\ell_S^2-1)\sqrt{\ell_R^2-\ell_S^2}}{\left[\sqrt{\ell_R^2-\ell_S^2}-\sqrt{\ell_R^2-1}\right]\left[(3\ell_R^2-2)\sqrt{\ell_R^2-\ell_S^2}-\ell_R^2\sqrt{\ell_R^2-1}\right]} = 2s_\parallel^{(R)}$$

$$S_R^z(0) = \frac{\pi}{2} \ \frac{(\ell_R^2-\ell_S^2)\sqrt{\ell_R^2-1}}{\left[\sqrt{\ell_R^2-\ell_S^2}-\sqrt{\ell_R^2-1}\right]\left[(3\ell_R^2-2)\sqrt{\ell_R^2-\ell_S^2}-\ell_R^2\sqrt{\ell_R^2-1}\right]} = s_\perp^{(R)} \quad . \tag{2.165}$$

The numbers $s_\parallel(\sigma)$ and $s_\perp(\sigma)$ are listed in Table 2.7 for different values of ℓ_S together with the equivalent weights for longitudinal and transverse phonons in an isotropic bulk Debye model, namely,

$$s^{(L)} = \ell_S^3/3 \quad \text{and}$$

$$s^{(T)} = 2/3 \quad , \tag{2.166}$$

We note that in the simplified Debye model, for an infinite solid the average density of states is assumed to be the same for all polarizations and is given by

Table 2.7. Relevant data for bulk and surface phonons such as frequencies $\omega_D(\sigma)$, weights $s_\perp(\sigma)$ and $s_\parallel(\sigma)$, and surface Debye frequencies $\omega_{D\perp}$ and $\omega_{D\parallel}$. Note that $s_\perp^{(H)} = 0$ and $s_\parallel^{(H)} = 1$, independent of ℓ_s. (Goldys et al.1982)

ℓ_s	$\dfrac{\omega_D^{(T)}}{\omega_D}$	$\dfrac{\omega_D^{(L)}}{\omega_D}$	$s^{(T)}$	$s^{(L)}$	ℓ_R	$\dfrac{\omega_D^{(H,GR)}}{\omega_D}$	$\dfrac{\omega_D^{(\pm)}}{\omega_D}$	$\dfrac{\omega_D^{(R)}}{\omega_D}$
0.1	0.874	8.737	2/3	0.00033	1.0475	0.795	7.946	0.759
0.2	0.875	4.374	2/3	0.00267	1.0496	0.796	3.978	0.758
0.3	0.877	2.925	2/3	0.00900	1.0537	0.798	2.660	0.757
0.4	0.883	2.207	2/3	0.02133	1.0606	0.802	2.007	0.757
0.5	0.891	1.783	2/3	0.04167	1.0724	0.811	1.621	0.756
0.6	0.904	1.507	2/3	0.07200	1.0939	0.822	1.370	0.752
0.7	0.921	1.316	2/3	0.11430	1.1391	0.836	1.196	0.735

ℓ_s	$s_\perp^{(B)}$	$s_\perp^{(GR)}$	$s_\perp^{(R)}$	$s_\parallel^{(B)}$	$s_\parallel^{(GR)}$	$s_\parallel^{(R)}$	$\omega_{D\perp}/\omega_D$	$\omega_{D\parallel}/\omega_D$
0.1	0.0007	0.593	0.725	0.0051	0.205	0.108	0.797	0.797
0.2	0.0565	0.582	0.753	0.0213	0.197	0.116	0.793	0.794
0.3	0.0204	0.562	0.805	0.0517	0.183	0.132	0.787	0.791
0.4	0.0528	0.529	0.894	0.102	0.160	0.161	0.775	0.785
0.5	0.114	0.477	1.048	0.179	0.126	0.214	0.756	0.775
0.6	0.216	0.398	1.338	0.291	0.078	0.324	0.723	0.758
0.7	0.353	0.290	1.974	0.411	0.039	0.599	0.668	0.725

$$d(\omega) = \frac{1}{3}\left(d_\ell(\omega) + 2d_t(\omega)\right)\Theta(\omega_D - \omega) \ , \tag{2.167}$$

where $d_{\ell,t}(\omega)$ are given by (2.146) with the Θ-functions removed. One thus gets the standard expression

$$d(\omega) = \frac{3\rho_0}{M_s\omega_D^3}\,\omega^2\,\Theta(\omega_D - \omega) \ . \tag{2.168}$$

Similarly, we can introduce the average surface densities of states in the surface Debye model as

$$d_\parallel^{(s)}(\omega) = \frac{1}{2}[d_{SH}^y(\omega,0) + d_+^x(\omega,0) + d_-^x(\omega,0) + d_{GR}^x(\omega,0) + d_R^x(\omega,0)]\Theta(\omega_{D\parallel} - \omega)$$

$$d_\perp^{(s)}(\omega) = [d_+^z(\omega,0) + d_-^z(\omega,0) + d_{GR}^z(\omega,0) + d_R^z(\omega,0)]\Theta(\omega_{D\perp} - \omega) \ , \tag{2.169}$$

for vibrations parallel and perpendicular to the surface, respectively, the latter contributing twice to the total density of states. The functions $d_\sigma^i(\omega,0)$ are given by (2.164) without the Θ-functions. Using (2.164,165,169) we get

$$d_{\parallel}^{(s)}(\omega) = \frac{3\rho_0}{M_s \omega_{D\parallel}^3} \omega^2 \Theta(\omega_{D\parallel} - \omega)$$

$$d_{\perp}^{(s)}(\omega) = \frac{3\rho_0}{M_s \omega_{D\perp}^3} \omega^2 \Theta(\omega_{D\perp} - \omega) \quad , \quad \text{where} \tag{2.170}$$

$$\omega_{D\parallel} = \left[\frac{3}{2+\ell_s^3} \left[s_{\parallel}^{(SH)} + s_{\parallel}^{(B)} + s_{\parallel}^{(GR)} + s_{\parallel}^{(R)} \right] \right]^{-1/3} \omega_D$$

$$\omega_{D\perp} = \left[\frac{3}{2+\ell_s^3} \left[s_{\perp}^{(B)} + s_{\perp}^{(GR)} + s_{\perp}^{(R)} \right] \right]^{-1/3} \omega_D \tag{2.171}$$

are called the surface Debye frequencies for vibrations parallel and perpendicular to the surface, respectively. Let us note that the same procedure
applied for $z=-\infty$ gives $\omega_D = \omega_{D\parallel} = \omega_{D\perp}$ and the result (2.168) for the infinite
solid. In the intermediate region, i.e., for $-\infty < z < 0$, one can define $\omega_{D\perp}$ and $\omega_{D\parallel}$
depending on ω through the product ωz; this, however, does not seem to lead
to any simplification. The surface Debye frequencies are listed in Table
2.7.

So far we have looked at the vibrational degrees of freedom of an elastic solid treating the latter as a continuum. Solids are, of course, aggre-
gates of molecules or ions arranged in regular lattice structures. A proper
treatment of vibrations starts from a collection of molecules interacting
via mutual two-body interactions. To simplify the problem, one usually
assumes a particular lattice structure, and expands the total potential
energy for small vibrations around the equilibrium positions, up to second
order in the displacements. In the full quasi-harmonic approximation, the
force constants, i.e., the second derivatives of the lattice potential energy
with respect to the displacements, are evaluated at the mean positions of
particles, with thermal expansion taken into account, rather than at the
positions of static equilibrium. In the strict harmonic approximation, thermal expansion is neglected and the force constants are evaluated at the
static equilibrium positions. Even in the harmonic approximation, as defined
above, the static relaxation of particles near the surface is to be taken
into account.

Allen et al. (1971 a and b) formulated a variety of harmonic and quasi-
harmonic theories to study surface vibrations:

(1)   In a "partial harmonic theory", the force constants are determined
       for particles in the bulk, i.e., deep within the crystal, at their
       positions of static equilibrium. The force constants for particles
       near the surface are then taken to be the same as those for parti-
       cles in the bulk.

(2)   In a "full harmonic theory", the force constants are evaluated in-
      dependently for each pair of particles in the crystal, with the
      mean positions taken to be the positions of static equilibrium.
      The relaxation, static displacements of the surface particles, and
      the resulting changes in the force constants near the surface, are
      taken into account.

(3)   In a "partial quasi-harmonic theory", the force constants are eva-
      luated with the mean positions taken to be the positions corres-
      ponding to uniform thermal expansion throughout the crystal. The
      picture is as follows: one starts with a static crystal in which
      the interparticle spacings near the surface are different from
      those in the bulk owing to static relaxation. One then allows the
      particles to vibrate, and the crystal will consequently expand. In
      this approximation, one assumes that the expansion is uniform, so
      that the spacing between any two particles in the crystal, includ-
      ing those near the surface, increases in proportion to the spacing
      between any other two particles.

Calculations (Allen et al. 1971b) have been performed for crystal slabs
of 3 to 21 layers with atoms interacting via Lennard-Jones 12-6 potentials
(2.5). Typical results for dispersion curves are given in Fig. 2.35 for a
(111)-oriented fcc slab. As the number of slabs increases, the bulk modes
thicken, forming a certain number of bands. These bands become continuous
in the limit of an infinitely thick slab; in this limit the band edges and
the gaps appearing between different bands for each p are exactly those of
the distribution of the frequencies $\omega_\sigma(p)$ of the cyclic infinite lattice,
where σ = 1, 2,...,3s, and p =(P,p_z) are the branch index and the three-
dimensional wave vector, when σ and p_z are allowed to assume all the possible values and P is kept constant. However, a finite number of slab modes,
labeled S_1, S_2, etc., remain outside the bands. Since their eigenvectors are
found to be large near the surface and rapidly decreasing with increasing
distance from the surface, they are interpreted as surface modes. In particular, the acoustic surface mode S_1 is localized below the acoustic band
in the long-wave limit. This mode is just the surface Rayleigh wave discussed above. The other surface modes have no analogue in the continuum
theory, since they appear in the dispersive region of the Brillouin zone.

In the long-wave limit, the monatomic slab admits only acoustic surface
waves, while the diatomic slab may have surface modes of optical character,
whose frequencies do not vanish as $P \to 0$. In addition to the macroscopic
modes, microscopic optical surface modes exist, such as the Lucas modes.

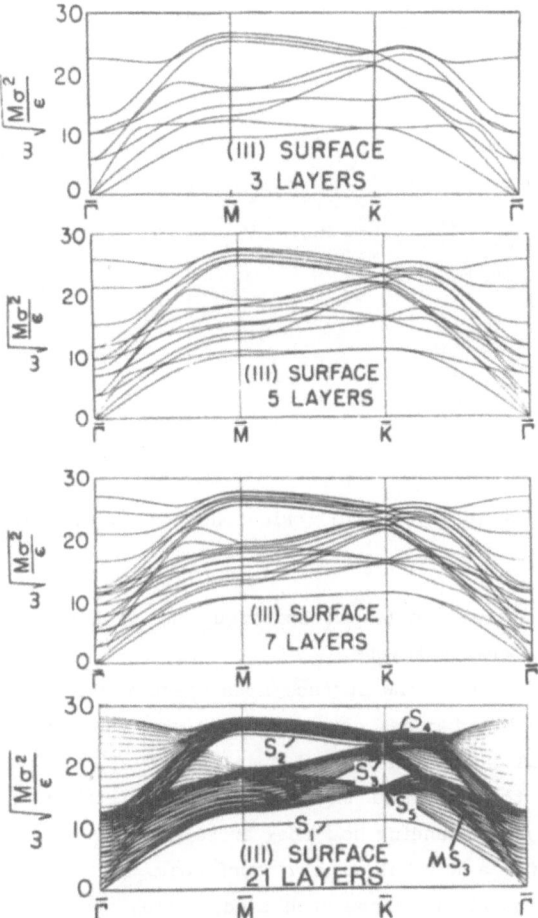

Fig.2.35. Dispersion curves of a monatomic fcc(111)-oriented slab along the boundaries of the irreducible part of the surface Brillouin zone, for different thicknesses. (Allen et al. 1971b)

Figure 2.36 shows the dispersion curves for a 15-layer slab of NaCℓ with (001) orientation, calculated from a 11-parameter shell model by Chen et al.(1977). The modes S_5 and S_4 are examples of Lucas modes with polarizations parallel and perpendicular to the surface, respectively. An interesting feature occurring for slabs with three-dimensional inversion symmetry (as in the case considered here) is that the surface modes occur in nearly degenerate pairs. The degeneracy occurs first in the short-wave region at the boundary of the two-dimensional Brillouin zone, but, as the number of layers N_L increases, it involves modes of larger and larger wavelength. This is due to the fact that each long-wave surface mode is deeply penetrating and involves the opposite surface. However, for $N_L \to \infty$, all surface modes are degenerate everywhere. These modes should be regarded as single surface modes of a semi-infinite lattice. Actually, quite precise information on the

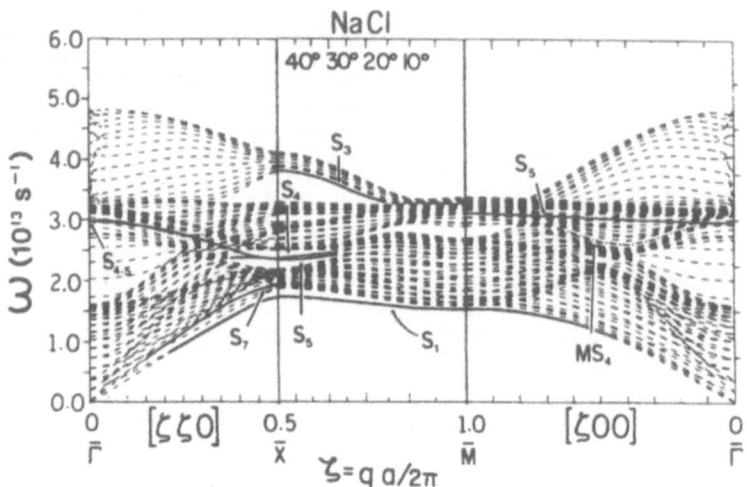

Fig.2.36. Dispersion curves of a 15-layer (001)-oriented NaCl slab. (Chen et al. 1977)

vibrations of the semi-infinite lattice can be extrapolated from slab calculations with reasonably large numbers of layers.

Despite the considerable complexity of the surface-mode spectra discussed
above, all the surface modes and mixed modes can be understood in the context of a simple phenomenological scheme developed by Allen et al. (1971).
This scheme does not allow one to determine the surface modes without a
calculation, but it does help in understanding how they arise.

We begin with the bulk bands for a crystal without surfaces. For a monatomic crystal, there should ordinarily be three such bands (which may overlap), corresponding roughly to two groups of transverse modes and one of
longitudinal modes. We then introduce the perturbation represented by the
surface, which actually consists of two parts - a "first-order" perturbation
due simply to the truncation of the crystal, and a "second-order" perturbation due to changes in the force constants near the surface. The second
part should not be important in monatomic crystals at large wavelengths,
where the surface modes penetrate deeply, but it is important at small
wavelengths.

The strength of the total perturbation depends on the point in the Brillouin zone, i.e., the value of the two-dimensional wave vector P. If the
perturbation is strong enough for a given value of P, it will peel one or
more surface-like modes off a given bulk band. Ordinarily, the perturbation
should correspond to a softening of the vibrations, since the truncation of
the crystal allows the surface atoms to vibrate more freely, and one

expects the surface atoms to relax outward, producing a decrease in the surface force constants. For such a softening perturbation, the surface modes should be peeled off the bottom of the bulk band. If for some reason, e.g., a change in the interaction between the particles near the surface, the total perturbation leads to a stiffening of the lattice vibrations, then the surface modes should be peeled off the top of the bulk band. Such high-frequency surface modes were produced in the calculations of Musser and Rieder (1970), when the surface force constants were stiffened, but it does not seem likely that they will occur naturally in monatomic crystals.

Ordinarily, the total perturbation due to the surface should first peel off, from a given bulk band, a mode primarily localized in the first layer. If strong enough, it should then peel off a mode primarily localized in the second layer, and so on. The n-th layer mode in this series has the same character in the n-th layer as the first-layer mode has in the first layer. Sometimes the pertubation will not succeed in completely peeling off the mode, in which case the mode remains within the bulk band as a mixed mode.

When a mode is peeled off, one of four things will happen:

(a) It may fall under all of the bulk bands, in which case it will necessarily be a surface mode.

(b) It may fall into a gap between two bulk bands, in which case it again will necessarily be a surface mode.

(c) Along a symmetry line associated with a reflection plane, it may fall into a region occupied only by bulk modes to which it is automatically orthogonal. In this case, once more, it will necessarily be a surface mode.

(d) It may fall into a region occupied by bulk modes to which it is not automatically orthogonal. In this case it will not be able to survive as a pure surface mode and will be a mixed mode instead.

Occasionally, two surface-mode branches will attempt to cross each other. In such a case there will be hybridization, with the hybrid branches exhibiting a mutual repulsion and interchange of character. The only case where two surface-mode branches could in fact cross, is along a symmetry line associated with a reflection plane, with the two modes belonging to mutually orthogonal classes and therefore being invisible to one another. Such a situation has been found in calculations for NaCℓ (Chen et al. 1971) and for an adsorbed layer (Alldredge et al. 1971). Similarly, a surface-mode branch which enters a bulk band will be repelled by the band and tend to bend away from it. The only exception is along a symmetry line when the band is invisible to the surface mode because its modes and the surface mode are automatically orthogonal.

The results of Figs. 2.35,36 can be interpreted as follows, according to the above scheme: For the (111) surface, S_2 is peeled off the "longitudinal band" of bulk modes along MK, S_4 is peeled off the uppermost band along TK, S_3 and its extension MS_3 is peeled off the upper "transverse band", and S_1 is peeled off the lower "transverse band". The perturbation is just strong enough to peel off a mode near the edge of the Brillouin zone, S_5, which is the second-layer analog of S_1.

Lattice dynamical calculations demonstrate clearly the emergence of new surface modes, in addition to the generalized Rayleigh waves of the continuum theory, which are characterized by the fact that they do not persist into the long-wavelength limit, and therefore cannot be obtained in the continuum approximation. The dependence of the amplitude of these modes upon the distance from the surface is generally rather complex; for details see Allen et al.(1971). Benedek (1982 and earlier references) has developed a Green's function method to calculate dispersion curves and densities of states for surface phonons of a semi-infinite solid, mainly for alkali-halide crystals. The results agree very well with large slab calculations. An example is given in Fig. 2.37 with a comparison with data obtained from inelastic atom scattering (Benedek 1982).

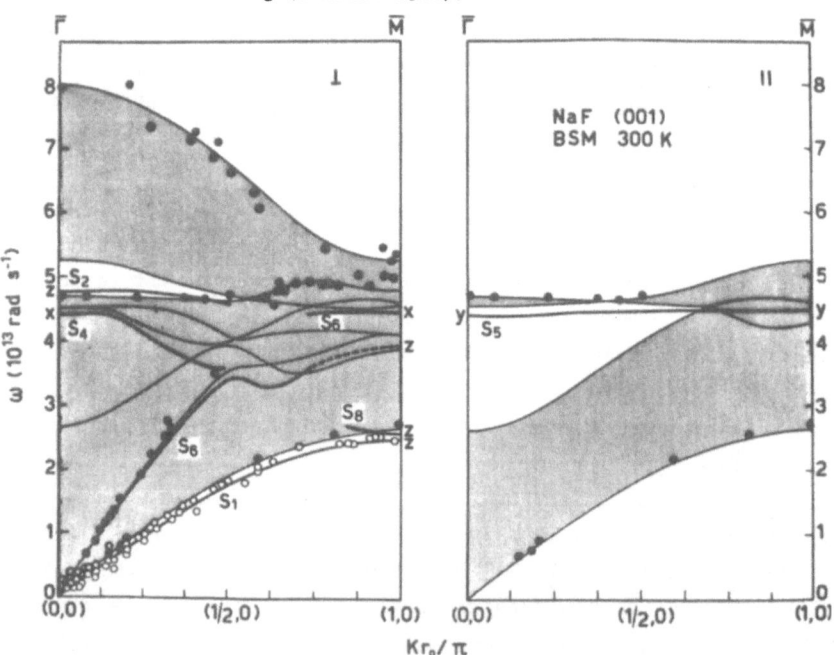

Fig.2.37. Surface phonon dispersion curves of NaF (001) along the TM direction. Curves calculated. Open circles from atomic beam scattering and dots from neutron scattering. (Benedek 1982)

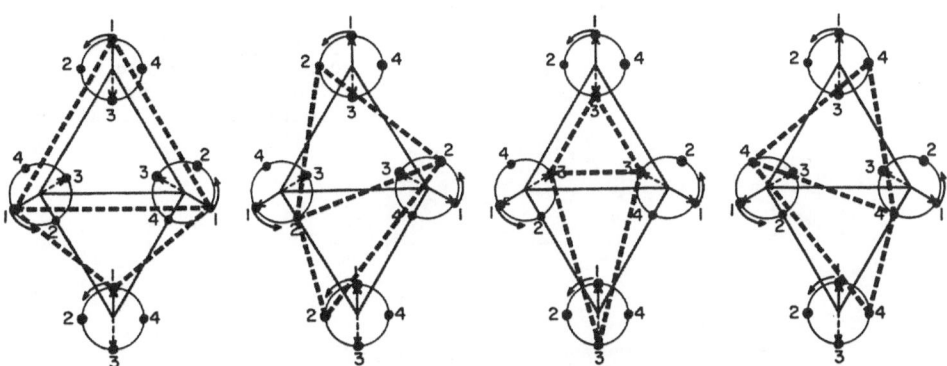

Fig.2.38. Counterclockwise rotation of atoms in a (111) plane for a wave moving to the right of K. The times 1,2,3, and 4 are 90° apart. The upward-pointing triangle is breathing in, while the downward-pointing triangle is showing a scissors motion. (Black, Shanes, and Wallis 1983).

Black, Shanes, and Wallis (1983) have examined the actual motion of atoms in the surface modes for the (111) surface of a fcc crystal. At the point K of the Brillouin zone, where the wavelength is comparable with the interatomic distance, the atoms in some planes move in circular paths about their equilibrium positions, while those in other planes move up and down perpendicular to the surface through their equilibrium positions. In Fig. 2.38 we show the counterclockwise rotation of atoms for a wave travelling to the right. It can be seen that the atoms forming the upward triangle are "breathing", that is alternately moving away from and toward the triangle centre. The atoms forming a downward-pointing triangle are exhibiting a "scissors" motion. In the case of a clockwise wave moving to the right, the triangles would be reversed. All upward triangles in a (111) plane exhibit the breathing for the counterclockwise wave and so on. The modes exhibit the following pattern with depth at K. For S_1 the surface atoms move perpendicularly to the surface, the atoms in the layer below rotate counterclockwise, the atoms in the third-layer rotate clockwise, and the pattern repeats. For S_3 the top-layer atoms rotate counterclockwise, the second-layer atoms rotate clockwise, and the third-layer atoms move perpendicularly to the surface; then the pattern repeats.

We next review briefly the vibrational properties of a surface covered by an adsorbate. Alldredge et al.(1971) have performed slab calculations assuming that the adsorbate forms a regular two-dimensional lattice. One finds additional surface modes of which three exist well into the small wave number region. In the light adatom case, they lie above the bulk bands, and below for the heavy adatom case, Figures 2.39,40.

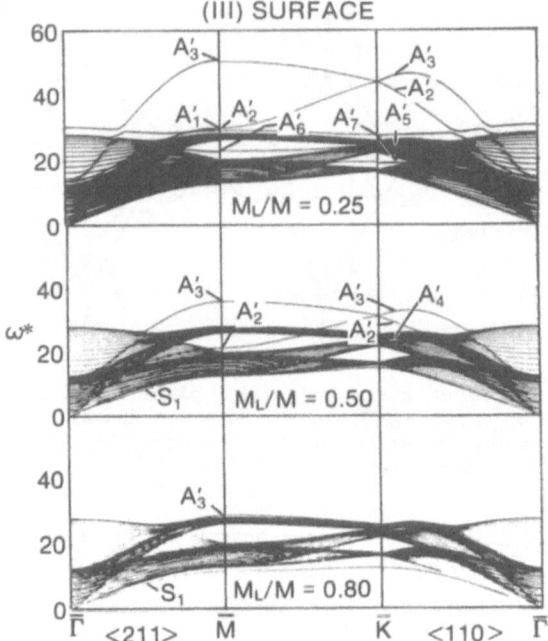

Fig.2.39. Surface dispersion curves of the fcc(111) slab with a light adsorbate of atomic mass M_L. (Alldredge et al. 1971)

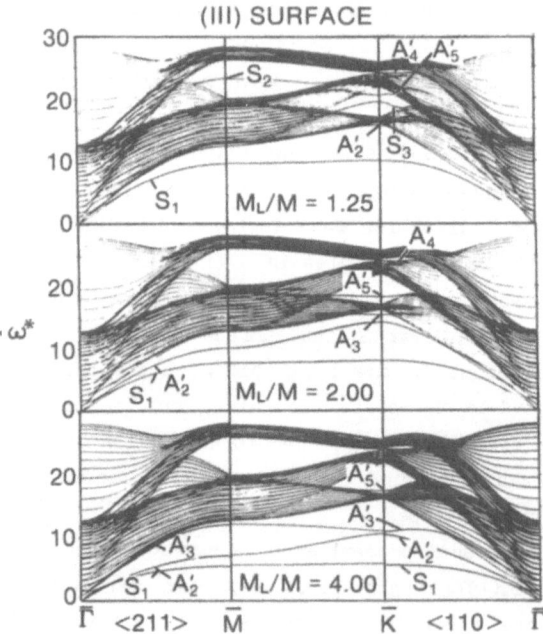

Fig.2.40. Surface dispersion curves for a fcc(111) slab with a heavy adsorbate. (Alldredge et al. 1971)

There is a certain lack of symmetry between the effects of light adatoms and the effects of heavy adatoms as a manifestation of dynamic decoupling (or lack thereof). For heavier adatoms, the three principal adsorbed-layer surface-wave branches S_1, A_2', and A_3', associated in a sense with the three degrees of freedom of the adsorbed monolayer, exist over most of the two-dimensional Brillouin zone. At the larger P values, the structure of this triplet of surface branches and the magnitude of the frequencies are somewhat similar to that of a single layer with atomic mass of adsorbed atoms M_L. Here the adatoms, moving in their surface modes with a long period due to their heavy mass, tend to drag along the first layer or so of the lighter substrate atoms, but with increasingly less response from the deeper substrate layers, since the period is much longer than that for the lowest natural period of the substrate. Thus, the localization of S_1, A_2', and A_3' increases with increasing M_L. When P is nearer to the center of the Brillouin zone, the first two branches make a smooth transition to the behaviour appropriate to the lowest surface, and pseudosurface branches S_1 and S_4 appropriate to the pure surface. In this sense, the principal adsorbed-layer surface branches in the heavy adatom case can be considered as acoustical.

In the light adatom extreme, this dynamic decoupling yields principal adsorbed-layer surface branches which can be considered as optical; here, the adatoms moving in their surface modes with a short period due to their light mass, tend to see the substrate of heavier atoms as an approximately static array of interaction centers, such that the principal adsorbed-layer surface branches A_1, A_2, and A_3 cannot tend to zero frequency as P → 0. For the lightest adatoms reported by Alldredge et al. (1971), one branch A_3 exists over the whole Brillouin zone; a second A_2 exists over most of the Brillouin zone as a pure surface branch and may be observed near $\overline{\Gamma}$ to continue into the bulk bands as a locus of mixed modes, which is called a pseudosurface branch; and the third A_1 also exists in a region around $\overline{\Gamma M}$ as a pure surface branch, which continues into the bulk band as a pseudosurface branch nearer $\overline{\Gamma}$. The P → 0 limit of all three surface and pseudosurface branches can be seen to have nonzero frequency , hence they are optical.

The penetration of adsorbed-layer modes into the substrate bulk bands even for M_L/M = 0.25 means that the dynamic coupling between light layer and substrate is still quite appreciable at long wavelengths. At much greater mass mismatches between adatoms and substrate atoms, one can expect the three surface branches associated with the monolayer to be completely separated from the bulk bands of the substrate. In such extreme cases, the adiabatic decoupling between the motions of the adsorbed layer and the substrate is more complete, and the three adsorbed-layer branches are approximately what would be expected of the atoms of the adsorbed layer moving in a potential field consisting of a contribution from the substrate atoms in a static configuration, in addition to the mutual interaction of the adatoms. Thus at P = 0, the frequencies of the adsorbed-layer branches will be something like that of a single adatom interacting with a stationary substrate, with motion normal to the surface having a frequency somewhat greater than that for motion parallel to the surface. As P increases, dispersion appears. The parts of the branches associated principally with vertical motion have relatively small dispersion. The parts of the branches associated principally with horizontal motion transverse to P exhibit little dispersion when the direction of motion of an adatom is between its nearest neighbors in the substrate [for P along <100> for the fcc(100) surface and along <100> for the bcc(100) surface], and exhibit quite large dispersion when the adatom motion is toward its nearest neighbors in the substrate.

The surface branches having peak amplitude beneath the outer layer, are observed to be relatively insensitive to the presence of heavy adatoms; in the opposite case of light adatoms, these subsurface branches are quickly extinguished, with a few notable exceptions. On the other hand, the surface branches having peak amplitude in the first layer are always shifted signi-ficantly by the presence of a monolayer, except in the long-wavelength limit for the acoustical branches. In particular, the Rayleigh wave S_1 is extin-guished for sufficiently light adatoms, except in a small region near P = 0. Alldredge et al. (1971) have performed slab calculations for graphite slabs with (1000) surfaces and also for xenon overlayers with $(30^\circ \sqrt{3}x\sqrt{3})$ symme-try.

2.5 The Gas-Solid Hamiltonian

To formulate the statistical mechanics of adsorption and desorption, we will need the Hamiltonian of the coupled gas-solid system which we write as

$$H = H_g + H_s + H_{g-s} \quad , \tag{2.172}$$

where H_g is the Hamiltonian of the gas particles including their kinetic energy and two-body interaction, H_s describes the vibrational degrees of freedom of the solid in the harmonic approximation, and H_{g-s} contains the static and dynamic atom-solid interaction. Because we are dealing with a many-body system, it is advantageous to use the formalism of second quanti-zation. We then get

$$H_S = \sum_J \hbar\omega_J \, b_J{}^\dagger \, b_J \quad , \tag{2.173}$$

where the b_J and $b_J{}^\dagger$ are the expansion coefficients in the displacement vector (2.78, 125). They are now interpreted as annihilation and creation operators of phonons of mode J with energy $\hbar\omega_J$. Here $J = (P,c,\sigma)$ or $J = (p,\sigma)$ for a semi-infinite or infinite solid, respectively (2.79-82).

To write down H_g, we must expand the field operators of gas particles in a complete set of single-particle wave functions. They are usually, in an infinite system, chosen as plane waves. For our surface problem, it is advantageous to choose the set of eigenfunctions of the static one-particle problem which are solutions of the Schrödinger equation

$$\left[-\frac{\hbar^2}{2m} \nabla^2 + V_S(r) - E_\iota \right] \phi_\iota(r) = 0 \quad , \tag{2.174}$$

where $V_S(r)$ is the static surface potential and ι a set of quantum labels for its bound states and continuum states. Expanding the gas particle field operators,

$$\psi^\dagger(r,t) = \sum_\iota \phi_\iota(r) \, \alpha_\iota{}^\dagger(t) \quad , \tag{2.175}$$

we introduce creation operators $\alpha_\iota{}^\dagger$ that generate a gas particle in state ι; likewise, α_ι is an annihilation operator. We can then write (2.172) as

$$H = H_g{}^{(0)} + H_S + H_{g-s}{}^{(1)} \quad , \qquad \text{where} \tag{2.176}$$

$$H_g{}^{(0)} = \sum_\iota E_\iota \, \alpha_\iota{}^\dagger \, \alpha_\iota + \sum_{\iota_1 \iota_2 \iota_3 \iota_4} V_{\iota_1 \iota_2 \iota_3 \iota_4}{}^{(g-g)} \, \alpha_{\iota_1}{}^\dagger \, \alpha_{\iota_2}{}^\dagger \, \alpha_{\iota_3} \, \alpha_{\iota_4} \quad , \tag{2.177}$$

with

$$V_{\iota_1 \iota_2 \iota_3 \iota_4}{}^{(g-g)} = \int \phi_{\iota_1}{}^*(r_1) \, \phi_{\iota_2}{}^*(r_2) \, V^{(g-g)}(r_1,r_2) \, \phi_{\iota_3}(r_1) \, \phi_{\iota_4}(r_2) \, d^3r_1 \, d^3r_2 \quad , \tag{2.178}$$

being the two-body interaction between gas particles. It is usually negligible in the gas phase proper and also in the adsorbate at very low coverage, but becomes dominant as a monolayer builds up. The last term in (2.176) is the dynamic atom-solid interaction that mediates the energy transfer between the adsorbate and the phonon system of the solid. From (2.68, 78) it can be written as

$$H_{g-s}{}^{(1)} = \sum_{\iota,\iota',J} X_{\iota\iota'}(J) \, \alpha_\iota{}^\dagger \, (b_J{}^\dagger + b_J) \, \alpha_{\iota'} \quad . \tag{2.179}$$

It provides the mechanism for gas particles to make transitions from state ι' to state ι within the static surface potential, aided by emission or absorption of phonons of mode J. The matrix element is given by

$$X_{\iota\iota'}(J) = -\sqrt{\frac{\hbar}{2\rho_0\omega_J}} \langle\iota| \; u^{(J)}(R,z=0)\cdot\nabla V_S(r) \;|\iota'\rangle \quad , \tag{2.180}$$

where $u^{(J)}$ are the normal modes of vibration which, for the bulk elastic model, are given by (2.79), and for the surface elastic model, by (2.82). If the surface potential depends on z only, (2.180) simplifies to

$$X_{\iota\iota'}(J) = -\sqrt{\frac{\hbar}{2\rho_0 A\omega_J}} \; \pi_z^{(J)} \; \delta_{K,K'+P} \; \langle\iota_z \;\Big|\; \frac{dV_S(z)}{dz} \;\Big|\iota'_z\rangle \quad , \tag{2.181}$$

where

$$\iota = (K,\iota_z) \quad \text{and} \quad p = (P,p_z) \quad . \tag{2.182}$$

In (2.181) A is the surface area of the solid, while $\pi^{(J)}$ is related to the polarization of the phonons and is given for bulk phonons by, see (2.79),

$$\pi^{(J)} = L_s^{-1/2} \; e_{p\sigma} \quad , \tag{2.183}$$

where L_s is the depth of the solid. For surface phonons we have, from (2.82),

$$\pi^{(J)} = g(P,c,\sigma; \; z = 0) \quad . \tag{2.184}$$

The particle annihilation operators satisfy Heisenberg's equation of motion

$$\frac{\partial}{\partial t} \; \alpha_\iota = -\frac{i}{\hbar} \; [\alpha_\iota \; , \; H] \quad , \tag{2.185}$$

which , with (2.176,177,179), can be written as

$$\frac{\partial}{\partial t} \; \alpha_\iota(t) = -\frac{i}{\hbar} \; E_\iota \; \alpha_\iota(t) \; -\frac{i}{\hbar} \; \sum_{\iota',J} \; X_{\iota,\iota'}(J) \; [b_J^\dagger(t) + b_J(t)]\alpha_{\iota'}(t) \quad . \tag{2.186}$$

A similar equation holds for the phonon operators b_J.

3. The Master Equation

3.1 The Mesoscopic Approach to the Master Equation

Physisorption kinetics must be based on kinetic equations that reflect the microscopic dynamics imposed by the Hamiltonian (2.176) and that take advantage of the statistical properties of the coupled gas-solid system. Theories have been worked out that start from the master equation, the Fokker-Planck equation, the Kramers equation, or the Langevin equation. In this and the next four chapters we will concentrate on the master equation approach. In Chap. 8 we will then derive the simpler kinetic equations from the master equation and discuss the work based on them.

There are two quite distinct approaches to the master equation. In a microscopic approach, to be reviewed in Sect. 3.3, one starts from the Liouville-von Neumann equation for the density matrix incorporating the Hamiltonian (2.176) for the coupled gas-solid system. Using, e.g., Zwanzig's projection operator method, one derives a generalized master equation which, for a Markov process, can be reduced to a master equation.

In a mesoscopic approach, which we outline presently, one first identifies a set of stochastic variables for the system and, relying on the general theory of stochastic processes, derives a master equation, again invoking, of course, the Markov nature of the adsorption-desorption process. We start with this approach and recall (see, e.g., van Kampen, 1981) that a multidimensional set of stochastic variables $\mathbf{X}=(X_1, \ldots, X_n)$ is defined by specifying

(i) a set of possible values $x_1, \ldots, x_n$ (called "range" or "set of states" or "phase space"), and

(ii) a probability distribution given by a nonnegative function

$$P(x_1,\ldots,x_n) \geq 0 \qquad\qquad (3.1)$$

that can be normalized

$$\int P(x_1,\ldots,x_n)dx_1\ldots dx_n = 1 \quad . \qquad\qquad (3.2)$$

Here $P(x_1,...,x_n)dx_1...dx_n$ is the probability that X_i assumes values between x_i and $(x_i + dx_i)$. For a given physical system one, of course, only specifies stochastic variables for those degrees of freedom that take part in the specific phenomenon under investigation. For example, for a molecular solid one could specify the momentum and polarization of phonons together with a Bose-Einstein distribution function as stochastic variables, rather than the positions and momenta of individual lattice constituents as long, that is, as local properties such as diffusion and defects are not important. For the gas phase consisting of N_g molecules, one could likewise specify momenta and positions, together with their equilibrium probability distributions, as stochastic variables. The theory of probability, then, provides the framework to transform a given set of stochastic variables into another one. In particular, if such a map depends on an additional variable, such as time, we speak of a stochastic process.

For the most part, we will consider the simple situation where the interaction between gas molecules, either in the gas phase or in the adsorbate, can be neglected so that it is sufficient to deal with a single gas molecule at a time. Its energy E depends on time t and, via equations of motion, on the initial conditions of the solid and the gas phase. If we can treat the coupled gas-solid system with classical mechanics, then these initial conditions are the positions r_i^0 and momenta p_i^0 at time $t = t_0$ of all particles involved so that

$$E(t) = f(r_1^0, r_2^0, ... ; p_1^0, p_2^0, ...; t) \qquad (3.3)$$

is a stochastic process, provided we specify the initial probability distribution $P(r_1^0, r_2^0, ...; p_1^0, p_2^0 ...)$. If quantum mechanics is used to describe the microscopic dynamics of the gas-solid system, then initial positions and momenta are replaced by a suitable set of quantum numbers.

We next define the probability density for the stochastic process E(t) to take the value E at time t,

$$P_1(E,t) = \int \delta(E-f(r,p,t)) P(r,p) \, dr dp \quad . \qquad (3.4)$$

Here we denote by r the set $(r_1^0, r_2^0, ...)$ and by p the set $(p_1^0, p_2^0, ...)$ or a suitable set of initial quantum numbers. Similarly, joint probabilities

$$P_n(E_1,t_1;E_2,t_2;...;E_n,t_n) = \int \delta(E_1-f(r,p,t_1)) \, \delta(E_2-f(r,p,t_2)) ...$$

$$... \, \delta(E_n-f(r,p,t_n))P(r,p)dr dp \qquad (3.5)$$

give the probability that $E(t)$ takes on values E_1 at t_1, E_2 at t_2, etc. Obviously we have

(i) $\quad P_n \geq 0$,

(ii) $\quad \displaystyle\int P_n(E_1,t_1;E_2,t_2;\ldots;E_n,t_n)dE_n = P_{n-1}(E_1,t_1;\ldots;E_{n-1},t_{n-1})$,

(iii) $\quad \displaystyle\int P_1(E,t)dE = 1$,

(iv) $\quad P_n(E_1,t_1;\ldots;E_i,t_i\ldots;E_j,t_j;\ldots;E_n,t_n)$

$\qquad = P_n(E_1,t_1,\ldots,E_j,t_j;\ldots;E_i,t_i;\ldots;E_n,t_n) \qquad\qquad (3.6)$
$\qquad$ for any pair (i,j).

Averages or correlations are then given by

$$\langle E(t_1)\, E(t_2)\, \ldots\, E(t_n)\rangle = \int E_1 E_2 \ldots E_n\, P_n(E_1,t_1;\ldots;E_n,t_n)dE_1\ldots dE_n \quad . \qquad (3.7)$$

Next, the conditional probability

$$P_{\ell|k}(E_{k+1},t_{k+1};\; \ldots\; ; E_{k+\ell},t_{k+\ell}\, | E_1,t_1;\ldots;E_k,t_k)$$

$$= \frac{P_{k+\ell}(E_1,t_1;\ldots;E_k,t_k;E_{k+1},t_{k+1};\ldots;E_{k+\ell},t_{k+\ell})}{P_k(E_1,t_1;\ldots;E_k,t_k)} \qquad\qquad (3.8)$$

is the probability density that $E(t)$ takes the values E_{k+1} at t_{k+1} to $E_{k+\ell}$ at $t_{k+\ell}$, given that its value at t_1 was E_1, at t_2 it was E_2, etc., to E_k being the given (fixed) value at t_k.

If the conditional probability density at time t_n is uniquely determined, given the value E_{n-1} at t_{n-1}, and thus does not necessitate any knowledge of its values at earlier times, i.e., if

$$P_{1|n-1}(E_n,t_n\, |E_1,t_1;\ldots;E_{n-1},t_{n-1}) = P_{1|1}(E_n,t_n\,|\,E_{n-1},t_{n-1}) \qquad (3.9)$$

we call the stochastic process $E(t)$ a Markov process. Then $P_{1|1}$ is called the transition probability. Together with the probability density $P_1(E_1,t_1)$, it uniquely determines all joint probabilities, e.g., for $t_1 < t_2 < t_3$ we have

$$P_3(E_1,t_1;E_2,t_2;E_3,t_3) = P_1(E_1,t_1)P_{1|1}(E_2,t_2|E_1,t_1)\, P_{1|1}(E_3,t_3|E_2,t_2) \quad . \qquad (3.10)$$

Integrating (3.10) over E_2 and dividing both sides by $P_1(E_1,t_1)$, one obtains the Chapman-Kolmogorov equation for $t_1 < t_2 < t_3$,

$$P_{1|1}(E_3,t_3|E_1,t_1) = \int dE_2 \; P_{1|1}(E_3,t_3|E_2,t_2) \; P_{1|1}(E_2,t_2|E_1,t_1) \; . \tag{3.11}$$

Together with the relation

$$P_1(E_2,t_2) = \int P_{1|1}(E_2,t_2|E_1,t_1) \; P_1(E_1,t_1)dE_1 \; , \tag{3.12}$$

it uniquely defines a Markov process. The Chapman-Kolmogorov equation (3.11) is rather difficult to deal with and to solve. It can, however, be cast into a linear integro-differential equation. To this end, we expand $P_{1|1}$ for small time increments (t_3-t_2),

$$P_{1|1}(E_3,t_3|E_2,t_2) = [1-a_0(E_2,t_2) \; (t_3-t_2)] \; \delta(E_3-E_2) + (t_3-t_2)W(E_3,E_2,t_2) \; , \tag{3.13}$$

such that

$$W(E_3,E_2,t_2) \geqq 0 \tag{3.14}$$

is the transition probability per unit time (and unit energy) to go from E_2 to E_3 at time t_2. The factor in square brackets in (3.13) is the probability that no transition takes place during (t_3-t_2); hence

$$a_0(E_2,t_2) = \int W(E_3,E_2,t_2)dE_3 \; . \tag{3.15}$$

We insert (3.13) into (3.11), divide by (t_3-t_2), and take the limit $t_2 \rightarrow t_3$. We get

$$\frac{\partial}{\partial t} P_{1|1}(E,t|E_0,t_0) = \int dE' \; [W(E,E',t) \; P_{1|1}(E',t|E_0,t_0)$$

$$- W(E',E,t) \; P_{1|1}(E,t|E_0,t_0)] \tag{3.16}$$

This is the master equation for a Markov process. It determines, for given transition probabilities $W(E,E',t)$, how the conditional probability evolves from a fixed value E_0 at time t_0. Expanding the second $P_{1|1}$ in (3.11) similarly to (3.13), we get a backward or adjoint equation

$$-\frac{\partial}{\partial t_0} P_{1|1}(E,t|E_0,t_0) = \int dE' \; [P_{1|1}(E,t;E',t_0) \; W(E',E_0,t_0)$$

$$- P_{1|1}(E,t;E_0,t_0) \; W(E',E_0,t_0)] \; , \tag{3.17}$$

which determines how the joint probability to have E at t, given that E_0 was fixed at t_0, evolves back toward the initial conditions.

It is sometimes more convenient to work with the probability density (3.12) itself. We therefore multiply (3.16) by an initial distribution $P_1(E_0, t_0)$ and integrate over E_0 to get

$$\frac{\partial}{\partial t} P_1(E,t) = \int dE' \; [W(E,E',t) P_1(E',t) - W(E',E,t) P_1(E,t)] \; . \tag{3.18}$$

This master equation for the probability density of a Markov process has an intuitively appealing interpretation: the probability that the particle has energy E at time t changes in time because there are $W(E,E',t) P_1(E',t)$ transitions per unit time from states E', occupied with probability $P_1(E',t)$, into the state E and, likewise, a loss due to transitions out of E into all other states according to the second term in (3.18). We will from now on use the notation $P_1(E,t) = n(E,t)$ to indicate that we are dealing with occupation probabilities.

Let us next consider a Markov process that is homogeneous in time. The joint probability $P_{1|1}(E,t;E_0,t_0)$ is then a function of the time difference $\tau = t-t_0$ and does not depend on t and t_0 separately, so that the transition probabilities $W(E,E',t) = W(E,E')$ in (3.13) are independent of time. Equation (3.18) then reads with the notation $n(E,t) = P_1(E,t)$,

$$\frac{\partial}{\partial t} n(E,t) = \int dE' \; [W(E,E')n(E',t) - W(E',E)n(E,t)] \; . \tag{3.19}$$

If the $W(E,E')$ are such that (3.19) has a time independent solution $n^{eq}(E)$, we call the underlying Markov process stationary. This is, in particular, guaranteed in isolated physical systems because they eventually evolve towards, and reach, equilibrium. Indeed, the stationary solution is such that, not only is the right-hand side of (3.19) zero, but each term in the square brackets vanishes individually, so that

$$W(E,E')n^{eq}(E') = W(E',E)n^{eq}(E) \; . \tag{3.20}$$

This is called the principle of detailed balance; for discussions, see van Kampen (1981) or Kreuzer (1981).

We indicated at the beginning of this chapter that we take E to be the energy of a gas molecule interacting with a solid. As long as we can treat it classically, its energy ranges from the depth of the surface potential $(-V_0)$ to infinity. A particle with $-V_0 < E < 0$ is part of the adsorbate whereas, with $E > 0$, it belongs to the gas phase. If we have to treat the dynamics of the molecule quantum mechanically, then the negative energy states in the surface potential are actually discrete for completely localized adsorption, or have at least a discrete contribution for mobile adsorption. Indeed, if we consider a gas-solid system enclosed in a finite volume, all energy levels are discrete, so we would define

$$n(E) = \sum_i n_i \, \delta(E-E_i) \quad ,$$

$$W(E,E_i) = \sum_j W_{ji} \, \delta(E-E_j) \quad . \tag{3.21}$$

The index i or j above stands in general for a set of quantum numbers necessary to specify a given state. Equation (3.19) then takes on its discrete form

$$\frac{\partial n_i}{\partial t} = \sum_j [W_{ij} \, n_j - W_{ji} \, n_i] \quad . \tag{3.22}$$

Defining a column vector n(t) such that its transpose is

$$n^T(t) = (n_1, n_2, \ldots) \tag{3.23}$$

and a matrix $\hat{W}$ with elements

$$\hat{W}_{ij} = W_{ij} - \sum_k W_{ki} \, \delta_{ij} \quad , \tag{3.24}$$

we can write for (3.22)

$$\frac{dn}{dt} = \hat{W} \cdot n \quad . \tag{3.25}$$

Its solution is formally given by

$$n(t) = e^{t\hat{W}} \cdot n(0) \tag{3.26}$$

and can be constructed explicitly by diagonalizing $\hat{W}$. Note, however, that $\hat{W}$ is not symmetric. Each eigenvalue $(-\lambda_\kappa)$ is therefore connected with a right and a left eigenvector,

$$\hat{W} \cdot e^{(\kappa)} = -\lambda_\kappa \, e^{(\kappa)} \quad , \tag{3.27}$$

$$\boldsymbol{\varepsilon}^{(\kappa)} \cdot \hat{W} = -\lambda_\kappa \, \boldsymbol{\varepsilon}^{(\kappa)} \quad , \tag{3.28}$$

which satisfy

$$\sum_i \boldsymbol{\varepsilon}_i^{(\kappa)} \, e_i^{(\kappa')} = \delta_{\kappa'\kappa} \quad . \tag{3.29}$$

Note that the $e^{(\kappa)}$ and $\boldsymbol{\varepsilon}^{(\kappa)}$ are not orthogonal sets. We have introduced minus signs in (3.27,28) because we will show in a moment that all λ_κ are nonnegative, if the matrices W and $\hat{W}$ satisfy detailed balance.

94

The solution (3.26) then reads

$$n(t) = \sum_\kappa f_\kappa \, e^{-\lambda_\kappa t} \, e^{(\kappa)} \quad , \tag{3.30}$$

where the coefficients f_κ are most conveniently determined by inverting the imposed initial conditions,

$$n(t=0) = \sum_\kappa f_\kappa \, e^{(j)} \quad . \tag{3.31}$$

If the left eigenvectors $\bar{e}^{(\kappa)}$ are also known we have

$$f_\kappa = \sum_\ell n_\ell(t=0) \, \bar{e}_\ell^{(\kappa)} \quad . \tag{3.32}$$

If the W_{ij} and $\hat{W}_{ij}$ satisfy detailed balance

$$W_{ij} \, n_j^{eq} = W_{ji} \, n_i^{eq} \quad , \tag{3.33}$$

then the left eigenvectors are simply given by

$$\bar{e}_i^{(\kappa)} = \frac{e_i^{(\kappa)}}{n_i^{eq}} \tag{3.34}$$

as one can easily check by inserting (3.34) into (3.28). Similarly, noting that the sum of $\hat{W}_{ij}$ over i vanishes, one shows that n^{eq} is nothing but the right eigenvector for $\lambda_0 = 0$, i.e.,

$$n_i^{eq} = e_i^{(0)} \tag{3.35}$$

so that according to (3.34),

$$\bar{e}_i^{(0)} = 1 \tag{3.36}$$

for all i.

With detailed balance satisfied, one can immediately symmetrize W_{ij} and consequently $\hat{W}_{ij}$ by a real transformation

$$S_{ij} = W_{ij} \sqrt{\frac{n_j^{eq}}{n_i^{eq}}} = S_{ji} \quad . \tag{3.37}$$

Thus the eigenvalues of $\hat{W}$ are real and its eigenvectors can be chosen real as well.

To show that all λ_κ are nonnegative, we multiply (3.27) from the left by $\bar{e}^{(\kappa)}$ and, using (3.29,34,35), get

$$-\lambda_\kappa = \sum_{i,j} \frac{e_i^{(\kappa)}}{e_i^{(0)}} \left[W_{ij} - \delta_{ij} \sum_\ell W_{\ell i} \right] e_j^{(\kappa)} \quad . \tag{3.38}$$

Calling the factor before the bracket X_i, and using detailed balance on the first term in the brackets, we get

$$-\lambda_\kappa = \sum_{i,j} e_i^{(0)} W_{ji} [X_i X_j - X_i^2]$$

$$= - \frac{1}{2} \sum_{i,j} e_i^{(0)} W_{ji} (X_i - X_j)^2 < 0 \quad , \tag{3.39}$$

where, to get the second line, one renames indices and applies detailed balance once more. For systems satisfying detailed balance, it is obvious that one should use the symmetrized master equation with (3.35,37) to take advantage of the simplicity of symmetric matrices.

For completeness of the discussion, we note that the adjoint or backward master equation (3.17) can be obtained by transposing (3.25)

$$\frac{d\tilde{n}}{dt} = \tilde{n} \cdot \tilde{W}^T \quad , \quad \text{i.e.,} \tag{3.40}$$

$$\frac{d\tilde{n}_i}{dt} = \sum_j [\tilde{n}_j - \tilde{n}_i] W_{ji} \quad . \tag{3.41}$$

With the help of detailed balance, one shows that the solution of (3.41) can be expressed in terms of the solution of (3.22) as

$$\tilde{n}_i(t) = \frac{n_i(t)}{n_i^{eq}} \quad . \tag{3.42}$$

The backward equation is sometimes advantageous to work with in first passage time problems (van Kampen 1981, Chap. VI).

3.2 The Master Equation for Physisorption Kinetics

In this section, we first present a more specific justification why adsorption-desorption in physisorbed gas-solid systems can indeed be modeled as a Markov process. We then introduce the appropriate conditions to calculate the relevant quantities to describe adsorption and desorption. Lastly, we survey the various theories that are based on the master equation.

We recall that desorption in physisorbed gas-solid systems is found experimentally to be a first-order process that is phenomenologically described by a rate equation (1.2) where the desorption time is given by (1.3). This implies that the time evolution in the adsorbate during the desorption process is essentially exponential, which strongly suggests an underlying homogeneous Markov process. Let us concentrate on a single gas molecule interacting with a solid. As it approaches the surface, it experiences an attractive (van der Waals) potential. Its long-range character minimizes the influence of the thermal motion of individual constituents in the solid, so that it is essentially static. Eventually, the gas molecule will be reflected at the repulsive wall of the surface potential. Because the latter is very steep, the particle will spend little time there. Pagni and Keck (1973) have argued that the time of collision τ_C is short compared to the time it takes the particle to traverse the attractive surface well, which one may equate with the classical oscillation period $\tau_0(E)$. The collision itself will, of course, involve the gas particle and one or at most two atoms in the surface of the solid. Depending on the (thermal) motion of these surface atoms, the gas particle might gain or lose energy during τ_C. Because $\tau_C \ll \tau_0$, we can thus assume that the gas particle is in a more or less well-defined energy state (for at least a time τ_0), occasionally undergoing transitions to other energy states. This, then, is the justification to treat the energy of the gas particle as a stochastic process. Next we note that the lattice constituents of the solid undergo random thermal motion, which will typically equilibrate the struck surface atom in a time τ_R that is again short compared with $\tau_0(E)$. The quantity τ_R is essentially the thermal relaxation time of the solid or the phonon-phonon collision time. Thus we observe that a gas particle returning to the repulsive part of the surface potential will again find a solid in equilibrium with no memory of the previous collision. This suggests that we are dealing with a Markov process, implying that adsorption and desorption can be treated within the master equation approach.

If the dynamics of the adsorbing gas particle has to be treated quantum mechanically, then the above (classical) arguments by Pagni and Keck (1973) translate into statements about level widths and lifetimes. Starting from a static situation, the gas particle will be found in any one of the (at least partially) discrete energy levels E_i of the surface potential. Due to the dynamic coupling to the solid, it can undergo transitions into other states with rates W_{ij} that result in a broadening of the energy levels by $\Gamma_{ij} = \hbar W_{ij}$. To talk about well-defined energy levels, one needs $\Gamma_{ij} \ll (E_i - E_j)$ or $1/W_{ij} \gg \hbar/(E_i-E_j)$. To keep the solid in equilibrium, one furthermore needs $1/W_{ij} \gg \tau_R$. These conditions can, and must, be checked a posteriori once the transition rates W_{ij} are determined, which we will do in the next chapter.

Let us now define the relevant quantities necessary to describe adsorption and desorption and specify the corresponding initial conditions to be imposed on the master equation. Let us, for simplicity, describe the dynamics of the gas-solid system by quantum mechanics. In this case we must label the energy states of a gas molecule by three quantum numbers, assuming that none of its internal degrees of freedom is effected by the adsorption-desorption process. Enclosing the gas phase in a finite volume of size L^3, we can label a continuum state by a discrete wave vector

$$\mathbf{k} = (k_x, k_y, k_z) \frac{\pi}{L} \qquad k_x, k_y, k_z = 0,1,2,\ldots \qquad (3.43)$$

such that

$$E_{\mathbf{k}} = (k_x^2 + k_y^2 + k_z^2) \frac{\hbar^2 \pi^2}{2mL^2} \ . \qquad (3.44)$$

In the limit $L \to \infty$, $E_{\mathbf{k}}$ becomes continuous. Likewise, the bound states are labelled by i. For localized adsorption, $i = (i_x, i_y, i_z)$ has three discrete indices, whereas on a completely flat surface, for which the surface potential V_s depends on z only, one has

$$E_i = E_i + \frac{\hbar^2 \pi^2}{2mL^2} (i_x^2 + i_y^2) \qquad i_x, i_y = 0,1,2,\ldots \qquad (3.45)$$

where E_i, $i=0,\ldots,N_{max}$ is an eigenvalue of one of the $(N_{max}+1)$ bound states in $V_s(z)$, and the remainder is the quasi-continuous translational energy of a mobile particle along the surface. Labeling from now on continuum states by $\mathbf{k}$ and bound states by i, we can rewrite the master equation (3.22) as

$$\frac{dn_i}{dt} = \sum_{i'} [W(i,i') n_{i'}(t) - W(i',i) n_i(t)]$$

$$+ \sum_{\mathbf{k}} W(i,\mathbf{k}) n_{\mathbf{k}}(t) - \sum_{\mathbf{k}} W(\mathbf{k},i) n_i(t) \ . \qquad (3.46)$$

Here, the two terms in square brackets account for bound state – bound state transitions. The next term deals with continuum-bound state transitions responsible for adsorption, and the last term entails bound state-continuum transitions leading to desorption. Equation (3.46) must be supplemented by equations for the continuum occupation functions

$$\frac{dn_{\mathbf{k}}}{dt} = \sum_{\mathbf{k}'} [W(\mathbf{k},\mathbf{k}') n_{\mathbf{k}'}(t) - W(\mathbf{k}',\mathbf{k}) n_{\mathbf{k}}(t)]$$

$$+ \sum_i W(\mathbf{k},i) n_i(t) - \sum_i W(i,\mathbf{k}) n_{\mathbf{k}}(t) \qquad (3.47)$$

where the terms in square brackets correspond to inelastic continuum-continuum transitions, the last two terms accounting for those particles coming from or going into the adsorbate, respectively.

We can define the fractional coverage θ by

$$\theta = \frac{N_g}{N_s A} \sum_i n_i \quad , \tag{3.48}$$

where N_g is the total number of gas particles and N_s is the maximum number of particles per unit area that can be adsorbed into a monolayer; A is the area of the exposed surface, i.e., $A = L^2$. For localized adsorption, the sum over i in (3.48) enumerates all bound states of all available adsorption sites; if different adsorption sites are mutually independent, one can alternatively replace $N_s A$ by 1 and run the sum over the bound states of a single adsorption site. Recall that n_k and n_i are probabilities so that

$$\sum_i n_i(t) + \sum_k n_k(t) = 1 \quad . \tag{3.49}$$

Summing (3.46) over all bound states we thus get

$$\frac{d\theta}{dt} = \frac{N_g}{N_s A} \sum_{i,k} [W(i,k) \, n_k(t) - W(k,i) \, n_i(t)] \quad . \tag{3.50}$$

Obviously bound state-bound state transitions do not alter the coverage. To reduce (3.50) further to the phenomenological equation (1.2), approximations must be invoked; they will be discussed below.

To calculate the isothermal desorption rate from (3.46), we recall that in an isothermal desorption experiment, starting from equilibrium at temperature T, one rapidly pumps away the gas phase, keeping the system's temperature constant. This implies for (3.46)

(i) the initial conditions $n_i(t<0) = n_i^{eq}$ \hfill (3.51)

and

(ii) $n_k(t>0) = 0$. \hfill (3.52)

The bound state occupations $n_i(t)$ will thus evolve for $t > 0$, due to bound state – bound state transitions, and due to a net loss out of the surface potential according to the last term in (3.46). For clarity, we rewrite (3.46) with this modification for isothermal desorption

$$\frac{dn_i}{dt} = \sum_{i'} [W(i,i') \, n_{i'} - W(i',i) \, n_i] - \sum_k W(k,i) \, n_i$$

$$= \sum_{i'} W^{iso}(i,i') \, n_{i'} \quad . \tag{3.53}$$

The transition probabilities of this truncated equation no longer satisfy detailed balance [because the $W(i,k)$ terms are missing], so that (3.53) has only the trivial solution $n_i(t \to \infty) = 0$ as a stationary solution. Note, however, that within a subspace spanned by the bound states, the matrix $W^{iso}(i,i')$ still satisfies detailed balance, but, unlike $\tilde{W}$, the sum over the first index i, does not vanish. If we therefore solve it by matrix diagonalization, we find that the lowest eigenvalue of W^{iso} is nonzero but for $W(k,i) \ll W(i',i)$ it is much smaller than the remaining eigenvalues. Solution (3.30) is still valid. This suggests to split it according to

$$ n_i(t) = f_0 \, e_i^{(0)} \, e^{-\lambda_0 t} + \sum_{\kappa > 0} f_\kappa \, e^{-\lambda_\kappa t} \, e_i^{(\kappa)} \quad . \tag{3.54} $$

If $\lambda_\kappa \gg \lambda_0$, then the terms with $\kappa > 0$ evolve much faster and represent initial transients, whereas the first term accounts for the actual desorption process. We can therefore identify

$$ R_d = t_d^{-1} = \lambda_0 \tag{3.55} $$

as the desorption rate constant. This will be shown to be valid in numerous numerical results.

The fact that during the desorption process there are two vastly different time scales, namely the fast one, λ_κ^{-1} for transients, and the slow one, λ_0^{-1} for desorption, suggests that the bound state – bound state transitions in (3.53) maintain more or less an equilibrium distribution that decreases monotonically. With an ansatz

$$ n_i(t) = \frac{N_s A}{N_g} \frac{n_i^{eq}}{\sum\limits_{i'} n_{i'}^{eq}} \, \theta(t) \tag{3.56} $$

suggested by (3.48), one then gets from (3.53), after summing over i to cancel the bound state – bound state transitions,

$$ \frac{d\theta}{dt} = - R_d^{PT} \, \theta \quad , \qquad \text{where} \tag{3.57} $$

$$ R_d^{PT} = \frac{\sum\limits_{i,k} W(k,i) \, n_i^{eq}}{\sum\limits_{i'} n_{i'}^{eq}} \tag{3.58} $$

is the quasi-equilibrium approximation for the desorption rate constant which results from applying perturbation theory to the master equation. This quantity was first calculated by Strachan (1935) and Lennard-Jones and Devonshire (1936a,b), without explicitly using a master equation; they called it the evaporation rate. We will see in Chap. 5 that (3.58) is a reasonable estimate, typically faster than the true rate constant by less than an order of magnitude.

At low coverage, adsorbed particles obey Maxwell-Boltzmann statistics. Due to the normalization (3.49), the probabilities are given in equilibrium by

$$n_\iota^{eq} = \frac{1}{N_g} \frac{1}{e^{\beta(E_\iota - \mu)} \pm 1} \tag{3.59}$$

with the plus (minus) sign holding for Fermi-Dirac (Bose-Einstein) statistics. According to the convention introduced below (2.150) ι denotes a set of quantum labels for both the adsorbate (bound states) and the gas phase (continuum states). To replace (3.59) by

$$n_\iota^{eq} = \frac{1}{N_g} e^{-\beta(E_\iota - \mu)} \quad , \tag{3.60}$$

we must ensure that

$$e^{\beta(E_\iota - \mu)} \gg 1 \quad , \tag{3.61}$$

where the chemical potential μ is given by the ideal gas expression for a dilute gas phase, i.e.,

$$e^{\beta\mu} = P \frac{(2\pi\hbar)^3}{k_B T (2\pi m k_B T)^{3/2}} \quad , \tag{3.62}$$

so that

$$P \ll e^{\beta E_\iota} (\frac{m}{2\pi\hbar^2})^{3/2} (k_B T)^{5/2} \quad . \tag{3.63}$$

Take $E_\iota = -V_0$ as a lower limit and a light particle, namely He. Then for $V_0/k_B \approx 140$ K, the well depth for Helium on graphite, we find that $P \ll 5.5$ Pa for a temperature $T \approx 10$ K, a limit well above standard high vacuum experimental conditions.

Let us then return to (3.58), use detailed balance (3.33) and (3.60) to get

$$R_d^{PT} = \frac{\sum_{i,k} W(i,k) e^{-\beta E_k}}{\sum_i e^{-\beta E_i}} \quad . \tag{3.64}$$

We can approximate the sum in the denominator by its lowest contribution, e.g., for localized adsorption by $\exp(\beta V_0)$ so that we get (the mobile case will be considered in detail in Chap. 5)

$$R_d^{PT} \approx \nu\, e^{-\beta V_0} \quad , \tag{3.65}$$

where the preexponential factor

$$\nu = \sum_{i,k} W(i,k) e^{-\beta E_k} \tag{3.66}$$

is typically weakly temperature dependent because the continuum states reached in desorption have energy of the order of k_BT. Recall that $W(\mathbf{i},\mathbf{k})$ mediates transitions from the continuum into bound states with the excess energy transferred to the solid. Such processes, dominated by spontaneous emission of phonons, are usually temperature independent, at least at low temperature. We have thus arrived, via some plausible approximations, at the Frenkel–Arrhenius parametrization of the rate constant of thermally activated desorption. This "derivation" clearly shows that the prefactor ν contains all the information about the dynamics of the desorption process. Identifying it with the attempt frequency of the particle to escape the surface potential well, is clearly not warranted.

Let us return to (3.50) and study the adsorption term. In the phenomenological approach (1.2) it would be identified as

$$\frac{N_g}{N_sA} \sum_{\mathbf{i},\mathbf{k}} W(\mathbf{i},\mathbf{k})\, n_{\mathbf{k}}(t) = S\, \frac{P}{N_s(2\pi m k_B T)^{1/2}}\, (1-\theta) \quad, \tag{3.67}$$

so that one gets the initial sticking coefficient at $\theta = 0$ as

$$S(\theta=0) = \frac{(2\pi m k_B T)^{1/2}}{P}\, \frac{N_g}{A} \sum_{\mathbf{i},\mathbf{k}} W(\mathbf{i},\mathbf{k})\, n_{\mathbf{k}}(t) \quad. \tag{3.68}$$

Now take for $n_{\mathbf{k}}(t)$ its equilibrium value $n_{\mathbf{k}}^{eq}$, assuming that adsorption at lower coverage does not deplete a sufficiently large gas phase. Next applying detailed balance and (3.60,62) we get

$$S(\theta=0) = \frac{h}{k_BT}\, \frac{h^2}{2\pi m k_B T}\, A^{-1} \sum_{\mathbf{i},\mathbf{k}} W\,(\mathbf{k},\mathbf{i})\, e^{-\beta E_{\mathbf{i}}}$$

$$= \frac{h}{k_BT}\, \frac{h^2}{2\pi m k_B T}\, R_d^{PT}\, A^{-1} \sum_{\mathbf{i}} e^{-\beta E_{\mathbf{i}}} \quad, \tag{3.69}$$

where the last line follows by using (3.64). These expressions were obtained by Lennard-Jones and Devonshire (1936a) who called S the condensation coefficient. Let us briefly look at a mobile adsorbate in a surface potential that is a function only of the distance z from the surface. In this case, $E_{\mathbf{i}} = E_i + \hbar^2 K_i^2/2m$ where K_i is the wave vector of an adparticle moving along the surface. Performing the sum over K_i in (3.69) yields for mobile adsorption

$$S(\theta=0) \approx \frac{h}{k_BT}\, R_d^{PT} \sum_{i} e^{-\beta E_i}$$

$$\approx \frac{h}{k_BT}\, R_d^{PT}\, e^{-\beta V_0} \quad. \tag{3.70}$$

Using (3.65) we thus have

$$S(\theta=0) = \frac{h}{k_B T} \nu \quad . \tag{3.71}$$

This shows that ν can only be identified with the thermal attempt frequency $k_B T/h$ if $S=1$.

3.3 The Microscopic Approach to the Master Equation

In Sect. 3.1, we have seen that the master equation is appropriate for the description of Markov processes. Because adsorption and desorption are kinetic phenomena in physical systems, it is instructive to derive it from the underlying microscopic dynamics. This can be done within the framework set forth, e.g., by van Hove or Prigogine and his collaborators [see, e.g., Kreuzer (1981) Chap. 10 for details] using in particular Zwanzig's projection operator techniques (Efrima et al. 1983). In such a general derivation, one unfortunately never uses any system-specific assumptions or arguments so that little more insight is gained that is not already apparent in Pauli's original heuristic approach to the master equation.

Let us assume that a (single) gas particle interacting with a solid via the time-dependent Hamiltonian (2.176) is in a state $\psi(\mathbf{r},t)$, whose time evolution is controlled by the Schrödinger equation

$$H \psi(\mathbf{r},t) = i\hbar \frac{\partial \psi(\mathbf{r},t)}{\partial t} \quad . \tag{3.72}$$

Rather than follow this wave-function approach, we prefer to develop here the equivalent operator formalism. Thus we treat $\psi(\mathbf{r},t)$ as field operators for gas particles for which we have an expansion in annihilation operators (2.175), which in turn are subject to Heisenberg's equation of motion (2.186). It can be formally integrated to yield

$$\alpha_l(t) = \alpha_l(t_0) \, e^{-iE_l(t-t_0)/\hbar} - \frac{i}{\hbar} \int_{t_0}^{t} dt' e^{-iE_l(t-t')/\hbar}$$

$$* \sum_{l',J} X_{ll'}(J) [b_J^\dagger(t') + b_J(t')] \alpha_{l'}(t') \quad , \tag{3.73}$$

where the matrix elements $X_{ll'}$ are given in (2.180) or (2.181). If we now assume that the phonon degrees of freedom of the solid are not affected by the adsorption-desorption process, then the time evolution of their creation and annihilation operators is determined by H_s as

$$b_J^\dagger(t) = b_J^\dagger(t_0) \, e^{i\omega_J(t-t_0)} \quad . \tag{3.74}$$

If we furthermore assume that $H_{g-s}^{(1)}$ is a small perturbation, we can iterate (3.73). Lowest-order perturbation theory thus gives

$$\alpha_\iota(t) \approx \alpha_\iota(t_0)\, e^{-iE_\iota(t-t_0)/\hbar} - \frac{i}{\hbar}\, e^{-iE_\iota(t-t_0)/\hbar} \int_{t_0}^{t} dt'$$

$$* \sum_{\iota' J} X_{\iota\iota'}(J)[b_J{}^\dagger(t_0)\, e^{-i(E_{\iota'}-E_\iota-\hbar\omega_J)(t'-t_0)/\hbar}$$

$$+ b_J(t_0)\, e^{-i(E_{\iota'}-E_\iota+\hbar\omega_J)(t'-t_0)/\hbar}]\, \alpha_{\iota'}(t_0) \quad . \tag{3.75}$$

We recall that

$$n_\iota(t) = <\alpha_\iota{}^\dagger(t)\,\alpha_\iota(t)> = \text{Tr}[\alpha_\iota{}^\dagger(t)\alpha_\iota(t)\hat{\rho}_0] \tag{3.76}$$

are the time-dependent occupation numbers of states ι, with $\hat{\rho}_0$ being the statistical operator at time t_0. In most applications it will correspond to the initial equilibrium state of the system before external conditions were changed to induce adsorption or desorption. We therefore take

$$\hat{\rho}_0 = \frac{e^{-\beta H_0}}{\text{Tr}[e^{-\beta H_0}]} \quad . \tag{3.77}$$

with $H_0 = H_g^{(0)} + H_S$ given by (2.176,177).

Let us first look at the expectation values in (3.76), with respect to the phonon degrees of freedom, and note that for $b_J = b_J(t_0)$ we have

$$< b_J b_{J'} > = < b_J{}^\dagger b_{J'}{}^\dagger > = 0$$

$$< b_J{}^\dagger b_{J'} > = \frac{\delta_{JJ'}}{e^{\beta\hbar\omega_J} -1} = \delta_{JJ'}\, n^{(ph)}(\omega_J)$$

$$< b_J b_{J'}{}^\dagger > = \delta_{JJ'}\, [n^{(ph)}(\omega_J) + 1] \quad , \tag{3.78}$$

where $n^{(ph)}(\omega_J)$ is the thermal occupation function (2.126) for phonons. We also have

$$< \alpha_\iota{}^\dagger(t_0)\alpha_{\iota'}(t_0) > = \delta_{\iota\iota'}\, n_\iota(t=t_0) \quad . \tag{3.79}$$

The various terms in (3.76) oscillate as a function of time with frequencies

$$\hbar\Omega = E_\iota - E_{\iota'} \pm \hbar\omega_J \quad . \tag{3.80}$$

For times $(t-t_0) \gg \Omega^{-1}$, terms like $\Omega^{-1}\cos\Omega(t-t_0)$, average to zero, whereas the long-time limits

$$\lim_{t\to\infty} \frac{\sin^2\Omega(t-t_0)}{\Omega^2} = \pi(t-t_0)\delta(\Omega) \tag{3.81}$$

are nonzero. This gives for (3.76),

$$n_1(t) \approx n_1(t_0) + (t-t_0) \sum_{1'} [W(1, 1') n_{1'}(t_0) - W(1', 1) n_1(t_0)] \quad , \qquad (3.82)$$

where

$$W(1,1') = \frac{2\pi}{\hbar} \sum_J \frac{1}{\omega_J} |X_{11'}(J)|^2 \{n^{(ph)}(\omega_J) \delta(E_1 - E_{1'} - \hbar\omega_J)$$

$$+ [n^{(ph)}(\omega_J) +1] \delta(E_1 - E_{1'} + \hbar\omega_J)\} \qquad (3.83)$$

is the transition probability from 1' to 1, according to Fermi's golden rule. We postpone its discussion for now and continue with (3.82) by recalling that on the one hand $(t-t_0) \gg h/|E_1-E_{1'}|$, to validate the use of (3.81), and on the other hand $(t-t_0) \ll [W(1,1')]^{-1}$, to guarantee that n_1 has changed but little. This we can use to define a time derivative, coarse-grained on the time scale $h/|E_1 - E_{1'}|$,

$$\left(\frac{dn_1}{dt}\right)_{t=t_0} = \lim_{t \to t_0} \frac{n_1(t) - n_1(t_0)}{t-t_0} = \sum_{1'} [W(1,1') n_{1'}(t_0) - W(1',1) n_1(t_0)] \quad .$$

$$(3.84)$$

This equation so far controls only the initial time evolution from an equilibrium state specified by $\hat{\rho}_0$. In general, as time goes on correlations will emerge. This is most easily seen by formally deriving an equation for $n_1(t)$ from Heisenberg's equation of motion for the density operator $\hat{\rho}(t)$ for the coupled gas-solid system

$$\frac{\partial \hat{\rho}}{\partial t} = -\frac{i}{\hbar} [H, \hat{\rho}(t)] \quad . \qquad (3.85)$$

Assuming again that the phonon system remains unchanged in the course of adsorption and desorption, we split off its contribution to $\hat{\rho}$ by defining a pair of orthogonal projection operators

$$\hat{\rho}(t) = \hat{\rho}_S \hat{\rho}_p(t) + [\hat{\rho} - \hat{\rho}_S \hat{\rho}_p(t)]$$

$$= P\hat{\rho}(t) + Q\hat{\rho}(t) \quad , \quad \text{where} \qquad (3.86)$$

$$\hat{\rho}_S = \frac{e^{-\beta H_S}}{Tr_S[e^{-\beta H_S}]} \quad \text{and} \qquad (3.87)$$

$$\hat{\rho}_p(t) = Tr_S[\hat{\rho}(t)] \quad , \qquad (3.88)$$

with Tr_S denoting the trace over all phonon states. The single-particle probabilities are now given by

$$n_1(t) = \langle 1| \hat{\rho}_p(t) |1\rangle \quad , \qquad (3.89)$$

where $\langle r|\iota\rangle = \phi_\iota(r)$ are the eigenstates (2.174) of $H_g^{(0)}$. Inserting the full Hamiltonian (2.176) into (3.85), one gets formally (Efrima et al. 1983)

$$i\hbar\,\frac{\partial\hat{\rho}_p(t)}{\partial t} = [H_g^{(0)},\hat{\rho}_p(t)]$$

$$- i \int_0^t dt'\,Tr_s[\hat{L}\,\exp[-i(1-P)\hat{L}t']\,[H_{g-s}^{(1)},\hat{\rho}_s\hat{\rho}_p(t-t')]]] \qquad (3.90)$$

Here $\hat{L}\hat{O} = (1/\hbar)\,[H,\hat{O}]$ is Liouville's operator acting on the operator $\hat{O}$ and $[\ ,\ ]$ is a commutator. The first term gives the evolution of the gas in the absence of the coupling to the solid, i.e., with $H_{g-s}^{(1)} = 0$. The second term is due to the perturbation of the particle system by the thermal fluctuations in the phonon system of the solid.

This equation is usually called the generalized master equation (e.g., Kreuzer 1981). To make the connection with Pauli's master equation, we must take quantum-mechanical expectation values according to (3.89). We will get terms on the right-hand side involving off-diagonal matrix elements $\langle\iota|\rho_p(t)|\iota'\rangle$ of the density operator. To extend the validity of (3.84) beyond the initial time period, one must invoke a repeated random phase approximation, setting

$$\langle\,\iota|\,\rho_p(t)\,|\iota'\,\rangle = \delta_{\iota\iota'}\,n_\iota(t) \qquad (3.91)$$

throughout the time evolution of the system. This, together with the elimination of additional memory effects from the time integral in (3.90), enforces Markov character on the adsorption-desorption process. It implies that (3.84) is valid at any time t and not just initially. In addition to having made more explicit the underlying physical assumptions, we managed to get a microscopic expression for the transition probabilities $W(t,t')$ per unit time, given in (3.83) in terms of the dynamical coupling between gas and solid. The first term, proportional to the phonon occupation number $n^{(ph)}(\omega_J)$, accounts for transitions of a gas particle from a lower energy $E_{\iota'}$ to a higher one E_ι, the energy difference $(E_\iota - E_{\iota'})$ being supplied by a phonon. The second term takes particles from higher into lower energy states, accompanied by stimulated [proportional to the number of phonons $n^{(ph)}(\omega_J)$] and spontaneous emission of phonons. We recall that the matrix elements $X_{\iota\iota'}(J)$, given in (2.180,181), involve the gradient of the surface potential which is largest in the repulsive region. In any case, plots of $\phi_\iota^*(r)\,\nabla V_s\phi_{\iota'}(r)$ suggest that this quantity is largest in a narrow region between the minimum of $V_s(r)$ and the solid, corroborating the arguments put forward by Pagni and Keck (1973) in their classical analysis.

106

3.4 The Master Equation at Finite Coverage

We have so far considered the master equation for a gas-solid system at such low coverage that the interaction between particles in the gas phase, as well as in the adsorbate, could be neglected. This allowed us to consider the dynamics of a single gas particle with the solid. This approach obviously becomes invalid as the particle density in the adsorbate increases. In the phenomenological rate equation (1.2), finite coverage effects up to a monolayer are taken into account by the blocking factor $(1-\theta)$ in the adsorption term. We briefly review two microscopic approaches to take coverage effects into account, namely a simple model by Kreuzer and Summerside (1981) and the mean field theory of physisorption kinetics by Summerside et al. (1982) and Sommer and Kreuzer (1982a-d).

Let us first point out the difficulties into which a straightforward generalization of lower-coverage theories can run. Let us assume that the surface potential (i.e., the net interaction between the solid and a physisorbed gas particle) develops a number of bound states, say at energies $E_0, \dots,$ $E_{N_{max}}$. In equilibrium, these bound states are occupied with probabilities (3.59) in which the +(-) sign has to be taken if the gas particles obey Fermi-Dirac (Bose-Einstein) statistics, $\beta = 1/k_B T$ is the inverse temperature and the chemical potential μ is determined by the gas phase, whose pressure P and temperature T are such that it satisfies Maxwell-Boltzmann statistics (3.62). In such situations, μ is large and negative, implying that as long as all $E_1 > \mu$, one has $n_1 \ll 1$. If at fixed pressure we now lower the temperature, the n_1 increases, eventually reaching a point where $N_g \Sigma n_1 > 1$. This implies that for localized adsorption, more than one gas particle is trapped into one and the same adsorption site; an impossible situation. Of course, the hard core repulsion between gas particles will ensure single occupancy by forcing, after completion of a monolayer, additional particles to form a second layer. Any microscopic theory that wants to describe gas-solid systems with monolayer coverage and beyond, must therefore include the two-body interaction between gas particles from the onset. This can be done in the Hartree-Fock approximation as shown by Summerside et al. (1982) and Sommer and Kreuzer (1982a-d). Kreuzer and Summerside (1981) developed a simple model for physisorption up to monolayer coverage by quantifying in a microscopic theory some ideas that underlie phenomenological theories. One argues typically that physisorbed gas particles are trapped into a potential well of depth V_0, roughly equal to the heat of adsorption. To avoid the difficulties with many bound states, they consider a model gas-solid system in which the surface potential develops just one shallow bound state at energy $E_0 = -V_0$. To ensure that only one gas particle can adsorb per site, one uses Fermi-Dirac statistics. In equilibrium we thus have

$$0 < n_0 N_g = \frac{1}{\exp[\beta(E_0-\mu)] + 1} \leq 1 \quad . \tag{3.92}$$

Note that for temperatures and pressures such that $\mu > E_0$, the occupancy $n_0 N_g$ of the bound state becomes of order, but does not exceed, one. Because the gas particles trapped into the bound state of the surface potential constitute the adsorbate, one can identify the occupancy of the bound state with the coverage ($n_0 N_g = \theta$), i.e., the fraction of available adsorption sites occupied by gas particles. Indeed, if the gas phase is dilute enough so that the chemical potential is given by the ideal gas law (3.62), then (3.92) is nothing but the Langmuir isotherm

$$\theta = \frac{P}{P +[(2\pi m)^{3/2}(k_B T)^{5/2}/h^3] \exp(-Q/k_B T)} \quad . \tag{3.93}$$

The use of Fermi-Dirac statistics, while of no consequence as far as the gas phase is concerned, ensures that only one gas particle can adsorb per bound state. By restricting the model to gas-solid systems, where the (localized) surface potential develops only one bound state, one thus ensures that only one gas particle can adsorb per adsorption site. The model then becomes quite reminiscent of spin-type lattice gas models of the liquid state (e.g., Schick 1981).

To derive kinetic equations for this model, we start from a Hamiltonian (for notation see Sect. 3.2)

$$H = E_0 \alpha_0^\dagger \alpha_0 + \sum_k E_k \alpha_k^\dagger \alpha_k$$

$$+ \sum_J \hbar\omega_J b_J^\dagger b_J + \sum_{k,J} [X_{k0}(J)\alpha_k^\dagger b_J^\dagger \alpha_0 + h.c.] \quad . \quad . \tag{3.94}$$

Heisenberg's equation of motion reads in integral form, see (3.73),

$$\alpha_0(t) = e^{-iE_0 t/\hbar} \alpha_0(0) - \frac{i}{\hbar} \int_0^t dt' e^{-iE_0(t-t')/\hbar}$$

$$* \sum_{k,J} X_{k0}(J) (b_J^\dagger(t') + b_J(t')) \alpha_k(t') \quad \text{and} \tag{3.95}$$

$$b_J(t) = e^{i\omega_J t} b_J(0) - \frac{i}{\hbar} \int_0^t dt' e^{-i\omega_J(t-t')} \sum_k X_{k0}(J)\alpha_k^\dagger(t')\alpha_0(t') \quad . \tag{3.96}$$

In deriving the master equation, we iterated an equation like (3.95) to second order, keeping only the first term of (3.96). We now insert (3.96) into (3.95) and iterate the resulting equation to second order. In taking the trace

$$N_g n_0(t) = < \alpha_0^\dagger(t)\alpha_0(t)> = \text{Tr}[\alpha_0^\dagger(t)\alpha_0(t)\hat{\rho}_s\hat{\rho}_g] \qquad (3.97)$$

over the initial equilibrium we encounter quartic terms

$$<\alpha_{\mathbf{k}_1}^\dagger(0)\alpha_{\mathbf{k}_2}^\dagger(0)\alpha_{\mathbf{k}_3}(0)\alpha_{\mathbf{k}_4}(0)> = N_g^2 n_{\mathbf{k}_1}(0)n_{\mathbf{k}_2}(0)\ (\delta_{\mathbf{k}_1\mathbf{k}_4}\delta_{\mathbf{k}_2\mathbf{k}_3} -\delta_{\mathbf{k}_1\mathbf{k}_3}\delta_{\mathbf{k}_2\mathbf{k}_4})\ , \qquad (3.98)$$

where

$$N_g n_{\mathbf{k}}(0) = \frac{1}{\exp[\beta(E_{\mathbf{k}}-\mu)]+1}\ . \qquad (3.99)$$

It is important to note here that the extraction of relations between the adsorption and desorption rates, and hence the equilibrium properties of the physisorption system presented in the next section, hinges on the terms of the form (3.98). These result from the terms in (3.95) cubic in the particle operators, arising from the complete treatment of the time evolution of the phonon operators (3.96). Physically, this introduces a phonon-mediated two-body interaction whereby a gas particle approaching the surface of the solid feels the influence on the phonon bath of the other gas particles competing for the same adsorption site. Note that a random phase approximation on the cubic terms in (3.95) leads to the phonon-type dressing of adsorbing gas particles as studied by Knowles and Suhl (1977).

To arrive at a master equation, we again extract the long-time behaviour according to Fermi's golden rule, via the asymptotic formula (3.81), and get

$$n_0(t) = n_0(0) + \frac{2\pi t}{\hbar} \sum_{\mathbf{k},J} |X_{\mathbf{k}0}(J)|^2\ \delta(E_0 - E_{\mathbf{k}} + \hbar\omega_J)$$
$$*[n_0(n_J^{(ph)} + 1) - n_0 n_{\mathbf{k}} - n_0 n_J^{(ph)}]\ , \qquad (3.100)$$

or with (3.84)

$$N_g \frac{dn_0(t)}{dt} = \lim_{t\to 0} N_g[n_0(t) - n_0(0)]/t = r_a - r_d\ , \qquad (3.101)$$

where the rates of adsorption and desorption are, respectively,

$$r_a = (A + B)\ (1 - N_g n_0)\ , \qquad (3.102)$$

$$r_d = (C-A)n_0\ , \quad \text{with} \qquad (3.103)$$

$$A = \frac{2\pi}{\hbar}\ N_g \sum_{\mathbf{k},J} |X_{\mathbf{k}0}(J)|^2 \delta(E_0 - E_{\mathbf{k}} + \hbar\omega_J)n_{\mathbf{k}}\ n_J^{(ph)} \qquad (3.104)$$

$$B = \frac{2\pi}{\hbar} N_g \sum_{\mathbf{k},J} \left|X_{\mathbf{k}0}(J)\right|^2 \delta(E_0 - E_\mathbf{k} + \hbar\omega_J)n_\mathbf{k} \tag{3.105}$$

$$C = \frac{2\pi}{\hbar} \sum_{\mathbf{k},J} \left|X_{\mathbf{k}0}(J)\right|^2 \delta(E_0 - E_\mathbf{k} + \hbar\omega_J)n_J{}^{(ph)} \quad . \tag{3.106}$$

We can rewrite (3.101) in the form of the master equation

$$\frac{dn_0}{dt} = \sum_\mathbf{k} W(0,\mathbf{k}) \, (1 - n_0 N_g) \, n_\mathbf{k} - \sum_\mathbf{k} W(\mathbf{k},0) \, (1 - n_\mathbf{k} N_g) n_0 \quad . \tag{3.107}$$

This model demonstrates nicely that coverage effects will show up in the master equation through an explicit dependence of the transition rates on the occupation probabilities of the final states n_0 and $n_\mathbf{k}$. Because they change as a function of time during adsorption and desorption, the underlying Markov process is no longer homogeneous at nonzero coverage. We will return to this model in Chaps. 5 and 7 in the context of the discussion of desorption times and sticking coefficients.

To go beyond this simple model, one has to honestly include the two-body interactions between the adsorbing gas particles. Adsorption and desorption thus becomes a full-fledged many-body problem which one can only hope to solve in some approximation. Trying to retain a single particle picture, Summerside et al. (1982) and Sommer and Kreuzer (1982a–d) argue that one should construct an effective coverage-dependent surface potential

$$V_S(\mathbf{r},\theta) = V_S(\mathbf{r}) + V_{MF}(\mathbf{r},\theta) \quad , \tag{3.108}$$

where $V_S(\mathbf{r})$ is the interaction (2.4) of a single gas particle with the solid, referred to from now on as the bare surface potential. In (3.108) $V_{MF}(\mathbf{r},\theta)$ is the potential arising from the mean field experienced by a gas particle in the presence of all other gas particles already in the surface region at a given coverage θ. As we outlined in Sect. 2.2., it can be calculated as a Slater average (2.55), employing the self-consistent solutions of the temperature-dependent Hartree-Fock equations (2.50). To get a master equation, we again iterate equations (3.95,96), suitably modified for many bound states. One gets

$$\frac{dn_\iota}{dt} = \sum_{\iota'} [W(\iota,\iota') \, (1 \pm N_g n_\iota) \, n_{\iota'} - W(\iota',\iota) \, (1 \pm N_g n_{\iota'}) \, n_\iota \,] \quad , \tag{3.109}$$

where the labels ι and ι' enumerate all bound states and continuum states. The plus (minus) sign must be taken for gas particles obeying Bose-Einstein (Fermi-Dirac) statistics. The transition rates are again given by (3.83), but note that the wave functions $\psi_\iota(\mathbf{r})$, being self-consistent solutions of

(2.50), are implicit functions of all occupation probabilities $n_{i'}$. Thus, to study the adsorption and desorption kinetics, including coverage effects, one must solve (2.47) and (3.109) concurrently. Results will be discussed in Chaps. 4 and 5.

4. Transition Probabilities in the Master Equation

In this chapter, we consider in detail the transition probabilities $W(\iota,\iota')$ in the master equation (3.46) that have been calculated or postulated in various theories of physisorption kinetics. In the first section we evaluate W for one-phonon processes based on (3.83). We consider the bulk and surface Debye models for the phonons and show how the theory reduces to a one-dimensional one for a highly mobile adsorbate. Sect. 4.2 deals with multiphonon processes and makes the connection with the correlation function approach. Sect. 4.3 reviews the classical soft cube model and phenomenological approaches. The final section then calculates W at finite coverage within the framework of mean field theory.

4.1 One-phonon Processes at Low Coverage

To calculate the transition probabilities $W(\iota',\iota)$ from (3.83), we must specify the phonon model and the static surface potential $V_s(\mathbf{r})$. Let us first look at a highly mobile adsorbate for which $V_s(\mathbf{r}) = V_s(z)$ is a function of z only. We then have

$$W(\iota_z,\mathbf{K}';\iota_z,\mathbf{K}) = \frac{\pi}{\rho_o A} \sum_J \frac{1}{\omega_J} \left| \pi_z^{(J)} \right|^2 \delta_{\mathbf{K}',\mathbf{K}+\mathbf{P}} \left| \langle \iota_z' \left| \frac{dV_s(z)}{dz} \right| \iota_z \rangle \right|^2$$
$$* [n^{(ph)}(\omega_J) \, \delta(E_{\iota'}-E_\iota-\hbar\omega_J) + [n^{(ph)}(\omega_J) + 1] \, \delta(E_{\iota'} -E_\iota + \hbar\omega_J)]$$

(4.1)

where, see (2.182),

$$E_\iota = E_\mathbf{K} + E_{\iota_z} = \hbar^2 K^2/2m + E_{\iota_z} \ .$$

(4.2)

The Kronecker delta ensures that, on a flat surface, parallel momentum is conserved in a transition, i.e., that the momentum $(\hbar\mathbf{K})$ of a gas particle moving parallel to the surface changes by the parallel component of the phonon momentum $(\hbar\mathbf{P})$ to $\hbar\mathbf{K}'=\hbar(\mathbf{K}+\mathbf{P})$. Momentum perpendicular to the surface is, of course, supplied to the gas particle by the surface force $-dV_s(z)/dz$.

We now note that at a given energy $k_BT = \hbar c|P| = \hbar^2 k^2/2m$, phonon and particle momenta $\hbar p$ and $\hbar k$ are vastly different,

$$\left|\frac{p}{k}\right| = \frac{1}{c}\sqrt{\frac{k_B T}{2m}} \ll 1 \quad . \tag{4.3}$$

Here c is the sound velocity. The ratio (4.3) is only 10^{-2} for He at $T = 10$ K and of the same order for heavier particles at higher temperatures. This suggests that the phonon momentum plays no major role in the desorption process of physisorbed particles. An intriguing consequence of (4.3) can, however, show up in a flash desorption experiment in which one suddenly raises the temperature of the solid from an initial T_i to a final T_f. Whereas the motion of the adparticle normal to the surface adjusts within picoseconds, its motion along a flat surface can only equilibrate to T_f due to collisions with phonons. Because their momentum, and thus the momentum transfer, is so small, thermalization times become very long (Gortel and Kreuzer 1985), so that desorbing particles emerge with a lateral average energy $k_B T_i$ and a normal energy $\frac{1}{2} k_B T_f$. This produces strong forward peaking of the flux of desorbing particles. But even in such a nonequilibrium situation, one can neglect the parallel phonon momentum P in the Kronecker delta in (4.1), choosing instead proper initial conditions as we will show explicitly in Chap. 5. Needless to say, for the description of beam scattering off a solid, one must not only keep P but also include surface corrugation, however small it may be (e.g., Toennies 198a,b).

We now look at a model in which we neglect the parallel phonon wave vector P in the Kronecker delta in (4.1), so that we get

$$W(\iota_z', K'; \iota_z, K) = \delta_{K',K} \, W(\iota_z', \iota_z) \quad . \tag{4.4}$$

Inserted in the master equation (3.46), this implies that the time evolution does not effect the energy E_K of the gas particles parallel to the surface. With an ansatz (and the change of notation $\iota = \iota_z$)

$$n_\iota(t) = n_\iota(t) \, n(E_K, t = 0) \quad , \tag{4.5}$$

the relevant part of the master equation then reads

$$\frac{dn_\iota}{dt} = \sum_{\iota'} [W(\iota, \iota')n_{\iota'} - W(\iota', \iota)n_\iota] \quad , \quad \text{where} \tag{4.6}$$

$$W(\iota, \iota') = \frac{\pi}{\rho_0 A} \sum_J \frac{1}{\omega_J} \left|\pi_z(J)\right|^2 \, [n^{(ph)}(\omega_J) \, \delta(E_\iota - E_{\iota'} - \hbar\omega_J)$$
$$+ [n^{(ph)}(\omega_J) + 1] \, \delta(E_\iota - E_{\iota'} + \hbar\omega_J) \,] \, \left|\langle \iota| \frac{dV_S(z)}{dz} |\iota'\rangle\right|^2 \quad . \tag{4.7}$$

113

This defines the one-dimensional cascade model of physisorption kinetics. In the form (4.4) it, of course, still allows us to calculate such three-dimensional quantities as the flux of desorbing particles. To evaluate (4.7) further, we must specify the phonon model. For the bulk Debye model we replace the summation over J by a summation over the polarization vector σ and an integration over phonon momentum $\mathbf{p}$. We note that only the polarization vectors depend on the angle of the phonon wave vector $\mathbf{p}$. Choosing a coordinate system for the $\mathbf{p}$-integration such that its polar axis is along the z-direction, we find for the polarization vectors (2.183)

$$\mathbf{e}_{\mathbf{p},\sigma} = L_s^{1/2} \pi(J)$$

$$\mathbf{e}_{\mathbf{p},L} = \mathbf{p}/p = (\cos\phi_p \sin\theta_p \, , \, \sin\phi_p \sin\theta_p \, , \, \cos\theta_p \,)$$

$$\mathbf{e}_{\mathbf{p},T} = (- \cos\phi_p \cos\theta_p \, , \, - \sin\phi_p \sin\theta_p \, , \, \sin\theta_p \,)$$

$$\mathbf{e}_{\mathbf{p},T'} = (- \sin\phi_p \, , \, \cos\phi_p \, , \, 0) \quad . \tag{4.8}$$

This allows us to do the angular integrations so that the δ-functions can be used to perform the remaining integral over $|\mathbf{p}|$ or over $\omega_J = c_\sigma|\mathbf{p}|$. We get

$$W(\iota,\iota') = \frac{1}{2\pi\hbar^2\rho_0 c_t^3} \, | \, \langle\iota| \, \frac{dV_s(z)}{dz} \, |\iota'\rangle|^2 \, (E_\iota \leftrightarrow E_{\iota'})$$

$$* \; \frac{1}{\exp[\beta(E_\iota - E_{\iota'})]-1} \; \frac{3}{2+\ell_s^3} \sum_{\sigma=L,T} s^{(\sigma)} \, \theta(\hbar\omega_D^{(\sigma)} - |E_\iota - E_{\iota'}|) \quad , \tag{4.9}$$

where the weights

$$s^{(L)} = \tfrac{1}{3} \, \ell_s^3 = \tfrac{1}{3} \, (c_t/c_\ell)^3$$

$$s^{(T)} = \tfrac{2}{3} \tag{4.10}$$

were introduced in (2.166). For $(E_\iota-E_{\iota'})>0$, this describes a transition from a lower to a higher energy accompanied by the absorption of a thermal phonon. Note that for $(E_\iota-E_{\iota'})<0$ we can use

$$n^{(ph)}(-\omega) = -[n^{(ph)}(\omega) + 1] \quad , \tag{4.11}$$

so that (4.9) describes downward transitions aided by stimulated and spontaneous emission of a phonon.

Turning next to the surface Debye model, we replace the summation over J in (4.7) according to (2.143) and note that the ϕ_p integration is only effected by the third row of the rotation matrix in (2.127). Using the δ-functions in (4.7) to do the P-integration, one arrives again at (4.9) with $s^{(\sigma)}$ for $\sigma = +, -,$ GR, and R given by $s^{(\sigma)} = S_\sigma^z(0)$ in (2.165). Note that the shear horizontal mode (σ = SH) does not contribute because it has no polarization perpendicular to the surface.

114

To continue with the evaluation of (4.9), we must specify the surface potential $V_s(z)$. The choices have been discussed in Sect. 2.1. For any one of them, one can readily calculate the eigenstates on a computer. There are a few potentials for which the calculations can be performed analytically, such as a separable potential (Gortel, Kreuzer and Spaner 1980), a Hulthen potential (Gortel, Kreuzer and Teshima 1980a) and the Morse potential (2.37) first considered in this context by Lennard-Jones and collaborators (Lennard-Jones et al. 1935-37). For the latter the bound state energies are given in (2.38) as

$$E_n = -\frac{(\hbar\gamma)^2}{2m}(\sigma_0-n-\tfrac{1}{2})^2 \ ; \ n = 0,1,\ldots N_{max} \quad , \quad \text{where} \tag{4.12}$$

$$\sigma_0^2 = \frac{2mV_0}{\hbar^2\gamma^2} \tag{4.13}$$

with V_0 and γ^{-1} as the depth and range of the Morse potential. Here N_{max} is the integer part of $(\sigma_0+1/2)$. The corresponding wave functions are

$$\phi_n(z) = \sqrt{\gamma} \ f_n(\xi-\xi_0) \quad , \quad \text{with} \tag{4.14}$$

$$f_n(\xi) = (-1)^n \sqrt{\frac{2\sigma_0-2n-1}{n! \ \Gamma(2\sigma_0-n)}} \ \frac{1}{\sqrt{2\sigma_0 e^{-\xi}}} \ W_{\sigma_0,\sigma_0-n-1/2}(2\sigma_0 e^{-\xi})$$

$$= \sqrt{\frac{n!(2\sigma_0-2n-1)}{\Gamma(2\sigma_0-n)}} \ (2\sigma_0 e^{-\xi})^{\sigma_0-n-1/2} \ \exp(-\sigma_0 e^{-\xi}) L_n^{2\sigma_0-2n-1}(2\sigma_0 e^{-\xi}) \tag{4.15}$$

where $\xi=\gamma z$ and $\xi_0=\gamma z_0$ with z_0 as the position of the minimum of $V_s(z)$. $W_{\sigma_0,\sigma_0'}(x)$ is a Whittaker function which can be expressed in terms of Laguerre polymials $L_n^\alpha(x)$, see, e.g., Abramowitz and Stegun (1972). The continuum wave functions are given by

$$\phi_k(z) = \frac{1}{\sqrt{2L}} \ f(\eta,\xi-\xi_0) \quad , \quad \text{with} \tag{4.16}$$

$$f(\eta,\xi) = \left[\frac{2\pi\eta\sinh(2\pi\eta)}{\cos^2(\pi\sigma_0)+\sinh^2(\pi\eta)} \ \frac{2\sigma_0 e^{-\xi}}{\left|\Gamma(\sigma_0+i\eta+\tfrac{1}{2})\right|^2}\right]^{\frac{1}{2}} W_{\sigma_0,i\eta}(2\sigma_0 e^{-\xi}) \quad , \tag{4.17}$$

where $\eta = k/\gamma$. The Whittaker function can also be expressed as a combination of two confluent hypergeometric functions that vanishes for $\xi \to -\infty$. Note that $\phi_k(z)$ is real and normalized in a finite box $-L < z < L$. Normalization is somewhat tricky, because it cannot be done analytically. However, we note that

     (i) the wave function decays with a double exponential $\exp(-\sigma_0 e^{-(\xi-\xi_0)})$ for $(\xi-\xi_0) < 0$ so that this region barely contributes to the normalization,

115

and

(ii) the region where $\phi_k(z)$ differs from a standing wave is only of order $\gamma^{-1} \ll L$ so that only its asymptotic behaviour for $\gamma^{-1} \ll z \leq L$, namely

$$\phi_k(z) \approx A \frac{\pi}{\sin\pi(2i\eta+1)} \left[\frac{(2\sigma_0)^{i\eta}}{\Gamma(1+2i\eta)\,\Gamma(\frac{1}{2}-\sigma_0-i\eta)} \, e^{i\eta\xi} - c.c. \right] , \qquad (4.18)$$

need be considered to determine the normalization constant A. Normalizing the wave functions in a box $\frac{1}{2}L < z < L$ seems somewhat contradictory because the gas phase is obviously restricted to $z > 0$. However, we note that for reasonable potential parameters $\gamma^{-1} \approx 1$ Å and $z_0 > 2$ Å, $V_S(z=0) > 40V_0$ implying that the wave functions are miniscule for $z < 0$. Thus, in evaluating the matrix elements of dV_S/dz, we can extend the lower integration limit from 0 to $-\infty$ and also the upper one from L to $+\infty$ because dV_S/dz decays exponentially. After a change of the integration variable from z to $(z-z_0)$, we find that the matrix elements become independent of z_0. These approximations have been tested numerically as to their effects on desorption times (Gortel, Kreuzer, and Teshima 1980c), see Sect. 5.1.

The matrix elements between bound states are given by

$$\left\langle n \left| \frac{dV_S}{dz} \right| n' \right\rangle = -\frac{\hbar^2\gamma^3}{4m} (-1)^{n-n'} (2\sigma_0-n-n'-1) |n-n'|$$

$$* \sqrt{(2\sigma_0-2n-1)(2\sigma_0-2n'-1) \frac{n_>!\,\Gamma(2\sigma_0-n_>)}{n_<!\,\Gamma(2\sigma_0-n_<)}} , \qquad (4.19)$$

between a bound state and a continuum state by

$$\left\langle k \left| \frac{dV_S}{dz} \right| n \right\rangle = -\sqrt{\frac{\pi}{\gamma L}} \frac{\hbar^2\gamma^3}{4m} (2\sigma_0)^{-i\eta} (-1)^n \left| \Gamma(\sigma_0+\tfrac{1}{2}+i\eta) \right| \left[(\sigma_0-n-\tfrac{1}{2})^2 + \eta^2 \right]$$

$$* \sqrt{\frac{(2\sigma_0-2n-1)\eta\sinh(2\pi\eta)}{n!\,\Gamma(2\sigma_0-n)\,(\cos^2(\pi\sigma_0)+\sinh^2(\pi\eta))}} , \qquad (4.20)$$

and between continuum states by

$$\left\langle k \left| \frac{dV_S}{dz} \right| k' \right\rangle = -\frac{\pi}{\gamma L} \frac{\hbar^2\gamma^3}{4m} (2\sigma_0)^{i(\eta-\eta')} (\eta^2-\eta'^2) \frac{\sqrt{\eta\eta'\sinh(2\pi\eta)\,\sinh(2\pi\eta')}}{\sinh(\pi(\eta+\eta'))\,\sinh(\pi(\eta-\eta'))}$$

$$* \left[\left| \frac{\Gamma(\frac{1}{2}-\sigma_0+i\eta)}{\Gamma(\frac{1}{2}-\sigma_0+i\eta')} \right| + \left| \frac{\Gamma(\frac{1}{2}-\sigma_0+i\eta')}{\Gamma(\frac{1}{2}-\sigma_0+i\eta)} \right| \right] . \qquad (4.21)$$

For the transition probabilities per unit time, we get for bound state – bound state transitions from (4.9)

$$W(n,n') = \omega_D \frac{3\pi}{2r^4} \frac{m}{M_S} (2\sigma_0-2n-1)(2\sigma_0-2n'-1)(2\sigma_0-n-n'-1)(n-n')^3$$

$$* \frac{n_>!}{n_<!} \frac{\Gamma(2\sigma_0-n_>)}{\Gamma(2\sigma_0-n_<)} \left[\exp\left[\frac{\delta}{r}(n^\pm n')(2\sigma_0-n-n'-1)\right] - 1 \right]^{-1}$$

$$* \frac{3}{2+\ell_S^3} \sum_\sigma s(\sigma)\, \theta\left[\frac{\omega_D(\sigma)}{\omega_D} - \frac{1}{r}|n-n'|(2\sigma_0-n-n'-1)\right] \quad . \tag{4.22}$$

Here

$$r = \frac{2m\omega_D}{\hbar\gamma^2} \quad, \tag{4.23}$$

$$\delta = \frac{\hbar\omega_D}{k_B T} \quad . \tag{4.24}$$

For transitions from bound states to continuum states we get

$$W(k,n) = \frac{\pi}{\gamma L}\,\omega_D \frac{3\pi}{2r^4}\frac{m}{M_S}\frac{2\sigma_0-2n-1}{n!\,\Gamma(2\sigma_0-n)}\frac{n\sinh(2\pi n)}{\cos^2(\pi\sigma_0)+\sinh^2(\pi n)}$$

$$* \left| \Gamma(\tfrac{1}{2}+\sigma_0+in) \right|^2 \left[(\sigma_0-n-\tfrac{1}{2})^2+n^2\right]^3 \left[\exp\left[\frac{\delta}{r}(\sigma_0-n-\tfrac{1}{2})^2 + \frac{\delta n^2}{r}\right]-1\right]^{-1}$$

$$* \frac{3}{2+\ell_S^3} \sum_\sigma s(\sigma)\, \theta(\frac{\omega_D(\sigma)}{\omega_D} - \frac{1}{r}(\sigma_0-n-\tfrac{1}{2})^2-\frac{n^2}{r}) \quad . \tag{4.25}$$

The transition probability $W(n,k)$ from the continuum into the bound states can be obtained from (4.25), using detailed balance (3.33) which amounts to adding '1' (for spontaneous emission of a phonon) to the Bose-Einstein factor.

Lastly, for continuum-continuum transitions we get

$$W(k,k') = (\frac{\pi}{\gamma L})^2\,\omega_D\frac{3\pi}{2r^4}\frac{m}{M_S}\frac{n n'\sinh(2\pi n)\sinh(2\pi n')}{\sinh^2[\pi(n+n')]\sinh^2[\pi(n-n')]}(n^2-n'^2)^3$$

$$*\{\exp\left[\frac{\delta}{r}(n^2-n'^2)\right]-1\}^{-1} \left[\left|\frac{\Gamma(\tfrac{1}{2}-\sigma_0+in)}{\Gamma(\tfrac{1}{2}-\sigma_0+in')}\right| + \left|\frac{\Gamma(\tfrac{1}{2}-\sigma_0+in')}{\Gamma(\tfrac{1}{2}-\sigma_0+in)}\right| \right]^2$$

$$* \frac{3}{2+\ell_S^3} \sum_\sigma s(\sigma)\, \theta(\frac{\omega_D(\sigma)}{\omega_D} - \frac{|n^2-n'^2|}{r}) \quad . \tag{4.26}$$

We note that these transition probabilities were first obtained by Lennard-Jones and collaborators (1935-37) for a bulk Debye model with a single cut-off frequency $\omega_D(\sigma) = \omega_D$, so that

$$\frac{3}{2+\ell_S^3} \sum_\sigma s(\sigma) = 1 \quad . \tag{4.27}$$

117

Before we present numerical values for these transition probabilities, we note again that detailed balance (3.33) allows us to relate, for example, the upward to the downward transitions

$$W(\eta',\eta) = W(\eta,\eta') \, e^{\beta(E_\eta - E_{\eta'})} \quad . \qquad (4.28)$$

The latter, being dominated by spontaneous emission of phonons, do not depend too strongly on temperature.

We first look at the ^{4}He-LiF system. The Morse potential parameters were chosen to get good agreement with the experimentally determined bound state energies (Derry et al. 1978). In Fig.4.1, we give the downward transition probabilities as a function of temperature. Note the typical order of magnitude for bound state - bound state transitions of 10^7-10^9 s^{-1} for the He-LiF system. To assess the likeliness of desorption, we plot in Fig. 4.2

$$R_{cn} = \sum_k W(k,n) \qquad (4.29)$$

as a function of inverse temperature; it is the transition probability per unit time to go from a bound state n to any continuum state. Because these processes are thermally activated, we see a more or less exponential temperature dependence like $\exp(\beta E_n)$. Note that below 10 K where adsorption becomes possible, the bound state-continuum transitions are rarer than downward bound state-bound state transitions by at least two orders of magnitude. This suggests, as indicated in the earlier discussion of the master equation, that fast bound state-bound state transitions will keep the adsor-

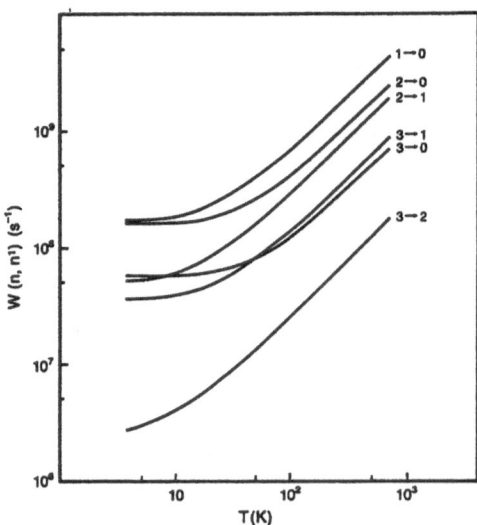

Fig.4.1. Downward transition probabilities as a function of temperature for the He-LiF system. $\sigma_0 = 4.023$, $r = 144.55$, $T_D = 730$ K.

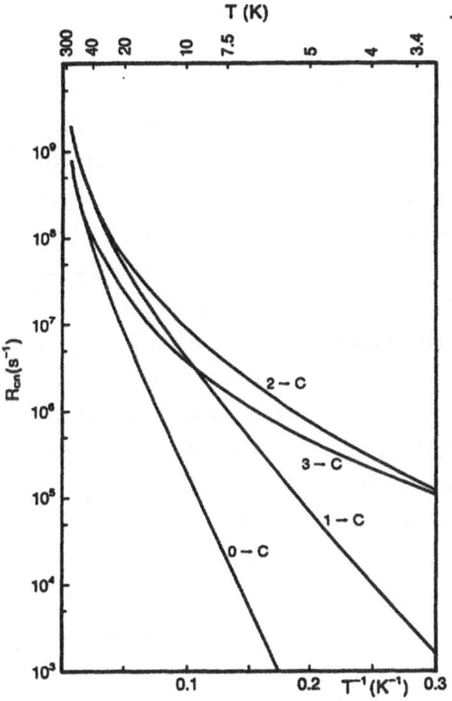

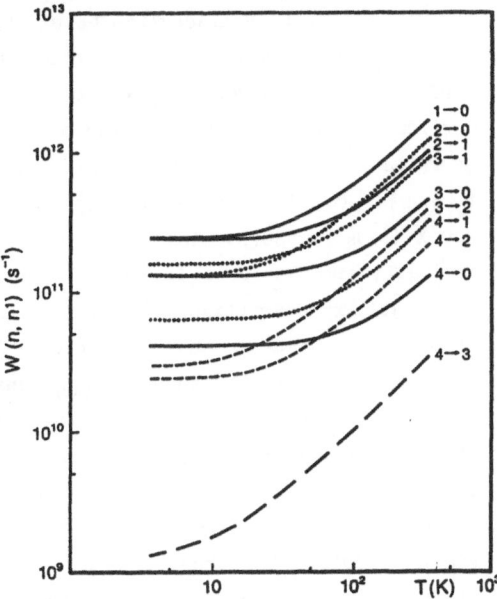

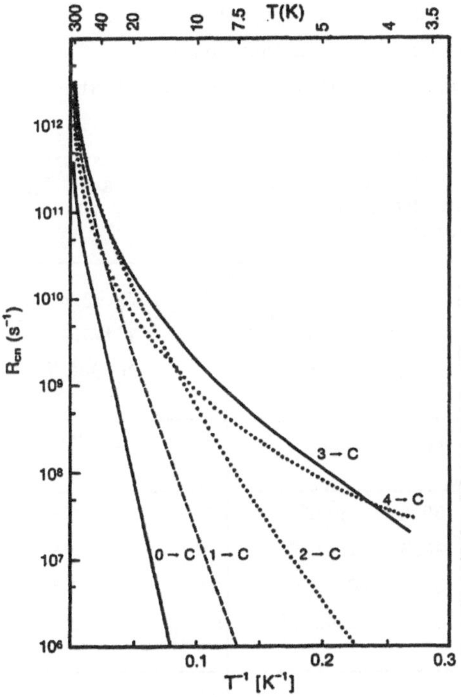

T (K)
300
40
20
10
7.5
5
4
3.4
$R_{cn}(s^{-1})$
10^9
10^8
10^7
10^6
10^5
10^4
10^3
2 – C
3 – C
1 – C
0 – C
0.1
0.2
$T^{-1}(K^{-1})$
0.3

◄ **Fig.4.2.** Transition probability R_{cn} in (4.29) from a bound state n into the continuum as a function of inverse temperature for He-LiF.

10^{13}
10^{12}
10^{11}
10^{10}
10^9
$W(n, n^1)$ (s^{-1})
1→0
2→0
2→1
3→1
3→0
3→2
4→1
4→2
4→0
4→3
10
10^2
T(K)
10^3

Fig.4.3. Downward transition probabilities as a function of temperature for the He-graphite system.

300
40
20
10
7.5
T(K)
5
4
3.5
R_{cn} (s^{-1})
10^{12}
10^{11}
10^{10}
10^9
10^8
10^7
10^6
3 – C
4 – C
0 – C
1 – C
2 – C
0.1
0.2
0.3
T^{-1} $[K^{-1}]$

Fig.4.4. Transition probability R_{cn} in (4.29) from a bound state n into the continuum as a function of inverse temperature for He-graphite.

bate in quasiequilibrium during the slow desorption process. Figures 4.3,4
give similar results for ⁴He-graphite.

The transition probabilities R_{cn} in (4.29) have actually been determined
experimentally by Bruisdeylins et al. (1980) from the angular halfwidths in
the cross section of helium atoms inelastically scattered off a LiF(001)
surface at a temperature T_s = 300 K. A comparison of experimental and theoretical values is given in Table 4.1 where we also list the bound state
energies in the surface potential. The choices of Morse potential parameters will be discussed around Table 5.1. The Debye temperature of LiF at
300 K is T_D = 630 K. With this choice the theoretical values for R_{cn} are
typically too low by a factor of five or so. However, we will see in Sect.
5.4 that for desorption from well-defined surfaces the Debye temperature
should be reduced by about a factor 0.75, see also Table 2.7. This increases the theoretical values for R_{cn} considerably. The remaining discrepancy
might be an indication that the effective Debye temperature is lower yet or
that the phonon coupling should be increased by reducing the width of the
surface potential. Further measurements will certainly clarify this point.

For a gas-solid system with hundreds of bound states such as Xe-W,
Fig.4.5 displays the transition probabilities W(E,E') plotted as perspective
views over a section of the (E+E',E-E') plane (Gortel et al. 1981). The pic-
tures are dominated by the bound state-bound state transition matrix elements for E+E'<0. The highest peaks correspond to transition rates of the
order of $3\omega_D m/M_s$. In Fig. 4.5, the transitions up to higher bound states

Table 4.1. Experimental and theoretical values of the bound state-continuum transition probabilities R_{cn} (from (4.29)) and bound state energies E_n
for He-LiF. The theoretical values are for different choices of Morse potential parameters and surface Debye temperatures.

	Experiment[a]		I: $V_0=89.0$ K $\gamma^{-1}=1.09$ Å $T_D=630$ K		II: $V_0=93$ K $\gamma^{-1}=0.9$ Å $T_D=630$ K		III: $V_0=93$ K $\gamma^{-1}=0.9$ Å $T_D=500$ K	
n	E_n[K]	$R_{cn}[s^{-1}]$	E_n[K]	$R_{cn}[s^{-1}]$	E_n[K]	$R_{cn}[s^{-1}]$	E_n[K]	$R_{cn}[s^{-1}]$
0	68.5	1.4×10^{11}	68.5	5×10^9	68.5	5×10^9	68.5	1.3×10^{10}
1	28.5	9×10^{10}	36.5	1.3×10^{10}	30.8	1×10^{10}	30.8	2.4×10^{10}
2	9.05	6×10^{10}	14.6	1.3×10^{10}	8.0	0.7×10^{10}	8.0	1.7×10^{10}
3	2.44	1.6×10^{10}	2.5	6.25×10^9	0.01	2.9×10^8	0.01	6.8×10^8

[a] Bruesdeylins et al. (1980)

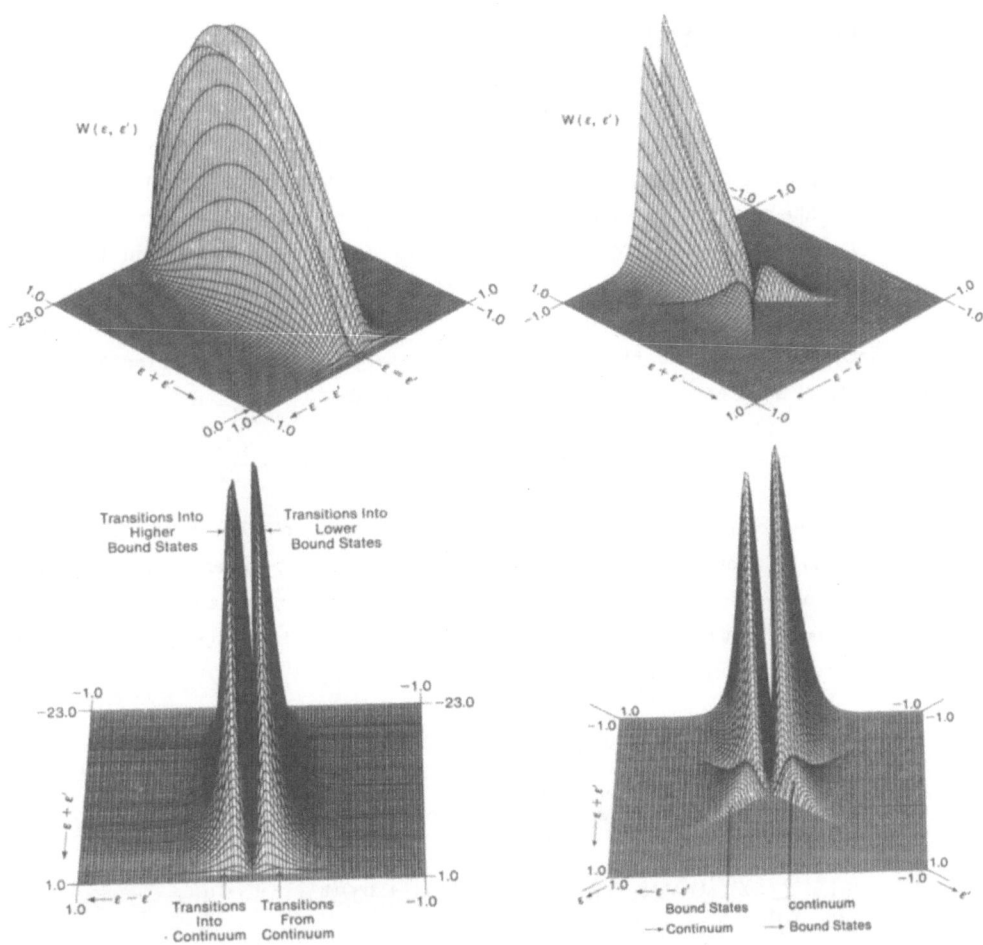

Fig.4.5. Perspective views of the transition probabilities $W(\epsilon,\epsilon')$, calculated from(4.22-26), plotted over the $(\epsilon+\epsilon',\epsilon-\epsilon')$ plane where $\epsilon=E/\hbar\omega_D$. The right-hand figures are sections around the origin. Other parameters: $m/M_S=0.714$ and $\omega_D=5.302\times10^{13}$ s^{-1} for the Xe-W system. (Gortel et al. 1981)

produce the lower peak on the left-hand side. The valley between the down and up transitions reflects the fact that $W(E,E)=0$ on account of the factor $(n-n')^3 \propto (\sqrt{E} - \sqrt{E'})^3$ in (4.22). To see the bound state-continuum and continuum-bound state transitions more clearly, we display on the right of Figure 4.5 two enlarged sections around the origin E=E'=0. Their matrix elements are, indeed, much smaller than those for bound state-bound state transitions, implying again that during the desorption process, the bound state occupation is reshuffled into a thermal distribution much faster than particles are actually desorbing.

4.2 Multi-phonon Processes and Correlation Functions

To go beyond the one-phonon approximation used in the preceeding section to calculate the transition probabilities, one can either carry the perturbation theory approach to, e.g., fourth order, thus including two-phonon processes, or formulate the theory in terms of correlation functions.

Gortel, Kreuzer, and Teshima (1980b) have calculated desorption times in a fourth-order time-dependent perturbation theory including all one-phonon and two-phonon processes. We note first that consistency requires that the expansion (2.64) of the time-dependent surface potential be carried to third order, i.e., in a one-dimensional model we include

$$V_S(z-u(t)) \approx V_S(z) - u(t) \frac{dV_S}{dz} + \tfrac{1}{2} u^2(t) \frac{d^2V_S}{dz^2} - \tfrac{1}{6} u^3(t) \frac{d^3V_S}{dz^3} \quad . \tag{4.30}$$

In principle, one should at this stage also include higher order Cole-Toigo corrections. Because they contribute little in lowest order, they can most likely also be neglected in higher orders. Using an expansion (2.78) for the displacements $u(t)$, the dynamic gas-solid Hamiltonian appropriate for a fourth-order calculation reads

$$
\begin{aligned}
H = {}& \sum_1 E_1 \alpha_1^\dagger \alpha_1 + \sum_J \hbar\omega_J b_J^\dagger b_J \\
& + L^{-1} \sum_{1,1'} \chi^{(1)}(1,1') \alpha_1^\dagger \sum_J \frac{1}{\sqrt{\omega_J}} (b_J^\dagger + b_J)\alpha_{1'} \\
& + L^{-1} \sum_{1,1'} \chi^{(2)}(1,1') \alpha_1^\dagger \sum_{J,J'} \frac{1}{\sqrt{\omega_J \omega_{J'}}} (b_J^\dagger + b_J)(b_{J'}^\dagger + b_{J'})\alpha_{1'} \\
& + L^{-1} \sum_{1,1'} \chi^{(3)}(1,1') \alpha_1^\dagger \sum_{J,J',J''} \frac{1}{\sqrt{\omega_J \omega_{J'} \omega_{J''}}} (b_J^\dagger + b_J)(b_{J'}^\dagger + b_{J'})(b_{J''}^\dagger + b_{J''})\alpha_{1'}
\end{aligned}
\tag{4.31}
$$

where $\chi^{(n)}(1,1')$ involves the n-th derivative of the surface potential. The importance of keeping higher-order derivatives in the Hamiltonian in going beyond second-order perturbation theory has also been stressed by Brenig and Schönhammer (1979). To render a fourth-order calculation feasible, Gortel et al. (1980b) chose a gas-solid system with only one bound state in which case the surface potential can be chosen to be separable. To briefly introduce the concepts of local, nonlocal, and separable potentials, let us look at the Schrödinger equation in Dirac notation,

$$(T + V_S) |\phi\rangle = E |\phi\rangle \quad , \tag{4.32}$$

where T is the kinetic energy and V_S the surface potential for a gas particle at energy E. Taking the coordinate representation we get

$$- \frac{\hbar^2}{2m} \nabla^2 \phi(\mathbf{r}) + \int \langle \mathbf{r} | V_S | \mathbf{r'} \rangle \phi(\mathbf{r'}) \, d^3\mathbf{r'} = E\phi(\mathbf{r}) \quad . \tag{4.33}$$

If the surface potential is chosen to be local, we set

$$\langle \mathbf{r} | V_S | \mathbf{r'} \rangle = V_S(\mathbf{r}) \, \delta(\mathbf{r}-\mathbf{r'}) \quad . \tag{4.34}$$

This, however, is not compelling. Indeed, quite often effective single-particle potentials turn out to be nonlocal, e.g., in mean-field theories as we have seen in Sect. 2.2. Moreover, it has been shown (Weinberg, 1963) that any potential operator can be expanded in Sturmian functions as

$$V_S = - \sum_{n=1}^{\infty} \lambda_n(E') \, |\psi_n(E')\rangle\langle\psi_n(E')| \quad , \tag{4.35}$$

where $|\psi_n(E')\rangle$ are solutions of the homogeneous Lippmann-Schwinger equation

$$V_S G_0(E') \, |\psi_n(E')\rangle = \lambda_n(E') \, |\psi_n(E')\rangle \tag{4.36}$$

at a fixed energy E' with

$$G_0(E') = \frac{1}{E'-T} \tag{4.37}$$

being the free particle Green's function. If necessary, E' may have an infinitesimal imaginary part. In particular, if V_S develops only one bound state at energy $E_1 < 0$ then keeping only one term in (4.35) and setting $E'=E_1$ yields the unitary pole approximation

$$\langle \mathbf{r} | V_S | \mathbf{r'} \rangle = -\langle \mathbf{r} | \psi_1(E_1) \rangle\langle\psi_1(E_1) | \mathbf{r'} \rangle$$
$$= - g \, v^*(\mathbf{r}) v(\mathbf{r'}) \tag{4.38}$$

because $\lambda_n(E_n)=1$. This is a separable potential. It has been used by Gortel, Kreuzer, and Spaner (1980) to calculate desorption times. A detailed discussion of local versus separable potentials has been given by Gortel, Kreuzer and Teshima (1980a). For the purpose of calculating all one- and two-phonon processes up to fourth order in time-dependent perturbation theory, they choose a form factor in (4.38) in a one-dimensional model

$$v(z) = \frac{1}{\sqrt{2}} \, e^{-\gamma z} \quad . \tag{4.39}$$

To simplify the fourth order calculations of the transition probabilities, including one- and two-phonon processes, Gortel, Kreuzer, and Teshima (1980b) perform a "dressing" transformation on the bound state part of the

Hamiltonian (4.31) (e.g., Schweber 1961) to avoid divergencies associated with diagonal elements of the perturbation Hamiltonian. It amounts to introducing a dressed bound state operator

$$\tilde{\alpha}_0 = e^{iS}\,\alpha_0 e^{-iS} = \exp\left[\frac{1}{\hbar L}\,X^{(1)}(0,0)\sum_J \omega_J^{-3/2}\,(b_J{}^\dagger - b_J)\right]\alpha_0 \tag{4.40}$$

corresponding to the dressed bound state with energy $\tilde{E}_0$, and dressed phonon operators

$$\tilde{b}_J = e^{iS}\,b_J e^{iS} = b_J + \frac{1}{\hbar L}\,\omega_J^{-3/2}\alpha_0{}^\dagger\alpha_0 \quad , \qquad \text{where} \tag{4.41}$$

$$S = iX^{(1)}(0,0)\,\frac{1}{\hbar L}\sum_J \omega_J^{-3/2}\,\alpha_0(b_J{}^\dagger - b_J)\alpha_0 \quad . \tag{4.42}$$

The actual calculation then amounts to iterating Heisenberg's equation of motion (2.185) for $\tilde{\alpha}_0$ to fourth order and taking expectation values (3.76) of which the long-time limit, similar to (3.81), eventually yields the transition probabilities. The rather involved details are given by Gortel et al. (1980b). In all, there are nine distinct contributions beyond the standard second-order results. Four of them are fourth order in the first derivative of the surface potential and differ in having one or two intermediate continuum states. In fourth order, one also gets one-phonon transitions as signaled by the appearance of an energy-conserving δ-function of the form $\delta(E_0 - E_k + \hbar\omega_J)$. As distinct from the second-order term, they involve two Bose-Einstein factors for the phonons with one real phonon of energy $\hbar\omega_J$ being absorbed, whereas the second phonon remains virtual, contributing to vertex and self-energy corrections. The appearance of the derivatives of the δ-functions reflects the fact that in a fourth-order calculation based on the dressed Hamiltonian, only self-energy corrections in $\tilde{E}_0$ are kept to all orders. Transferring the derivative from the δ-function onto the matrix elements, one gets the lowest-order vertex correction.

Apart from the one-phonon terms, one also gets two-phonon contributions in the rates with the factor

$$n_J(ph)n_{J'}(ph)\delta(\tilde{E}_0 - E_k + \hbar\omega_J + \hbar\omega_{J'}) \tag{4.43}$$

for the absorption of two phonons, and the factor

$$n_J(ph)(n_J(ph) + 1)\,\delta(\tilde{E}_0 - E_k + \hbar\omega_J - \hbar\omega_{J'}) \tag{4.44}$$

for the absorption of a phonon of energy $\hbar\omega_J$ and the simultaneous emission of a phonon of energy $\hbar\omega_{J'}$. Trivially, it is only terms with a factor (4.43) that survive for deep bound states with $-2\hbar\omega_D < \tilde{E}_0 < -\hbar\omega_D$.

Next come the contributions involving the second derivative of the surface potential. First there is a second-order term in $\chi^{(2)}$. In addition, there will be several terms in which second-order terms in $\chi^{(1)}$ interfere with the first-order terms in $\chi^{(2)}$. They are again classified according to having no or one intermediate continuum state. Lastly, the interference terms involving $\chi^{(1)}$ and $\chi^{(3)}$ both in first order contribute to the one-phonon processes via a rate — given here as an example —

$$W^{(10)}(k,0) = \frac{12\pi}{L^2\hbar} \; \chi^{(1)}(0,k) \; \chi^{(3)}(k,0) \sum_{J,J'} (\omega_J\omega_{J'})^{-1}$$
$$*[n_J(ph)n_{J'}(ph) + n_J(ph)[n_{J'}(ph) + 1]] \; \delta(\tilde{E}_0 - E_k + \hbar\omega_J) \quad . \quad (4.45)$$

We observe that the fourth order transition probabilities necessitate — even for the simple separable surface potential (4.38) — numerical evaluation of singular multiple integrals. This is described in detail by Gortel et al. (1980a).

A rather different approach to include multiphonon processes in the first-principles calculation of transition probabilities is accessed via correlation functions. Following Bendow and Ying (1973a,b), we define the transition probability per unit time for a particle to go from state ι to state ι' in the surface potential (see, e.g., Messiah, 1961)

$$W(\iota',\iota) = \frac{2\pi}{\hbar} \sum_{s,s'} e^{-\beta\varepsilon_s} \; |\langle s'; \iota'| \; T(\varepsilon_s + E_\iota) \; |s,\iota\rangle|^2$$
$$* \; \delta(\varepsilon_s - \varepsilon_{s'} - E_\iota + E_{\iota'}) / \sum_{s''} e^{-\beta\varepsilon_{s''}} \quad , \quad\quad\quad (4.46)$$

where ε_s and $\varepsilon_{s'}$ represent the initial and final lattice energies with corresponding states $|s\rangle$ and $|s'\rangle$ and

$$T(E) = H_{g-s} + H_{g-s} \; G(E) \; H_{g-s} \quad\quad\quad\quad\quad (4.47)$$

is the scattering operator with

$$G(E) = (E-H+i\varepsilon)^{\pm 1} \quad\quad\quad\quad\quad (4.48)$$

being the Green's function for the coupled gas-solid system. Replacing $T(E)$ by H_{g-s} in (4.46) yields Fermi's Golden rule (3.83). For H_{g-s} Bendow and Ying (1973a) take [see the first line of (2.83)]

$$H_{g-s} = \delta V_s(\mathbf{r}) = \sum_i [V_2(\mathbf{r}-\mathbf{r}_i) - V_2(\mathbf{r}-\mathbf{r}_i^{(0)})] \quad , \quad\quad\quad (4.49)$$

where

$$r_i = r_i^{(0)} + u_i(t) \quad . \tag{4.50}$$

Defining Fourier transforms

$$v_q = \int d^3r \; e^{-iq \cdot r} \; V_2(r) \quad , \tag{4.51}$$

and form factors

$$f_q(\iota,\iota') = \int d^3r \; \phi_\iota(r) \; e^{iq \cdot r} \; \phi_{\iota'}(r) \quad , \tag{4.52}$$

involving eigenstates $\phi_\iota(r)$ of the static Hamiltonian $H_g^{(0)}$ given by (2.177), one gets for a harmonic solid with H_S given in (2.173) and with T replaced by H_{g-s} in (4.46),

$$W(\iota',\iota) = \hbar^{-1} \sum_{q,q',\ell,\ell'} \exp[i(q' \cdot r_{\ell'}^{(0)} - q \cdot r_\ell^{(0)})] \; v_q \; v_{q'}^*$$

$$*f_q(\iota,\iota') \; f_{q'}^*(\iota,\iota') \int_{-\infty}^{\infty} dt \; \exp[-i(E_{\iota'}-E_\iota)t/\hbar - W_{\ell\ell'}(q,q';\beta) + q \cdot C_{\ell\ell'}(t;\beta) \cdot q']$$

(4.53)

Here we introduced a Debye-Waller factor [see also (2.101,102)]

$$W_{\ell\ell'}(q,q';\beta) = \tfrac{1}{2} \frac{Tr\{[(q \cdot u_\ell)^2 + (q' \cdot u_{\ell'})^2]e^{(-\beta H_S)}\}}{Tr\{e^{(-\beta H_S)}\}} \tag{4.54}$$

characterizing the mean square lattice displacements at finite temperature and reflecting the reduction of the scattering amplitude by the thermal motion of the lattice (e.g., Callaway 1974). One also defines the time- and position-dependent displacement-displacement correlation tensor

$$C_{\ell\ell'}(t;\beta) = \frac{Tr\{u_\ell(t)u_{\ell'}(0)\exp(-\beta H_S)\}}{Tr\{\exp(-\beta H_S)\}} \quad , \tag{4.55}$$

which can be written - using (2.78) for the displacement vectors - as

$$C_{\ell\ell'}(t,\beta) = \sum_J \frac{\hbar}{2\rho_0\omega_J} [(n_J^{(ph)}+ 1) \; e^{-i\omega_J t} \; u^{(J)}(r_\ell)u^{(J)*}(r_{\ell'})$$

$$+ n_J^{(ph)} \; e^{i\omega_J t} \; u^{(J)*}(r_\ell)u^{(J)}(r_{\ell'})] \quad , \tag{4.56}$$

To evaluate this further, Bendow and Ying (1973a) assume an isotropic bulk Debye model for which $C_{\ell\ell'}$ is a multiple of the 3x3 unit matrix. To incorporate the surface in an approximate way they take (with $J = (p,\sigma)$)

$$u^{(J)}(r_\ell) = \frac{1}{\sqrt{AL_s}} \; e_{p\sigma} \; \cos(p \cdot r_\ell) \tag{4.57}$$

with standing waves rather than with propagating waves as in (2.79). Bendow and Ying (1973a) proceed further by expanding the exponential involving the correlation function. The time integration in the lowest nonvanishing order then yields a term in (4.53)

$$\int_{-\infty}^{\infty} dt\, e^{i(E_{l'}-E_{l})t/\hbar} \sum_J \frac{\hbar}{2\rho_0\omega_J A L_s} \left[[n_J(ph)+1]\, e^{-i\omega_J t} + n_J(ph)\, e^{i\omega_J t}\right]$$

$$= \sum_J \frac{\hbar}{2\rho_0\omega_J A L_s} \left[[n_J(ph)+1]\, \delta(E_{l'}-E_l-\hbar\omega_J) + n_J(ph)\, \delta(E_{l'}-E_l-\hbar\omega_J)\right] \quad . \tag{4.58}$$

This reduces (4.53) to almost the one-phonon result (3.83) if we note that $q v_q$ is proportional to the Fourier transform of ∇V_2 where V_2 is still the two-body interaction between a gas and a solid particle, so that approximate summations over lattice sites as discussed in Sect. 2.1 are still needed to recover the ∇V_s dependence. Higher terms in the expansion of $\exp(q \cdot C \cdot q')$ yield multiphonon processes, but only those that result from second-order perturbation theory in the higher derivatives of V_2. To include all other multiphonon processes simultaneous iteration of the scattering equation (4.47) must be performed. This program is outlined by Bendow and Ying (1973a).

To evaluate (4.53) further, Bendow and Ying (1973a) consider highly localized adsorption in a single bound state for which the Bloch-like wave function

$$\phi_l(R,z) = \frac{1}{\sqrt{A}} \sum_\ell e^{i(K \cdot R_\ell)}\, W(R-R_\ell,z) \tag{4.59}$$

is expressed in terms of a localized Wannier function chosen as

$$W(R,z) = \frac{1}{(2\pi)^{3/4} r_2 \sqrt{r_3}} \exp\left[-\tfrac{1}{4} R^2/r_2^2 - \tfrac{1}{4} (z-z_0)^2/r_3^2\right] \quad . \tag{4.60}$$

The final continuum states are calculated in WKB approximation

$$\phi_k(R,z) = \frac{1}{\sqrt{A}} e^{iK \cdot R} \sqrt{\frac{2}{L \kappa_z}} (2mE_z)^{1/4} \sin\left[\int_a^{z_0} dz'\, k(z') + \kappa_z(z-z_0) + \frac{\pi}{4}\right] \quad , \tag{4.61}$$

where $E_z=\hbar^2 k_z^2/2m$, $k^2(z)=2m[-V_0(z)+E_z]/\hbar^2$, $\kappa_z^2=2m(\tilde{E}_0-E_z)/\hbar^2$, a is the inner turning point, and z_0 and $\tilde{E}_0$ are the position and depth of the minimum of the average surface potential $V_0(z)$, defined like (2.4,14) averaged over the surface plane, but with a two-body interaction taken to be a sum of Gaussians

$$V_2(r) = \tilde{V}_0 \left[\exp\left[-\tfrac{1}{4}(r^2-r_0^2)/r_1^2\right] - \exp\left[-\tfrac{1}{8}(r^2-r_0^2)/r_1^2\right]\right] \quad . \tag{4.62}$$

The Wannier function in (4.60) corresponds to the lowest eigenstate in an anisotropic harmonic approximation to $V_S(\mathbf{r})$, so that $r_z^{-2} = 2m(\partial^2 V_S(\mathbf{r})/\partial R^2)_0/\hbar^2$ and $r_3^{-2} = 2m(\partial^2 V_S/\partial z^2)_0/\hbar^2$, where the derivatives are evaluated at the minimum of $V_S(\mathbf{r})$.

A similar correlation function approach to the calculation of transition probabilities has been worked out by Efrima et al. (1980,1983) and Jedrzejek et al. (1981b,1983). They consider mobile adsorbates and treat the gas dynamics in a one-dimensional model. Rather than using (4.49) as the time-dependent perturbation that causes transitions of the gas particle between the states of the surface potential, they take the difference between the true Hamiltonian and its thermal average (2.112). This implies that the zero-order Hamiltonian is the thermally averaged one. By choosing a Morse potential, the thermal averaging can be done explicitly as shown in (2.109), where it is argued that the temperature-induced changes in the surface potential depth and width are negligible. Whereas Bendow and Ying (1973a,b) consider the thermal motion of the individual lattice particles explicitly in constructing the dynamic part of the Hamiltonian, see (4.56), Efrima et al. (1983) use the simplified approach (2.63) in which only the vibration in the z-direction of the surface just below the adparticle is taken into account, so that

$$\delta V_S = V_S(z-u(t)) - \langle V_S(z-u(t))\rangle \tag{4.63}$$

is the perturbation. In calculating transition probabilities Efrima et al. consider only terms quadratic in δV_S. However, they do not expand the exponential involving the lattice correlation function into a multiphonon series. To perform the numerical time integration, they include a damping factor $\exp[-\Delta(\omega_J)t]$ in the correlation function (4.56) which, for the present model, of course, has no spatial structure, i.e., it simplifies for a bulk Debye model to

$$\langle u(t)u(0)\rangle = \frac{\hbar}{2M_S N_S} \sum_{p\sigma} \frac{1}{\omega_{p\sigma}} \left|e_{p\sigma}^z\right|^2 \{[n^{(ph)}(\omega_{p\sigma}) +1] e^{-i\omega_{p\sigma}t}$$
$$+ n^{(ph)}(\omega_{p\sigma}) \exp(i\omega_{p\sigma}t)\}\exp(-\Delta(\omega_{p\sigma})t) \quad . \tag{4.64}$$

Physically, the damping factor is supposed to reflect phonon relaxation due to lattice anharmonicity and electron-hole pair excitations (in metals). This ad hoc modification unfortunately leads to a violation of detailed balance for the transition probabilities, so that the master equation does not relax towards equilibrium for an isolated system. To avoid this problem, Efrima et al. (1983) calculate the downward transition probabilities microscopically and then get the upward transitions from detailed balance. To achieve numerical stability of the time integration, the damping constant $\Delta(\omega_J)$ has to exceed about 50 cm^{-1}. For a Morse surface potential, the transition probabilities in a one-dimensional model read, for bound state-bound state transitions (Efrima et al.1980, 1983)

$$W(m,n) = \frac{2V_0^2}{\hbar^2} \, \text{Re}\left\{ \int_0^\infty dt \, \exp[i(E_n-E_m)t/\hbar] \, \{(B_{nm}^{(2)})^2 \, \exp(4\gamma^2\langle u^2\rangle) \right.$$

$$*\{\exp[4\gamma^2\langle u(t)u(0)\rangle]-1\} + 4(B_{nm}^{(1)})^2\exp(\gamma^2\langle u^2\rangle)\{\exp[\gamma^2\langle u(t)u(0)\rangle]-1\}$$

$$- 4B_{nm}^{(1)}B_{nm}^{(2)} \, \exp(2.5\gamma^2\langle u^2\rangle)\{\exp[2\gamma^2\langle u(t)u(0)\rangle]-1\}\}\} \quad . \qquad (4.65)$$

The coefficients are

$$B_{nm}^{(j)} = \int_{-\infty}^\infty \phi_n^*(z) \, e^{-j\gamma z} \, \phi_m(z)dz$$

$$= \{[n_> \, (2\sigma_0-n_>-1) - n_< \, (2\sigma_0-n_<-1) + 2\sigma_0]/2\sigma_0\}^{j-1}$$

$$* \frac{1}{2\sigma_0} \, [(2\sigma_0-2n-1) \, (2\sigma_0-2m-1)]^{1/2} \left[\frac{n_>! \, \Gamma(2\sigma_0-n_>)}{n_<! \, \Gamma(2\sigma_0-n_<)}\right]^{1/2} \quad , \qquad (4.66)$$

with $n_> = \max(n,m)$ and $n_< = \min(n,m)$, σ_0 is given in (4.13).

The transition probability from a bound state n to all continuum states is found to be, after some approximations such as using Stirling's formula for Γ-functions,

$$\sum_k W(k,n) = 4\pi(\frac{V_0}{2\sigma_0\hbar})^2 \left[\frac{2\sigma_0-2n-1}{n! \, \Gamma(2\sigma_0-n)}\right] \exp(2\sigma_0\ln\sigma_0-2\sigma_0)$$

$$*[\frac{1}{(2\sigma_0)^2} \, \exp(4\gamma^2\langle u^2\rangle) \, \text{Re}\{ \int_0^\infty dt \, \exp(-E_nt/\hbar)$$

$$*[\exp(4\gamma^2\langle u(t)u(0)\rangle)-1] \, (A^2\beta^{-1} + 2A\beta^{-2} + 2\beta^{-3})\}$$

$$+ 4 \, \exp(\gamma^2\langle u^2\rangle) \, \text{Re}\{ \int_0^\infty dt \, \exp(-iE_nt/\hbar) \, \{\exp[\gamma^2\langle u(t)u(0)\rangle] - 1\}\beta^{-1}\}$$

$$- \frac{4}{2\sigma_0} \, \exp(2.5\gamma^2\langle u^2\rangle) \, \text{Re}\{ \int_0^\infty dt \, \exp(-iE_nt/\hbar)$$

$$*\{\exp[2\gamma^2\langle u(t)u(0)\rangle] - 1\} \, (A\beta^{-1} + \beta^{-2})\}] \quad . \quad \text{Here,} \qquad (4.67)$$

$$A = (\sigma_0-n-\frac{1}{2})^2 + 2\sigma_0$$

$$\beta = \sigma_0^{-1} + i\hbar\gamma^2t/2m$$

$$E_n = -\hbar^2\gamma^2 \, (\sigma_0-n-\frac{1}{2})^2/2m \quad . \qquad (4.68)$$

Expanding (4.64,66) in powers of the correlation function $\langle u(t)u(0)\rangle$ yields a multiphonon expansion

$$W(m,n) = \left(\frac{V_0}{\hbar}\right)^2 \sum_{p=1}^{3} B_p^{nm} \, \delta_p \sum_{\ell=1}^{\infty} \frac{1}{\ell!} \left(\frac{\hbar}{2M_s}\right)^{\ell} \int \prod_j \frac{g\,\omega_j\,d\omega_j}{\omega_D^3}$$

$$* \sum_{\xi_j = \pm 1} a_\ell(\omega_\ell) \frac{2\ell\Delta}{((E_n - E_m)/\hbar + \omega_\ell)^2 + \ell^2\Delta^2} \quad \text{where} \tag{4.69}$$

$$B_1^{nm} = (B_{nm}^{(2)})^2 \exp(4\gamma^2\langle u^2\rangle)$$

$$B_2^{nm} = 4(B_{nm}^{(1)})^2 \exp(\gamma^2\langle u^2\rangle)$$

$$B_3^{nm} = -4B_{nm}^{(1)} B_{nm}^{(2)} \exp(2.5\gamma^2\langle u^2\rangle)$$

$$\delta_1 = 4, \ \delta_2 = 1, \ \delta_3 = 2$$

$$\omega_\ell = \sum_{j=1}^{\ell} \xi_j \omega_j$$

$$a_\ell = \prod_{j=1}^{\ell} \left[n^{(ph)}(\omega_j) + \tfrac{1}{2} + \tfrac{1}{2}\xi_j \right].$$

The ℓth term is the ℓ-phonon contribution. Likewise, the bound-continuum terms expand to give, after some rearrangement,

$$\sum_k W(k,n) = 2\pi \frac{1}{(2\sigma_0)^2} \frac{2\sigma_0 - 2n - 1}{n!\,\Gamma(2\sigma_0 - n)} \exp[2\sigma_0 \ln(\sigma_0) - 2\sigma_0]$$

$$* \sum_{r=1}^{3} B_r \sum_{\ell=1}^{\infty} \frac{1}{\ell!} \left(\frac{\hbar}{2M_s}\right)^{\ell} \int \prod_j \left[\frac{g\,\omega_j\,d\omega_j}{\omega_D^3}\right]$$

$$* \sum_{\xi_j = \pm 1} a_\ell(\omega_\ell) \, I_r\left(\frac{E_r}{\hbar} + \omega_\ell, \frac{\ell\Delta}{\lambda\sigma_0}\right) \quad , \quad \text{where} \tag{4.70}$$

$$I_r(\omega,d) = \frac{2(\lambda^{-r})}{(r-1)!} \int_0^{\infty} e^{-y} y^{r-1} \frac{d}{d^2 + (y-\omega)^2} \, dy$$

$$\lambda = \hbar\gamma^2/2m \quad .$$

We note that in the limit $\Delta \to 0$, the one phonon contributions to the bound state–bound state transitions (4.65) reduce to (4.22). However, this is not the case for the transitions to continuum states due to a series of approximations invoked in deriving (4.67).

The role of many-phonon contributions in the calculation of transition probabilities has also been examined by Böheim and Brenig (1981) in the context of inelastic atom-surface scattering. They also employ a correlation function approach, but in terms of the autocorrelation function of the force. Various approximations that reduce the problem to a one-dimensional one, allow them to evaluate the energy distribution P(ΔE) of atoms scattered inelastically from a solid surface. This function would seem to be related to the transition probability in the master equation by

$$P(\Delta E) = \frac{L}{v} \, W(E + \Delta E, E) \quad , \tag{4.71}$$

where $E = mv^2/2$ and L is the length of the box enclosing the gas phase. Contrary to Efrima et al. (1980, 1983), Böheim and Brenig (1981) do not need a phonon damping factor to achieve numerical convergence. Some of their results, namely for the rare gases scattering off a tungsten surface, are reproduced in Fig. 4.6. We note:

(i)   for helium the one-phonon approximation is acceptable even at temperatures as high as $T = T_D = 400$ K. We note in passing that for continuum-continuum transitions the transition probabilities are nonzero at zero energy transfer ΔE=0 whereas they are zero at ΔE=0 for bound state - bound state transitions;

(ii)   already for Ne-W significant multiphonon contributions appear at high temperatures, leading to a breakdown of the one-phonon approximation;

(iii) for the heavier rare gases the transition probabilities are dominated by multiphonon processes at all temperatures.

In such situations, they can be approximated by a Gaussian

$$W(E',E) = W_0 \, \frac{1}{\sqrt{4\pi k_B T \Delta}} \, \exp\left[- \frac{(E'-E+\Delta)^2}{4 k_B T \Delta} \right] \quad , \tag{4.72}$$

provided that $E,E' \gg V_0$. Böheim and Brenig (1981) give various estimates for Δ. Comparing their theory with experiments by Janda et al.(1980) on Ar-W scattering, they find good agreement. A comparison with classical trajectory calculations for the Ne-Ag system by Barker et al. (1980) shows that their "fine structure" in the classical energy distribution is completely masked by quantum effects. Taking recoil effects into account by simply replacing the mass m of the scattered particle by the reduced mass $\mu = mM_S/(m+M_S)$, where M_S is the mass of a unit cell of the solid, results in considerable narrowing of the energy distributions.

A phenomenological transition probability (4.72) has earlier been used by Müller and Brenig (1979) and also by Pagni and Keck (1973). Such a Gaussian and many other forms are discussed at great length by Cercignani (1975) in

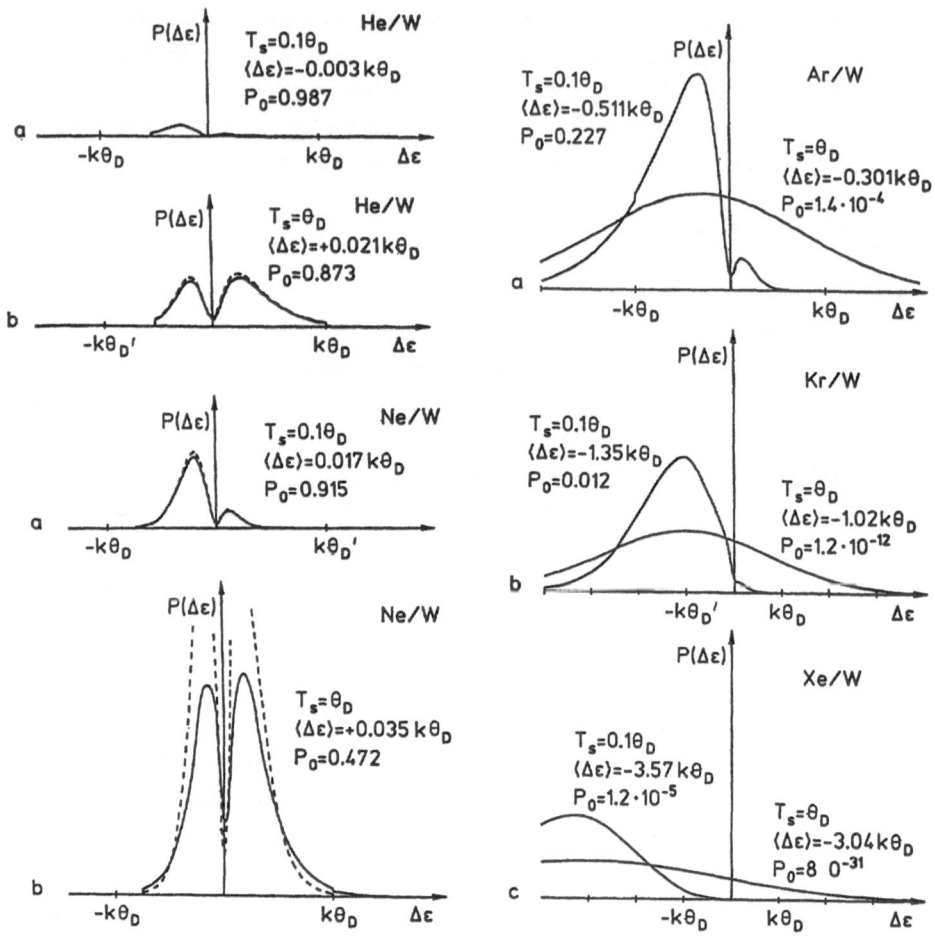

Fig.4.6. Energy distribution related via (4.71) to the transition probabilities for rare gases scattered off tungsten at initial energy $E_i = 0.1k_BT_D$ at two different solid temperatures T_s. $\langle\Delta\varepsilon\rangle$ is the average energy transfer. Solid lines: multiphonon theory; dashed lines: one-phonon approximation. Morse potential parameters:

	He	Ne	Ar	Kr	Xe
$V_d[K]$	161	250	1200	2000	3200
$\gamma[A^{-1}]$	1.3	1.7	1.5	1.5	1.5

(Böheim and Brenig 1981)

the context of establishing proper boundary conditions to be imposed on the solutions of the Boltzmann equation.

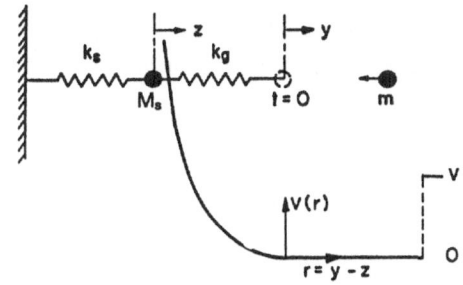

Fig.4.7. Soft cube model: the attractive potential is of arbitrary shape of height V at fixed position several lattice spacings from the surface. Repulsive quadratic potential follows the surface oscillator motion. (Pagni and Keck 1973)

4.3 Soft Cube Model and Phenomenological Models at Low Coverage

Pagni and Keck (1973) and Pagni (1973) have calculated the transition probabilities $W(E',E)$, based on the classical soft cube model of Logan and Keck (1968), illustrated in Fig. 4.7. It is assumed:

(i)   tangential momentum is conserved as appropriate for a mobile adsorbate.

(ii)   The surface atoms are independent simple harmonic oscillators with a single characteristic frequency obtained by approximating the lattice frequency spectrum with a delta function at the Debye frequency. It is also assumed that only a single surface atom is involved in any given collision. When the adatom velocity parallel to the surface is so large that the distance travelled during a collision is greater than the lattice constant, this one-on-one approximation fails.

(iii)   The gas-surface potential consists of a stationary arbitrarily shaped attraction, taken, e.g., as a step of height V, and an oscillating quadratic repulsion. The interaction is described classically and would not strictly apply to incident atoms whose de Broglie wavelength is of the order of the lattice spacing or to solid temperatures below the surface Debye temperature.

We denote by k_s and k_g, the lattice and gas-surface spring constants; M_s and m are the surface and gas atom masses. The time origin marks the initiation of a collision. For t<0, the gas atom is force free, and the surface oscillator is in simple harmonic motion. Thus, the surface atom displacement from its equilibrium position evolves according to

$$Z(t) = \frac{V}{\omega_s} \sin(\omega_s t + \phi) \tag{4.73}$$

and the gas atom displacement is given by

$$Y(t) = -ut + \frac{V}{\omega_s} \sin\phi \quad , \tag{4.74}$$

133

where u is the incident gas atom velocity in the well, ω_s is the angular frequency of the surface atom, and V and ϕ are the velocity amplitude and phase of the surface atom. The initial conditions are established as the gas atom impacts at the end of the gas-surface spring. At t=0

$$Y(0) = Z(0) = \frac{V}{\omega_s} \sin\phi$$

$$\dot{Y}(0) = -u$$

$$\dot{Z}(0) = V \cos\phi \quad . \tag{4.75}$$

The equations of motion during collision (t>0) are

$$m \frac{d^2Y(t)}{dt^2} = - k_g[Y(t) - Z(t)] \quad , \tag{4.76}$$

$$M_s \frac{d^2Z(t)}{dt^2} = -k_s Z(t) + k_g[Y(t)-Z(t)] \quad . \tag{4.77}$$

The collision ends at time $t=\tau_C$ when the gas-surface spring returns to its equilibrium extension, with the gas atom moving away from the surface, so that

$$Y(\tau_C) = Z(\tau_C) \quad , \tag{4.78}$$

with $\dot{Y}(\tau_C)$ positive and greater than $\dot{Z}(\tau_C)$. The collision duration τ_C is the smallest nonzero root of (4.78). The nondimensional energy transfer $\Delta(\varepsilon)$ is defined as the increase in the gas atom kinetic energy E during the collision ($\varepsilon = E/k_BT$)

$$\Delta(\varepsilon) = \varepsilon'-\varepsilon= \frac{m}{2k_BT} \left[\dot{Y}(\tau_C)^2-\dot{Y}(0)^2\right] \quad . \tag{4.79}$$

Defining ω_s^{-1} as a characteristic time and $u\omega_s^{-1}$ as a characteristic length, the system is specified by five dimensionless parameters: the mass ratio $\mu = m/M_s$, the frequency ratio $\nu = (\mu k_s/k_g)^{1/2} = \omega_s/\omega_g$, the surface well depth v $= V_0/k_BT$, the initial particle velocity $V_C = \dot{Z}(0)/u = V \cos\phi/u$ and the initial particle position $V_S = Z(0)\omega_s/u = V \sin\phi/u$. The solution of (4.76,77) then gives the gas atom trajectory

$$y(t) = \frac{Y(t)\omega_s}{u} = - \frac{\zeta V_S}{1-\zeta} \cos(\omega_1 t) - \frac{1+V_C-\omega_2^2\nu^2}{(1-\zeta)\omega_1^3\nu^2} \sin(\omega_1 t)$$

$$+ \frac{V_S}{1-\zeta} \cos(\omega_2 t) + \frac{1+V_C-\omega_1^2\nu^2}{(1-\zeta)\omega_1^2\omega_2\nu^2} \sin(\omega_2 t) \quad , \tag{4.80}$$

and the surface atom trajectory

134

$$z(t) = \frac{Z(t)\omega_S}{u} = -\frac{(1-\omega_1^2\nu^2)\,\zeta V_S}{1-\zeta}\cos(\omega_1 t)$$

$$-\frac{(1-\omega_1^2\nu^2)\,(1+V_C-\omega_2^2\nu^2)}{(1-\zeta)\omega_1^3\nu^2}\sin(\omega_1 t) + \frac{(1-\omega_2^2\nu^2)V_S}{1-\zeta}\cos(\omega_2 t)$$

$$+\frac{(1-\omega_2^2\nu^2)\,(1+V_C-\omega_1^2\nu^2)}{(1-\zeta)\omega_1^2\omega_2\nu^2}\sin(\omega_2 t) \quad, \tag{4.81}$$

where $\zeta = \omega_2^2/\omega_1^2$ with ω_1 and ω_2 being the normal mode frequencies of (4.76,77) given by

$$\omega_{1,2} = \frac{1}{4\nu}\sqrt{\nu^2+\mu+1 \pm \sqrt{(\nu^2+\mu+1)^2-4\nu^2}} \quad. \tag{4.82}$$

Using (4.78,80,81), the implicit expression for the collision duration τ_C is

$$\cos(\omega_1\tau_C+\phi_1) = -\left|\frac{\alpha_2}{\alpha_1}\right|\zeta\cos(\omega_2\tau_C+\phi_2) \quad, \tag{4.83}$$

where α_i and ϕ_i (i=1,2) are the amplitudes and phases of the normal modes, respectively. In terms of the initial conditions, these are

$$|\alpha_1|^2 = \frac{V_S^2\zeta^2}{4(1-\zeta)^2} + \frac{(1+V_C-\nu^2\omega_2^2)^2}{4(1-\zeta)^2\omega_1^6\nu^4}$$

$$|\alpha_2|^2 = \frac{V_S^2}{4(1-\zeta)^2} + \frac{(1+V_C-\nu^2\omega_1^2)^2}{4(1-\zeta)^2\omega_1^4\omega_2^2\nu^4} \quad\text{and} \tag{4.84}$$

$$\phi_1 = \tan^{-1}\left[-(1+V_C-\nu^2\omega_2^2)/V_S\omega_2^2\omega_1\nu^2\right]$$

$$\phi_2 = \tan^{-1}\left[-(1+V_C-\nu^2\omega_1^2)/V_S\omega_1^2\omega_2\nu^2\right] \quad. \tag{4.85}$$

Once the collision duration is known, the energy transfer in a collision is obtained directly from

$$\Delta = \varepsilon(\dot{y}(\tau_c)^2-1) \quad, \tag{4.86}$$

where $\varepsilon = mu^2/2k_BT$. Typical trajectories are given in Fig. 4.8.

To proceed to the calculation of transition probabilities, we note that Pagni and Keck (1973) consider a symmetrized version of the master equation different from the one introduced in (3.46), by defining a "specific" distribution

$$x(\varepsilon,t) = n(\varepsilon,t)/n^{eq}(\varepsilon) \quad, \tag{4.87}$$

so that the relevant symmetric transition probabilities, called one-way equilibrium transition rates, are

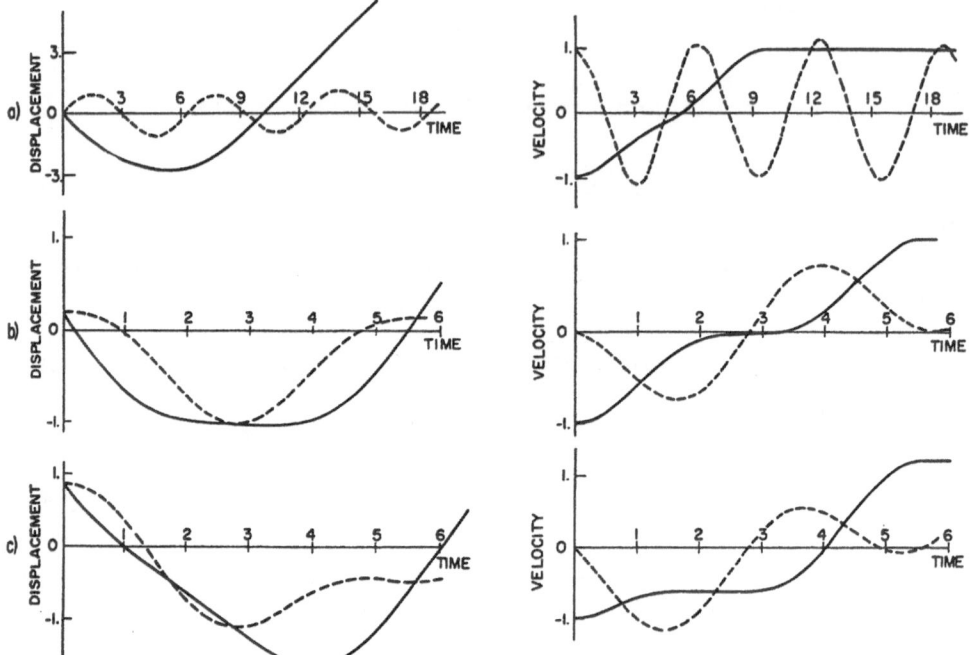

Fig.4.8. Gas y(t) - and surface z(τ) -- atom trajectories. **a** Adiabatic collision (ν=3.16, μ=0.5, V_S=0, V_C=1.0) with $τ_C$=3.06 and Δ/ε=-0.007. **b** Limiting case between single and second collisions (ν=μ=1, V_S=0.2, V_C=0) with $τ_C$=2.80 and Δ/ε=-1 at the end of the first collision. **c** second collision (ν=μ=1, V_S=0.86, V_C=0) with $τ_C$=1.71 and Δ/ε=-0.69 at the end of the first collision. Length and time scales are nondimensional with respect to $uω_S^{-1}$ and $ω_S^{-1}$, respectively. (Pagni and Keck 1973)

$$R(ε',ε) = W(ε',ε) n^{eq}(ε) \quad . \tag{4.88}$$

Followng Hughes (1959), they can be expressed as a flux in phase space from state ε to ε',

$$dR = ρ(\mathbf{v} \cdot \mathbf{n})ds \quad , \tag{4.89}$$

where ds is a differential element of a surface in phase space separating the states ε and ε' on which $(\mathbf{v} \cdot \mathbf{n}) > 0$, $\mathbf{v}$ is the generalized velocity of a point in phase space, $\mathbf{n}$ is the unit outward normal to ds, and $ρ(\mathbf{q},\mathbf{p})$ is the density of points representing the state of a system of n particles in a 6n-dimensional phase space, the axes of which are the conjugate momentum $\mathbf{p}$ and position $\mathbf{q}$ coordinates of the particles.

For the case of a gas atom-surface atom collision, the only nonignorable variables are the momentum of the gas atom, p_g, the momentum of the surface atom p_S, and the position of the surface atom Z. The one-way equilibrium transition rate may therefore be written in terms of the initial gas atom energy, ε, and the initial velocity V_C and position V_S of the surface atom.

$$R(V_C,V_S,\varepsilon) \; d\varepsilon dV_C dV_S = \rho(p_g,p_S,Z) \; (\dot{Y}-\dot{Z}) dp_g dp_S dZ \tag{4.90}$$

The surface in phase space across which the flux is monitored, is given by $Y-Z=0$ with the condition $\dot{Y}-\dot{Z}<0$.

With the definitions: $dp_g = dE_g/u$, $dp_S = M_S d\dot{Z}$, and $\dot{Y} = -u$, we get

$$R(V_C,V_S,\varepsilon) \; d\varepsilon dV_C dV_S = \rho(E_g,\dot{Z},Z) M_S \left[1+(\dot{Z}/u)\right] dE_g d\dot{Z} dZ \quad . \tag{4.91}$$

Since it is assumed that the gas atom and surface atom are independent prior to the collision, the phase space density in the initial state $\rho(E_g,\dot{Z},Z)$ can be separated as a product of the gas particle density $\rho_g(E_g)$ and the surface particle density $\rho_S(E_S)$

$$\rho(E_g,\dot{Z},Z) = \rho_g(E_g)\rho_S(E_S) = \frac{N_g \exp(-E_g/k_B T)}{Q_g} \frac{\exp(-E_S/K_B T)}{Q_S} \quad , \tag{4.92}$$

where N_g is the number of gas atoms, Q_g is the gas partition function, $Q_S = 2\pi k_B T/\omega_S$ is the solid atom partition function, and it is assumed that, for the equilibrium transition rate calculation, the phase space points have a Boltzmann distribution. Substituting (4.92) into (4.91) with $E_g = k_B T \varepsilon$, $E_S = 1/2 \; M_S u^2 (V_C^2 + V_S^2)$, $d\dot{Z} = u dV_C$, and $dZ = u \omega_S^{-1} dV_S$ yields

$$R(V_C,V_S,\varepsilon) \; d\varepsilon dV_C dV_S = \frac{N_g}{Q_g} e^{-\varepsilon} \frac{M_S u^2}{2\pi}$$
$$* \exp\left[-(\frac{\varepsilon}{\mu})(V_C^2 + V_S^2)\right] (1+V_C) \; d\varepsilon dV_C dV_S \quad . \tag{4.93}$$

This shows that the one-way equilibrium transition rate has a strong maximum in initial condition (V_C,V_S) space. In the limit $\varepsilon \to \infty$, $R(V_C,V_S,\varepsilon)$ resembles a delta function at

$$V_S = 0$$

$$V_C = \tfrac{1}{2} \left[\sqrt{1+2\mu/\varepsilon} - 1\right] \to 0 \quad . \tag{4.94}$$

This behavior is useful in making the variable transformation from $R(V_C,V_S,\varepsilon)$ to $R(\varepsilon',\varepsilon)$. The energy transfer, and therefore the final energy for a given initial energy, is given as an implicit but exact function of ε, V_C, and V_S by (4.86). It can be shown that the surface, $(\Delta/\varepsilon) \; (V_C,V_S)$, is approximately a plane in the neighborhood of the maximum of $R(V_C,V_S,\varepsilon)$. The equation of

137

that plane is obtained by expanding $(\Delta/\varepsilon)(V_C, V_S)$, given by (4.86), in a two-dimensional Taylor series about the (V_C, V_S) values given by (4.94), and truncating the expansion at the linear terms

$$(\Delta/\varepsilon)(V_C, V_S) = a - 1 + bV_S + cV_C \quad , \quad \text{where} \tag{4.95}$$

$$a = \dot{y}^2 \tau_C$$

$$b = 2\dot{y}(\tau_C)\,\dot{y}_{V_S}(\tau_C) \quad \text{and} \tag{4.96}$$

$$c = 2y(\tau_C)\,\dot{y}_{V_C}(\tau_C) \quad , \tag{4.97}$$

with subscripts indicating differentiation, and τ_C for $V_C = V_S = 0$ is obtained from (4.78). A rotation to two new coordinates, V_{Sr} and V_{Cr} whose axes are parallel and normal to the intersection of the ε' plane with the plane of the (V_C, V_S) axes, picks up the direction of the normal coordinate V_{Cr}. This transformation is

$$V_{Sr} = \frac{V_C}{\sqrt{1+c^2/b^2}} - \frac{V_S}{\sqrt{1+b^2/c^2}}$$

$$V_{Cr} = \frac{V_C}{\sqrt{1+b^2/c^2}} + \frac{V_S}{\sqrt{1+c^2/b^2}} \quad . \tag{4.98}$$

The Jacobian is unity. Inverting (4.98), substituting into (4.93), and integrating over V_{Sr} gives

$$R(\varepsilon, V_{Cr})dV_{Cr}d\varepsilon = \frac{N_g k_B T}{Q_g}\sqrt{\frac{\varepsilon}{\pi\mu}}\left[1 + \frac{V_{Cr}}{\sqrt{1+b^2/c^2}}\right]$$

$$*\exp[-(\varepsilon/\mu)V_{Cr}^2 - \varepsilon]dV_{Cr}d\varepsilon \quad . \tag{4.99}$$

The one-way equilibrium transition rate is then obtained using

$$R(\varepsilon', \varepsilon)d\varepsilon'd\varepsilon = R[\varepsilon, V_{Cr}(\varepsilon')]\frac{dV_{Cr}}{d\varepsilon'}d\varepsilon'd\varepsilon$$

$$V_{Cr}(\varepsilon', \varepsilon) = \frac{\varepsilon' - a\varepsilon}{\varepsilon\sqrt{b^2+c^2}} \quad , \tag{4.100}$$

with the desired result

$$R(\varepsilon', \varepsilon) = \frac{N_g k_B T}{Q_g}\frac{\varepsilon(c^2+b^2)+c(\varepsilon'-a\varepsilon)}{\sqrt{\pi\mu}\,[\varepsilon(c^2+b^2)]^{3/2}}\exp\left[-\frac{(\varepsilon'-a\varepsilon)^2}{\mu\varepsilon(b^2+c^2)} - \varepsilon\right] \quad . \tag{4.101}$$

This approximate rate does not satisfy the symmetry required by detailed balancing. A symmetric form can be constructed by taking the arithmetic mean of $R(\varepsilon, \varepsilon')$ and $R(\varepsilon', \varepsilon)$. Numerical comparisons show negligible distinction between the symmetric and unsymmetric rates.

138

Introducing the approximations

$$a \approx 1$$

$$\varepsilon \approx \bar{\varepsilon} = \frac{\varepsilon + \varepsilon'}{2} \quad , \tag{4.102}$$

and $\Delta/2\varepsilon \ll 1$ yields a simple symmetric kernel

$$R(\bar{\varepsilon},\Delta) = \frac{N_g k_B T}{Q_g} \frac{1}{\sqrt{\mu \pi \bar{\varepsilon}(b^2 + c^2)}} \exp\left[-\frac{\Delta^2}{\mu\bar{\varepsilon}(b^2 + c^2)} - \bar{\varepsilon}\right] \quad . \tag{4.103}$$

This approximate one-way equilibrium transition rate is used by Pagni and Keck (1973) and Pagni (1973). Figure 4.9 shows contours of constant dimensionless one-way equilibrium transition rate $R(\varepsilon',\varepsilon)$ in the range 10^0 to 10^{38} in $(\varepsilon,\varepsilon')$ space. The gradient of $R(\varepsilon',\varepsilon)$ along the Δ axis is much greater than along the $\bar{\varepsilon}$ axis. This difference allows an expansion of the population $x(\varepsilon,t)$ about $\Delta=0$ to obtain an equivalent diffusion equation for $x(\varepsilon,t)$ from the master equation. The exponential decay in $\bar{\varepsilon}$ follows from the Boltzmann factor. A typical well depth, $v = 13$, is indicated by the dashed line in Fig. 4.9. There are four transition regions separated by $\varepsilon'=v$ and $\varepsilon=v$: for ε and $\varepsilon'>v$, the gas atom is never trapped; when ε and $\varepsilon'<v$, transitions occur between states in the well; for $\varepsilon<v$ and $\varepsilon'>v$ desorption takes place; and in the region $\varepsilon'<v$ and $\varepsilon>v$ gas phase atoms are adsorbed.

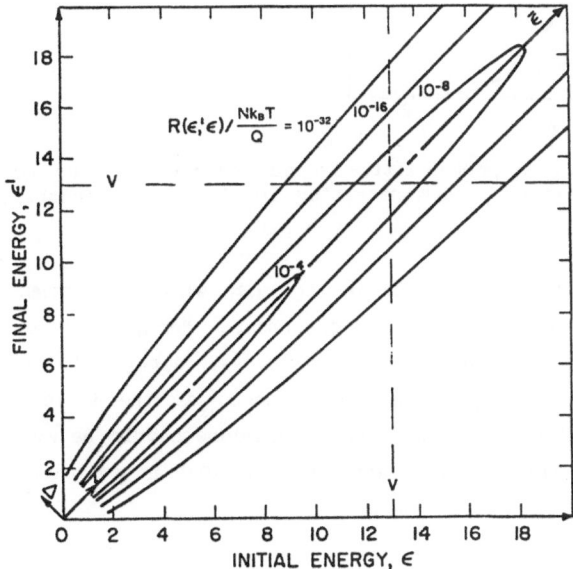

Fig.4.9. Contours of constant one-way equilibrium transition rate for $v=3,16$ and $\mu=0.5$. (Pagni and Keck 1973)

Once the equilibrium transition rate is known, the one-way equilibrium desorption rate $R(f,\varepsilon)$ is obtained as the integral of $R(\varepsilon',\varepsilon)$ over $v < \varepsilon' < \infty$. This is equivalent to integrating (4.103) over $\frac{1}{2}(v+\varepsilon) < \bar{\varepsilon} < \infty$ using $\Delta = 2(\bar{\varepsilon}-\varepsilon)$ with ε as a parameter. The result is

$$R(f,\varepsilon) = \frac{1}{2} \frac{N_g k_B T}{Q_s} e^{-4qp\varepsilon+8q^2\varepsilon} [erfc(p\ell-2q\varepsilon/\ell) + e^{8qp\varepsilon} erfc(p\ell+2q\varepsilon/\ell)] ,$$

(4.104)

where

$$q = \frac{1}{\sqrt{\mu(b^2+c^2)}}$$

$$p = \sqrt{4q^2+1} \quad \text{and} \quad (4.105)$$

$$\ell = \frac{1}{2}\sqrt{v+\varepsilon} . \quad (4.106)$$

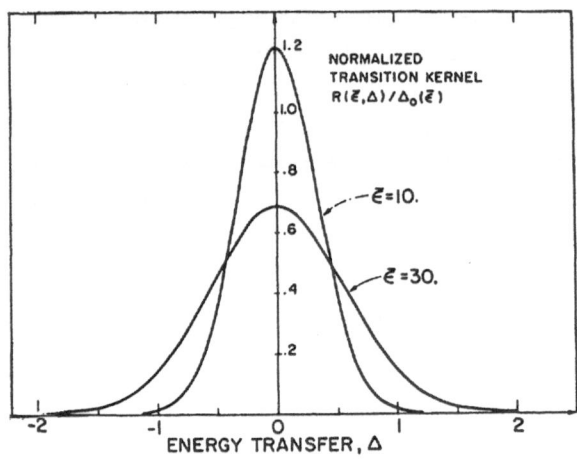

Fig.4.10. Normalized transition probabilities versus energy transfer for $v=3.16$ and $\mu=0.5$ with $\bar{\varepsilon}=10$ and $\bar{\varepsilon}=30$, demonstrating the Gaussian character. (Pagni and Keck 1973)

The Gaussian behavior of (4.103) is clear in Fig. 4.10 where the dimensionless transition rate is plotted versus energy transfer for $\mu = 0.5$ and $v = 3.16$. The two curves, equivalent to slicing Fig. 4.9. along $\bar{\varepsilon}=10$ and $\bar{\varepsilon}=30$, indicate that the mean energy transfer increases as the incident energy increases. This is reasonable since harder collisions should be more impulsive. The broadening of the halfwidth appears in (4.103) as a Gaussian scale factor of $\bar{\varepsilon}^{-1}$.

4.4 Mean-Field Theory at Finite Coverage

In this section, we very briefly review the mean field approach to the kinetics of physisorbed multilayers as developed by Summerside et al. (1982)
and Sommer and Kreuzer (1982a-d). In Sect. 2.2, we have outlined the calculation of an effective coverage-dependent surface potential (2.41) in which
the mutual interaction between gas particles is accounted for with a mean
field potential calculated from the temperature-dependent Hartree-Fock
equations (2.50). An extensive discussion is given in Sect. 2.2.

Sommer and Kreuzer (1982a-d) calculate the transition probabilities in
the master equation within the one-phonon approximation (4.7). Some clari-
fying remarks are necessary when the adsorbate reaches monolayer coverage.
First, the surface loading may change the phonon modes as has been observed
in gas-solid systems with chemisorption of heavy adsorbates (Ibach and
Bruchmann 1980); see also Fig. 2.38. However, for ^{3}He and ^{4}He adsorbates,
one does not expect too great an effect and one can therefore keep the unp-
erturbed phonon spectrum of the unloaded solid for the calculations of
desorption times. Second, at monolayer coverage desorbing particles might
draw the necessary energy not directly from the solid but via a collective
excitation in the adsorbate itself. Such a coupling could be incorporated
explicitly in the random phase approximation. The calculation of the transition probabilities $W(\iota,\iota')$ from (4.7) as a function of temperature and coverage proceeds by straightforward numerical integration of the matrix elements with the self consistent eigenfunctions of the Hartree-Fock equations
(2.50). Results will be presented when we discuss desorption times in the
next chapter.

5. Desorption Times

We briefly discussed isothermal and temperature programmed desorption experiments in the introductory chapter. We also collected data on the desorption time t_d for a number of physisorbed gas-solid systems in Table 1.1. Recently, Goodstein and his collaborators have performed a series of low-temperature experiments on the desorption kinetics of helium on metal surfaces. As these experiments yield time of flight spectra exhibiting a wealth of information in addition to the desorption time, we postpone a detailed discussion to the appropriate chapter, 6. In this chapter, we report on the calculation of desorption times from the master equation, with the various transition probabilities surveyed in Chap. 4. We begin with the simplest model, namely a mobile adsorbate on a uniform surface, with the further simplification that the parallel momentum of the desorbing particle is conserved according to (4.4). The master equation is then (4.6). Splitting the sums into bound state contributions, labeled with an index i, and continuuum contributions, labeled by k, it reads

$$\frac{dn_i}{dt} = \sum_{i'} W(i,i')n_{i'} - \sum_{i'} W(i',i)n_i - \sum_{k} W(k,i)n_i + \sum_{k} W(i,k)n_k \quad . \tag{5.1}$$

To calculate the isothermal desorption rate, we recall that in an isothermal desorption experiment, starting from a gas-solid system in equilibrium at temperature T, one rapidly pumps away the gas phase keeping the system's temperature constant. This implies the initial conditions

$$n_i(t \leq 0) = n_i^{eq} \quad , \tag{5.2}$$

but it also suggests that one should keep the continuum states unoccupied, i.e.,

$$n_k(t \geq 0) = 0 \quad . \tag{5.3}$$

This drops the last term in (5.1); the resulting truncated matrix of transition probabilities we called W^{iso} in (3.53). Diagonalizing it according to the procedure outlined in (3.25,32), we can write the time evolution of the

142

bound state occupation functions as

$$n_i(t) = \sum_\kappa A_{i\kappa} e^{-\lambda_\kappa t} \quad , \tag{5.4}$$

where the λ_κ are the eigenvalues of the matrix W^{iso}. Summing over all bound states

$$\frac{N(t)}{N(0)} = \frac{\displaystyle\sum_i n_i(t)}{\displaystyle\sum_i n_i(0)}$$

$$= \sum_\kappa S_\kappa e^{-\lambda_\kappa t} \quad , \tag{5.5}$$

we get the probability that particles are still found in the adsorbate at time t.

5.1 Model System with Two Bound States

We begin our discussion of (5.5) by examining its structure analytically for a simple model system with just two bound states (Gortel, Kreuzer, and Teshima, 1980c). In this case we get

$$N(t)/N(0) = S_0 e^{-\lambda_0 t} + S_1 e^{-\lambda_1 t} \quad , \tag{5.6}$$

with

$$\lambda_{0,1} = \tfrac{1}{2} [R_{C0} + R_{C1} + W(1,0) + W(0,1)]$$
$$\pm \tfrac{1}{2} [[R_{C0} + W(1,0) - R_{C1} - W(0,1)]^2 + 4 W(0,1) W(1,0)]^{1/2} \tag{5.7}$$

$$S_{0,1} = \pm \frac{1}{\lambda_1 - \lambda_0} \left[\lambda_{1,0} - \frac{n_0(0)}{N(0)} R_{C0} - \frac{n_1(0)}{N(0)} R_{C1} \right] \quad , \tag{5.8}$$

where $W(i,i')$ and R_{Ci} are given in (4.22,25,29) in the one-phonon approximation with the surface potential chosen as a Morse potential (2.37). Recall that $W(i,i')$ is the transition probability for an adsorbed particle to jump from bound state level i' to level i whereas R_{Ci} is the probability to go from the i-th bound state into any continuum state.

It is instructive to look at the low-temperature behavior of the expressions (5.7) and (5.8). One finds for temperatures such that

$$\exp[-\beta(E_1 - E_0)] \ll 1 \quad , \tag{5.9}$$

$$\lambda_0 \approx R_{C0} + \frac{\Gamma_{10}(R_{C1}-R_{C0})}{R_{C1}-R_{C0}+\Gamma_{10}} \exp[-\beta(E_1-E_0)] \tag{5.10}$$

$$\lambda_1 \approx R_{C1} + \Gamma_{10} + \Gamma_{10} \frac{R_{C1}-R_{10}+2\Gamma_{10}}{R_{C1}-R_{C0}+\Gamma_{10}} \exp[-\beta(E_1-E_0)] \tag{5.11}$$

$$S_0 = 1 - S_1 \tag{5.12}$$

$$S_1 \approx (\frac{R_{C1}-R_{C0}}{R_{C1}-R_{C0}+\Gamma_{10}})^2 \exp[-\beta(E_1-E_0)] \quad . \tag{5.13}$$

We also separated the temperature dependence from the transition probabilities by rewriting (4.9) as

$$W(i,j) = \frac{\Gamma_{ij}}{\exp\beta(E_i-E_j) - 1} \frac{3}{2+\ell_s{}^3} \sum_\sigma s^{(\sigma)} \theta(\hbar\omega_D{}^{(\sigma)} - |E_i-E_j|) \tag{5.14}$$

Note first that the coefficient S_1 becomes negligibly small in the low-temperature region implying that for $t > \lambda_1{}^{-1}$ the time evolution of the system is characterized by one time scale

$$t_d = \lambda_0{}^{-1} \quad . \tag{5.15}$$

Indeed, the difference $(\lambda_1-\lambda_0)$ becomes larger for lower temperatures, implying that the small transients controlled by λ_1 die out very quickly. For not too shallow states, E_1, one finds numerically that $\Gamma_{10} \gg R_{C1} > R_{C0}$ at low temperatures. If gas particles can desorb from both bound states by absorption of a single phonon, i.e., if $-E_0 < \hbar\omega_D$, then we have at low temperatures $\lambda_0 \approx R_{C0}$ and $\lambda_1 \approx \Gamma_{10}$. This can be understood by noting that (5.4) implies a negligible occupation of the upper bound state E_1. Because $W(0,1) \gg W(1,0)$, a particle in E_1 will first make a transition to E_0, emitting a nonequilibrium phonon before it desorbs from E_0 into the continuum. For $-E_0 < \hbar\omega_D$ and at low temperatures, the desorption in such a system with two bound states thus appears to be solely controlled by the lower bound state E_0. One finds numerically that over a limited temperature regime, t_d can be parametrized by a Frenkel-Arrhenius formula (1.3)

$$t_d = t_d^0 e^{E_d/k_B T} \quad , \tag{5.16}$$

where the desorption energy E_d is slightly larger than the lowest bound state energy $-E_0$. Turning now to a system, still at low temperatures, with $-E_1 < \hbar\omega_D < -E_0 < 2\hbar\omega_D$ and $|E_1-E_0| < \hbar\omega_D$, we obviously have $R_{C0} = 0$ and

$$t_d \approx \frac{R_{C1}+\Gamma_{10}}{R_{C1}\Gamma_{10}} \exp[-\beta(E_1-E_0)] \quad . \tag{5.17}$$

Desorption in this situation can only proceed via a phonon cascade $0 \to 1 \to$ continuum. If $\Gamma_{10} \gg R_{ci}$, as one can check numerically, one then gets

$$t_d \approx \frac{1}{R_{c1}} \exp[-\beta(E_1-E_0)] \quad . \tag{5.18}$$

Again R_{c1} can be parametrized as

$$R_{c1} \approx \nu_1 e^{-\tilde{E}_1/k_BT} \quad , \tag{5.19}$$

where, over a limited low-temperature region, ν_1 is independent of T and $\tilde{E}_1 \gtrsim -E_1$, so that the Frenkel-Arrhenius parametrization of t_d again holds true with $E_d=(E_1-E_0+\tilde{E}_1)\gtrsim-E_0$. This time, however, $t_d^0=\nu_1^{-1}$ is controlled by the matrix elements between the higher state E_1 and the continuum. Thus the prefactor t_d^0 is, in this case, controlled by the upper state, whereas the desorption energy E_d is close to the energy $-E_0$ of the lower state. Let us note in passing that the results of the above analysis hold independently of the particular form of the surface and interaction potentials and are equally valid for both localized and mobile adsorption. The condition $\Gamma_{01} \gg R_{ci}$ for not too shallow bound states is also generally satisfied, being a result of rapid oscillations of the continuum wavefunctions.

A situation with $-E_1<\hbar\omega_D<-E_0<2\hbar\omega_D$ is not very likely to be met in real systems, which will always have many bound states if $-E_0>\hbar\omega_D$. Before we turn our attention to such systems, we want to discuss briefly the dependence of t_d on the position z_0 of the minimum in the Morse potential (2.37). We have seen earlier that one has to choose $z_0 \gtrsim \gamma^{-1}$ in order to prevent gas particles from penetrating into the solid. In Fig. 5.1 we show λ_0^{-1} as a function of $\xi_0=\gamma z_0$ for two systems, each developing two bound states. For ease of reference, we recall that we characterized a Morse potential (2.37) by two parameters,

$$\sigma_0^2 = \frac{2mV_0}{(\hbar\gamma)^2}$$

$$r = \frac{2m\omega_D}{\hbar\gamma^2} \quad , \tag{5.20}$$

and that we measure all energies, including the temperature, in units of the Debye energy of the solid,

$$\varepsilon = \frac{E}{\hbar\omega_D}$$

$$\delta = \frac{\hbar\omega_D}{k_BT} \quad . \tag{5.21}$$

In Fig. 5.1. we also give t_d for the system with $\sigma_0 = 2.49$ and $\xi_0 = \infty$ as a function of δ . We see that in the system with $\sigma_0 = 1.55$ for $\xi_0 = 1$ and at

145

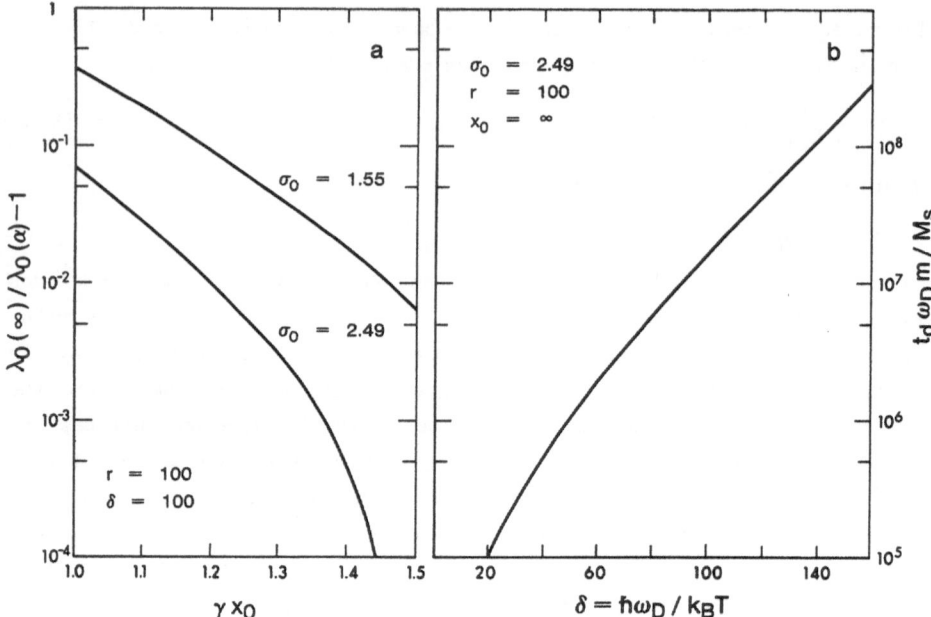

Fig.5.1. a Relative change in λ_0 as a function of the shift z_0 in the Morse potential for a gas-solid system with two bound states. $\alpha=2\sigma_0\exp(\gamma z_0)$. **b** Temperature dependence of the isothermal desorption time $t_d=\lambda_0^{-1}$. (Gortel, Kreuzer, and Teshima 1980c)

δ = 100, λ_0^{-1} is smaller by some 36% than its value at ξ_0 = ∞ which it, indeed, reaches within 0.6% at ξ_0 = 1.5. For the larger σ_0 = 2.49, the change in λ_0^{-1} from ξ_0=∞ to ξ_0=1 amounts to only 7% for δ>70 and to 11% at δ=30 and to 20% at δ=5. For a system with many bound states, , i.e., with larger σ_0, even less dependence is found on ξ_0, so that one can safely set it equal to ∞. This justifies the use of the much simplified transition probabilities (4.22,26).

5.2 Systems with a Few Shallow Bound States

In this section, we will present calculations of isothermal desorption times for gas-solid systems that have several shallow bound states (Gortel, Kreuzer, and Teshima 1980c). We call a bound state shallow if a gas particle can desorb from it by absorbing a single phonon. Systems in this category are He-LiF, He-NaF, H-LiF, H-NaF, He-graphite, etc. These systems are also known to show mobile adsorption, implying that the motion of particles in the adsorbate is more or less unhindered in the direction parallel to the surface, so that our one-dimensional model is adequate.

Table 5.1. Bound state energies $|E_n|/k_B$ [K], experimental and calculated for Morse potentials with different parameters. (Gortel et al. 1980c)

	⁴He-LiF			⁴He-NaF		⁴He-Graphite		
γ^{-1}[Å]	1.09	1.09		0.97		0.95	0.95	
U_0[K]	81.75	89.0	expt.	77.78	expt.	160.54	171.34	expt.
n=0	62.69	68.47	68.47	57.10	57.10	129.33	139.03	139.03
1	32.16	36.45	28.55	25.30	21.70	77.04	84.54	73.46
2	11.72	11.43	9.05	6.27	6.27	38.21	43.53	34.07
3	1.38	2.44	2.44			12.86	16.01	11.49
4						0.98	1.97	1.97

We first deal with the ⁴He-LiF system. Its surface potential has four bound states given in Table 5.1. To reproduce them in a Morse potential, Gortel et al. (1980c) followed two options leading to somewhat different potential parameters. One set they get by a least-squares fit to the bound state energies. Because we know from the previous section that the lowest bound-state energy basically defines the activation energy, we get a second set of potential parameters by fitting the deepest and shallowest bound states for this system. In the temperature region $T \leq 40$ K, the Debye temperature of LiF is $\hbar\omega_D/k_B \approx 730$ K, dropping sharply to a minimum of 610 K at $T \approx 60$ K (Scales 1958). In Table 5.2, we present the λ_K's and S_K's of (5.5) for some typical temperatures for the first potential. At high temperatures, , i.e., for $\delta \leq 20$ or $T \geq 40$ K, all λ_K's are of the same order although the coefficient S_0 is larger than S_1 to S_3. For $\delta \geq 20$, the lowest eigenvalue λ_0 splits off dramatically from λ_1 to λ_3 and S_0 approaches unity very closely, indicating that apart from very small and very fast transients, the system is indeed controlled by one exponential in analogy with the analytical results obtained for systems with two shallow states at low temperatures. For $\delta \geq 20$, we can then identify $t_d = \lambda_0^{-1}$. This isothermal desorption time is plotted in Fig.5.2. as a function of δ. Although $\ln(t_d)$ is not completely linear in δ, we can parametrize it by a Frenkel-Arrhenius formula over a limited temperature range,

$$t_d = 1.9 \times 10^{-9}\text{s} \exp(71\text{K}/T), \qquad 4 \text{ K} < T < 40 \text{ K} . \tag{5.22}$$

Again the desorption energy E_d in (5.22) is larger than the lowest bound state by some 5%. In Fig. 5.2, we also show the desorption times calculated from the second Morse potential in Table 5.1. We see that the two potential fits give comparable desorption times.

Table 5.2. Eigenvalues λ_K in s^{-1} and coefficients S_K in (5.5) for ^{4}He-LiF for $\gamma^{-1} = 1.09$ Å and $V_0/k_B = 81.75$ K as a function of δ. (Gortel, Kreuzer, and Teshima 1980c)

δ	λ_0	λ_1	λ_2	S_0	S_1	S_2
10	4.06×10^8	6.03×10^8	1.37×10^9	0.842	0.112	0.031
20	1.08×10^8	2.47×10^8	5.96×10^8	0.980	0.001	0.015
30	3.45×10^7	1.63×10^8	3.81×10^8	0.992	0.003	0.006
50	4.19×10^6	1.17×10^8	2.43×10^8	0.999	10^{-4}	0.0006
100	2.87×10^4	9.97×10^7	1.81×10^8	1.000	10^{-7}	10^{-7}
150	2.46×10^2	9.78×10^7	1.78×10^8	1.000	10^{-11}	10^{-10}

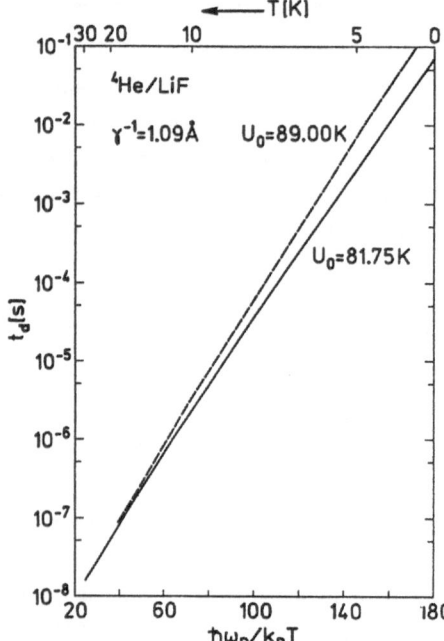

Fig. 5.2. Temperature dependence of the isothermal desorption time for ^{4}He-LiF for two sets of Morse potential parameters (Tables 5.1,3). (Gortel, Kreuzer, and Teshima 1980c)

The ^{4}He-LiF system, as mentioned above, has four shallow bound states from which gas particles can desorb by absorption of a single phonon. In this case one might argue that only bound state – continuum transitions are important with cascades like $0 \to 1 \to \ldots \to$ continuum being negligible at least as long as $k_B T$ is not too much smaller than $-E_0$. This idea has been checked numerically by setting all bound state–bound state transition probabilities $W(i,i')$ in (5.1) equal to zero. For ^{4}He-LiF, one finds that for 4 K $<T<$ 40 K λ_0 is reduced by about a factor of two. With uncertainties in t_d arising from different choices of potential parameters being of the same size, the approximation of setting bound state-bound state transitions equal to zero seems acceptable for the ^{4}He-LiF system with its rather shallow bound states. However, if the lowest bound state $-E_0$ gets much lower, for example, $-E_0 \lesssim \hbar\omega_D$, this approximation fails dismally, and, of course, is even qualitatively wrong for $-E_0 > \hbar\omega_D$, because one-phonon processes cannot empty such a deep state. As a numerical example, we consider a hypothetical gas-solid system with $r=60$, $\sigma_0=8.05$, , i.e., having eight bound states, the lowest of which is $\varepsilon_0 = E_0/\hbar\omega_D = -0.95$. In this case, $t_d^{-1} = 10^{-10}(M_S/m)\omega_D$ at $\delta=20$; without bound state-bound state transitions, this time is reduced by a factor of 10^{-3}. At high temperatures, say such that $\delta \approx 1$, the approximation is still unacceptable because then the coefficient S_0 becomes substantially smaller than one, with the other S_K increasing proportionately.

Gortel, Kreuzer, and Teshima (1980c) have also calculated isothermal desorption times, identified as $t_d = \lambda_0^{-1}$ for ^{4}He-NaF and ^{4}He-graphite. The relevant parameters and the predicted desorption times are given in Table 5.3. The "theoretical error bars", due to some arbitrariness in choosing γ^{-1} and V_0, are similar to those discussed for the He-LiF system.

We have repeatedly argued that the structure of the transition probabilities in the master equation is such that over the temperature range of experimental interest, the smallest eigenvalue λ_0 is much less than all others and occurs in the time evolution (5.5) with a weight S_0 of order one. This allowed us to identify λ_0 as the desorption time t_d. We should point out that there can be a pathological situation where this interpretation needs some thought. We are referring to a gas-solid system for which the

Table 5.3. Frenkel-Arrhenius parametrization (1.3) of the isothermal desorption times for various systems and Morse potential parameters. (Gortel, Kreuzer, and Teshima 1980c)

System	N	T [K]	t_d^0 [s]	Q [K]	γ^{-1} [Å]	V_0 [K]	T_D [K]
^{4}He-LiF	4	4-40	1.9×10^{-9}	71	1.09	81.75	730
	4	4-40	1.4×10^{-9}	77	1.09	89.0	730
^{4}He-NaF	3	4-15	8.5×10^{-10}	63.2	0.97	77.78	450
^{4}He-	5	4-15	2.1×10^{-12}	139.5	0.95	160.54	185
graphite	5	4-15	1.8×10^{-12}	149.3	0.95	171.34	185
H-NaCl	7	15-35	7.2×10^{-12}	370	1.78	399.3	280
^{4}He-Ar	6	4-10	2.3×10^{-12}	151	1.09	171.93	92
	10	4-10	4.1×10^{-12}	151	1.98	160.07	92

surface potential develops a very shallow bound state of energy E_N so that the associated time scale $|\hbar/E_N|$, or the classical oscillation period at that energy, becomes of the order or even larger than the desorption time. For a Morse potential, this would happen if σ_0 is slightly larger than half integer. Such a situation is depicted in Fig. 5.3. We note that the pathological lowest eigenvalue (pathological because it disappears for a slightly different choice of potential parameters) contributes little to the time evolution (5.5) as evidenced by its small, and, for $\delta > 50$, negligible, weight.

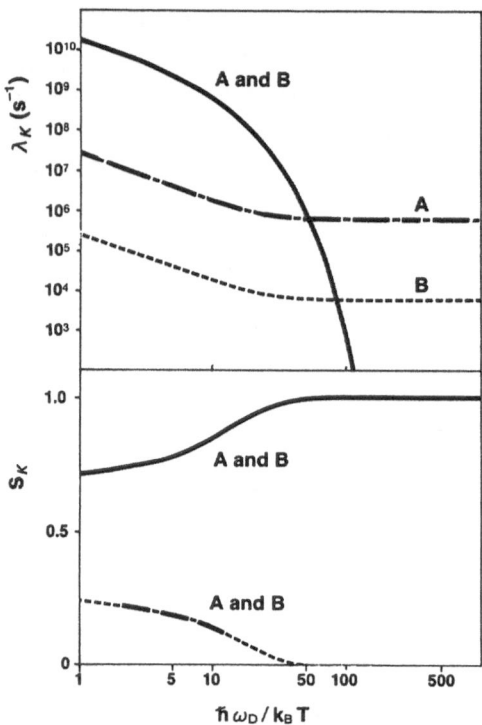

Fig.5.3. The two lowest eigenvalues and associated weights S_K in (5.5) for two gas-solid systems with a very shallow bound state. Morse potential parameters: r=70.496; for system A σ_0=3.501 and for B σ_0=3.50001. ω_D=5.892x10^{13} s^{-1}; m/M$_S$=0.0952 appropriate for ^{4}He-NaF.

5.3 Perturbation Theory of the Master Equation

In this section, we want to develop an appropriate perturbation theory for the calculation of the desorption time from the master equation (Kreuzer and Teshima 1981). We recall that the full matrix (4.7) of transition probabilities satisfies detailed balance (the Greek indices ι,ι' run over bound states and continuum states, whereas Latin indices n,n' run over N+1 bound states only):

$$W(\iota',\iota)\, e^{-\delta\varepsilon_\iota} = e^{-\delta\varepsilon_{\iota'}}\, W(\iota,\iota') \quad . \tag{5.23}$$

The matrix $W(\iota,\iota')$, thus far nonsymmetric, can therefore be symmetrized by a transformation (3.37)

$$S(\iota,\iota') = e^{\delta\varepsilon_\iota/2}\, W(\iota,\iota')\, e^{-\delta\varepsilon_{\iota'}/2} = S(\iota',\iota)$$

$$\chi_\iota(t) = n_\iota(t)\, e^{\delta\varepsilon_\iota/2} \quad . \tag{5.24}$$

Equation (5.1) then reads for the bound state occupation probabilities

$$\frac{d}{dt}\, \chi_n(t) = \sum_{n'\neq n=0}^{N} S(n,n')\chi_{n'}(t)$$

$$- \left[R_{cn} + \sum_{n'\neq n=0}^{N}{}' e^{-\delta\varepsilon_{n'}/2}\, S(n',n)\, e^{\delta\varepsilon_n/2} \right] \chi_n(t) \tag{5.25}$$

Instead of diagonalizing the nonsymmetric matrix $W^{iso} = W_0 + W_c$ with elements

$$W_0(n,n') = W(n,n') - \delta_{n,n'} \sum_{n''=0}^{N} W(n'',n) \tag{5.26}$$

$$W_c(n,n') = - \delta_{nn'}\, R_{cn} \quad , \tag{5.27}$$

we work now with the symmetric matrix

$$S^{iso} = S_0 + W_c \quad , \tag{5.28}$$

where the elements of S_0 are

$$S_0(n,n') = S(n,n') - \delta_{n,n'} \sum_{n''=0}^{N} e^{\delta(\varepsilon_n - \varepsilon_{n''})/2}\, S(n'',n)$$

$$= e^{\delta(\varepsilon_n - \varepsilon_{n'})/2}\, W(n,n') - \delta_{nn'} \sum_{n''=0}^{N} W(n'',n) \quad . \tag{5.29}$$

Let us rewrite (5.25) as

$$\frac{d\chi}{dt} = S \cdot \chi \quad , \tag{5.30}$$

where χ is a column vector with components χ_n, $n=0,\ldots,N$. Diagonalizing S, we get the eigenvalue equation for S

$$S \cdot \mu^{(\kappa)} = -\lambda_\kappa\, \mu^{(\kappa)} \quad , \tag{5.31}$$

with real eigenvalues λ_κ assumed to be ordered $\lambda_0 < \lambda_1 < \ldots < \lambda_N$ and an orthonormal set of eigenvectors $\mu^{(\kappa)}$. The solution of (5.30) then reads

$$\chi(t) = \sum_{\kappa=0}^{N} [\mu^{(\kappa)} \cdot \chi(0)] e^{-\lambda_\kappa t}\, \mu^{(\kappa)} \quad , \tag{5.32}$$

151

which yields (5.4) again,

$$n_n(t) = \sum_{\kappa=0}^{N} A_{n\kappa} \, e^{-\lambda_\kappa t} \quad , \tag{5.33}$$

where the coefficients are now given in terms of the eigenvectors of the symmetric matrix S,

$$A_{n\kappa} = e^{-\delta\epsilon_n/2} \, \mu_n(\kappa) \sum_{n'=0}^{N} \mu_{n'}(\kappa) \, n_{n'}(0) e^{\delta\epsilon_{n'}/2} \quad . \tag{5.34}$$

Here, $n_{n'}(0)$ is the initial occupation of the n'-th bound state, and $\mu_n(\kappa)$ is the n-th component of the κ-th eigenvector. We have shown in detail in the previous section, that, as the temperature in the gas–solid system is lowered, the lowest eigenvalue λ_0 splits off dramatically from the others, becoming much smaller. Let us look at another numerical example, choosing a gas–solid system whose surface potential develops many bound states. Recall from the explicit expressions (4.22,26) that for phonon-mediated gas-solid interactions, all transition probabilities $W(\iota,\iota')$ calculated in second-order time-dependent perturbation theory are proportional to the Debye frequency ω_D of the solid (if we measure energies in units of the Debye energy), and to the ratio m/M_S of the masses of a gas particle m and of a unit cell of the solid M_S. In the following examples (Gortel et al. 1981), we choose $\hbar\omega_D/k_B$=405 K for tungsten and m/M_S=0.714 for the Xe-W system. For the depth of the Morse potential, we choose $V_0/\hbar\omega_D$ = 11.56 so that for a range γ^{-1} = 1.5 Å, chosen arbitrarily, the lowest boundstate energy is E_0/k_B = -4662 K = $-Q/k_B$, which equals the heat of adsorption for the Xe-W system (Dresser et al. 1971, 1974). We thus have from (5.20) r = 4969.0 and σ_0 = 239.66, so that this particular Morse potential develops 240 bound states. Diagonalizing (5.28), we find that at δ = 1.0, λ_0 = 3.7895 x 10^6 s^{-1}, λ_1 = 1.7352 x 10^{10} s^{-1}, λ_2 = 2.5647 x 10^{10} s^{-1}, and λ^{240} = 1.8680 x 10^{13} s^{-1}. More generally, we find that, for $\delta \gtrsim$ 0.5, all λ_κ for κ>0 are much larger than the smallest eigenvalue λ_0. This confirms our earlier assertion, substantiated already in Table 5.2, that for times t>>$(\lambda_\kappa - \lambda_0)^{-1}$ with κ>0, all fast transients in (5.33) with rate constants λ_κ, κ>0 have died out, and the time evolution settles down to an exponential decay

$$n_n(t) \approx A_{n0} e^{-\lambda_0 t} \quad , \tag{5.35}$$

which is characterized by a single relaxation time $t_d = \lambda_0^{-1}$.

In Fig. 5.4, we plot the deviation of the nonequilibrium distribution function $n_n(t)$ from the initial equilibrium one for t>>$(\lambda_\kappa-\lambda_0)^{-1}$, i.e., we plot

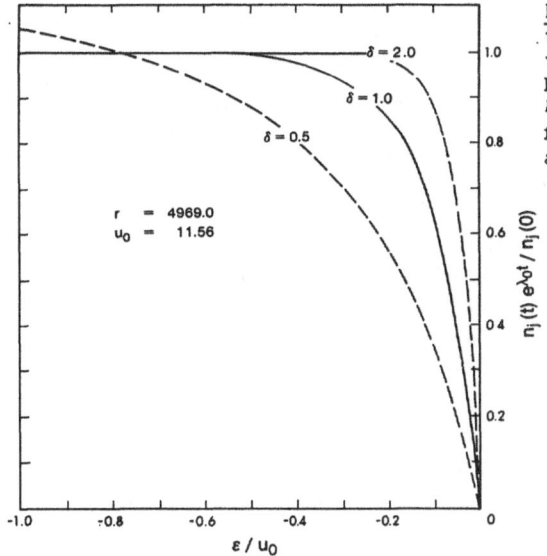

Fig.5.4. Steady-state occupation of the bound states for various temperatures $\delta = \hbar\omega_D/k_BT$. Parameters: $\hbar\omega/k_B = 405$ K, $m/M_s = 0.714$ appropriate for the Xe-W system. (Kreuzer and Teshima 1981)

$$\frac{n_n(t)e^{\lambda_0 t}}{n_n(0)} \cdot \tag{5.36}$$

Deviating markedly from the value one at higher bound-state energies indicates that the desorption process removes particles from these bound states faster than bound state – bound state transitions can manage to rearrange the bound state occupation into an equilibrium one. Decreasing the temperature to $\delta = 2.0$ yields $\lambda_0 = 74.691$ s^{-1}. The desorption process slows down considerably, allowing the bound state – bound state transitions to become more effective, so that (5.36) is closer to one up to higher bound state energies. Also note that at higher temperatures, e.g., at $\delta=0.5$, the lower bound states are relatively more occupied in the steady state than in equilibrium, simply because desorption depletes the higher levels too fast. The parameter dependence of these results can be seen by comparing Figs. 5.4 and 5.5; for the latter, we choose $r = 550$ and $\sigma_0 = 79.734$, so that this Morse potential, of the same depth V_0 as the previous one but of a range $\gamma^{-1} = 0.5$ Å, develops only 80 bound states. Because a smaller range implies a stronger coupling of the adparticle to the phonons of the solid, bound state-bound state transitions become more effective, maintaining the adsorbate closer to equilibrium during the desorption process. Similar conclusions were also reached by Pagni (1973) who based his calculations on the transition probabilities $R(\varepsilon,\Delta)$ given in (4.103), as calculated from the classical soft cube model. This certainly encourages one to try a perturbation

153

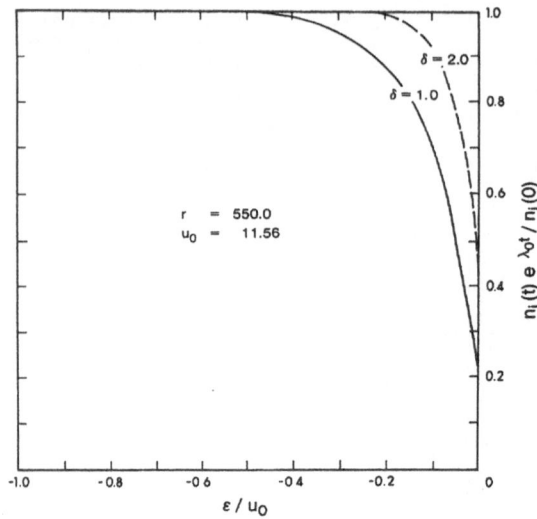

Fig.5.5. Steady-state occupation of the bound states for various temperatures $\delta = \hbar\omega_D/k_BT$. Parameters: $\hbar\omega/k_B = 405$ K, $m/M_s = 0.714$ appropriate for the Xe-W system. (Kreuzer and Teshima 1981)

calculation of λ_0 at low temperatures, based on the decomposition (5.28). Following Kreuzer and Teshima (1981), we diagonalize

$$\mathbf{S}_0 \cdot \mathbf{v}^{(\kappa)} = -\lambda_\kappa^{(0)} \, \mathbf{v}^{(\kappa)} , \qquad \kappa = 0,\ldots N \qquad (5.37)$$

and observe immediately that the lowest eigenvalue is $\lambda_0^{(0)} = 0$. The corresponding normalized eigenvector $\mathbf{v}^{(0)}$ has components

$$\nu_n^{(0)} = \frac{e^{-\delta\varepsilon_n/2}}{\sqrt{\displaystyle\sum_{m=0}^{N} e^{-\delta\varepsilon_m}}} \qquad (5.38)$$

corresponding to equilibrium occupation functions. Because $\mathbf{S}^{iso}$ is symmetric, we can, starting from $\mathbf{S}_0$, do straightforward perturbation theory on (5.28) and get

$$\lambda_\kappa = \lambda_\kappa^{(0)} + \mathbf{v}^{(\kappa)}\mathbf{T}\cdot\mathbf{W}_C\cdot\mathbf{v}^{(\kappa)} + \sum_{\kappa'\neq\kappa=0}^{N} \frac{\left|\mathbf{v}^{(\kappa')}\mathbf{T}\cdot\mathbf{W}_C\cdot\mathbf{v}^{(\kappa)}\right|^2}{\lambda_\kappa^{(0)}-\lambda_{\kappa'}^{(0)}} + \ldots, \qquad (5.39)$$

where $\mathbf{v}^{(\kappa)}\mathbf{T}$ is the transpose of $\mathbf{v}^{(\kappa)}$. Note that

$$\lambda_0^{(0)} = 0$$

$$\lambda_0^{(1)} = \frac{\displaystyle\sum_n R_{cn} e^{-\delta\varepsilon_n}}{\displaystyle\sum_n e^{-\delta\varepsilon_n}} = \frac{\displaystyle\sum_n R_{cn}n_n(0)}{\displaystyle\sum_n n_n(0)} . \qquad (5.40)$$

154

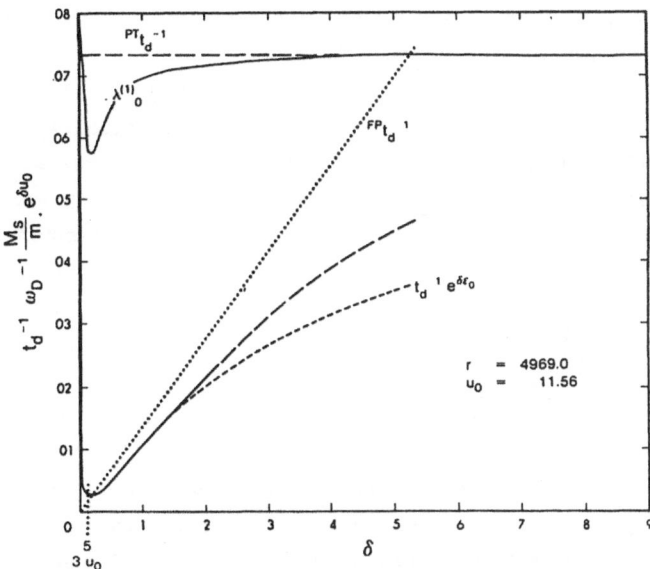

Fig.5.6. Inverse desorption time in various approximations. $^{PT}t_d$ is calcu-lated from (5.42). $^{FP}t_d$ is the Fokker-Planck result given in (8.5,30). Heavy lines mean first passage time. (Kreuzer and Teshima 1981)

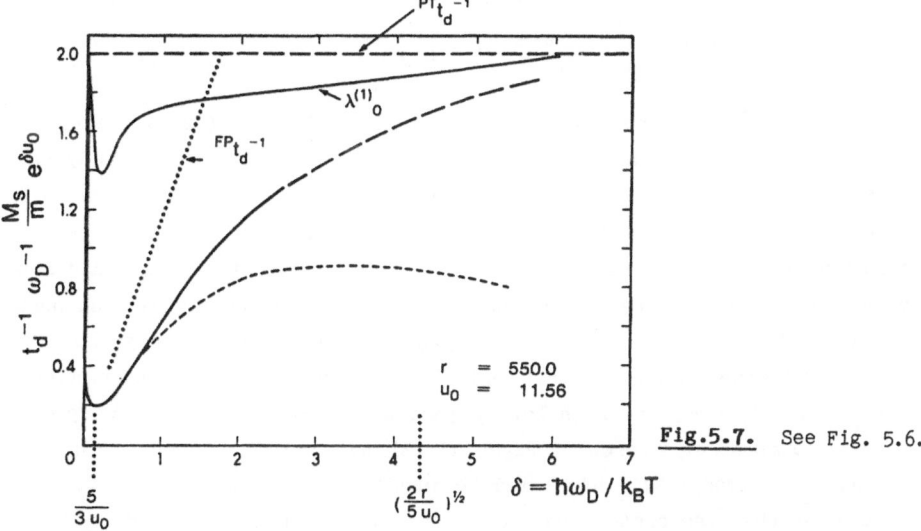

Fig.5.7. See Fig. 5.6.

The calculation of $\lambda_0^{(1)}$ thus presumes that the initial equilibrium distribu-tion $n_n(0)$ is maintained during the desorption process. Kreuzer and Teshima (1981) have calculated λ_0 by exact diagonalization of (5.28) and also by eva-luating the perturbation series (5.39) up to third order. The results are given in Figs. 5.6,7. It is fair to say that lowest-order perturbation theory, i.e., $\lambda_o^{(1)}$, is a poor approximation to t_d^{-1} throughout the temperature range depicted, in particular for $1 \gtrless \delta \gtrless$ 4, where t_d can actually be measured in the Xe-W system. For larger δ, λ_o^{-1} becomes immeasurably large, implying also that the ratios λ_o/λ_κ for $\kappa>0$ of the exact eigenvalues λ_κ of S^{iso} become very small, making exact diagonalization on most computers impossible. Fortunately, third-order perturbation theory works very well in this region, in particular after one resums (5.39) in a Padé-type fashion

$$\lambda_o = \lambda_o^{(1)} + \lambda_o^{(2)} + \lambda_o^{(3)} + \dots$$

$$= \frac{\lambda_o^{(1)}}{1 - \frac{\lambda_o^{(2)}/\lambda_o^{(1)}}{1 - \lambda_o^{(3)}/\lambda_o^{(2)} + \lambda_o^{(2)}/\lambda_o^{(1)} + \dots}} \ . \tag{5.41}$$

This last expression leads to the dashed curves in Figs. 5.6,7. Kreuzer and Teshima (1981) have obtained a simple analytical expression for the desorption time by starting from (5.40) and expanding the Γ-functions in R_{cn} for large r and δu_o. We will review this approach in Chap. 8 dealing with the Kramers equation. Here we merely quote their result

$$PT t_d = \omega_D^{-1} \frac{M_s}{m} \left[\frac{\hbar \omega_D}{V_o}\right]^4 \left[\frac{2mV_o}{\hbar^2 \gamma^2}\right]^{3/2} e^{V_o/k_B T} \tag{5.42}$$

We can therefore conclude with Kreuzer and Teshima (1981) that in the temperature region of most interest, namely for $1 \gtrless \delta \gtrless 4$, the exact diagonalization yields the only trustworthy desorption times. Lowest-order perturbation theory $\lambda_o^{(1)}$ is unacceptable. The latter is used in the so-called "equilibrium theory" of desorption where it is assumed that the bound state occupation function does not deviate from an equilibrium distribution during the desorption process (e.g., Jewsbury and Beeby 1975; Holloway and Beeby 1975 and Armand 1977). Figures 5.4-7 demonstrate that this assumption is not warranted, except at such low temperatures where the desorption process becomes unmeasurably slow. Pagni (1973) has made the same point very strongly. Armand (1977) has tried to justify the equilibrium approximation by quoting that the error thus introduced, is less than 15% if $V_o/k_B T>5$ and less than 10% if $V_o/k_B T>10$. Figures 5.6. and 5.7. do not confirm these numbers. To get a 15% agreement between λ_o and $\lambda_o^{(1)}$ one needs $\delta>9$ in Fig. 5.6. and $\delta>4.5$ in Fig. 5.7., implying with $V_o/\hbar\omega_D = 11.56$ that $V_o/k_B T>100$ and 50, respectively. Kreuzer and Teshima (1981) argue that the ratio $V_o/k_B T$ is not the crucial parameter to assess the validity of the equilibrium assumption at all. Rather, it is a question of time scales. If the bound state –

bound state transitions are much faster than the bound state - continuum transitions, then equilibrium will be approximately maintained during the desorption process. However, these transitions are caused by the energy-dissipating coupling of the gas to the solid, and not by the static surface potential alone, so that $V_0/k_B T$ cannot be the critical parameter that determines the validity of the equilibrium assumption.

We have seen above that in the examples studied so far, the time evolution of the adsorbate density (5.5) is for $\delta \gtrsim 0.5$ dominated by the lowest eigenvalue λ_0^{-1} of the matrix of transition probabilities, which was identified with the desorption time. Kreuzer and Teshima (1981) go on to show that in this situation, the mean first passage time (Montroll and Shuler 1958)

$$\tau = \int_0^\infty \frac{N(t)}{N(0)} \, dt$$

$$= \sum_{\kappa=0}^N S_\kappa \lambda_\kappa^{-1} \tag{5.43}$$

yields the same time scale. Indeed, if $\lambda_0 \ll \lambda_\kappa$ for $\kappa > 0$, we know from Table 5.2 that $S_\kappa \ll S_0 \approx 1.0$, so that $\tau \approx \lambda_0^{-1}$. This can also be substantiated straightforwardly by perturbation theory for $\delta \gg 1$, and confirms similar conclusions by Kim (1958). The situation is, of course, different at high temperatures where many eigenvalues contribute. The time evolution of the desorbing gas is then transient in nature, with many exponential terms adding up in (5.5). It might then be futile to characterize such a situation by a single time scale, such as the mean first passage time, in particular if its calculation involves an exact diagonalization in any event. The mean first passage time has also been calculated by Efrima et al.(1980,1983) and, in some approximations, by De et al. (1980).

5.4 Desorption Kinetics Mediated by Surface Phonons

We continue now to present a detailed comparative investigation of desorption times calculated

    (i)   using a bulk Debye model for the phonons, as done so far in this chapter, and

    (ii)   using a surface Debye model (Goldys et al. 1981, 1982).

As we saw, e.g., in (4.9), the relevant transition probabilities differ in the one-phonon approximation by the sums over the respective phonon modes which have different weights $s^{(\sigma)}$ and different Debye cutoffs $\omega_D^{(\sigma)}$. However, for $k_BT \ll \hbar\omega_D^{(\sigma)}$, these cutoffs can be safely extended to infinity, so that the transition probabilities, and thus the desorption times calculated in the bulk and surface models, are related as follows

$$\frac{t_d(\text{bulk})}{t_d(\text{surface})}\Bigg|_{\text{mobile}} \approx \frac{3}{2+\ell_s^3} \left[S_+{}^Z(0) + S_-{}^Z(0) + S_{GR}{}^Z(0) + S_R{}^Z(0) \right]$$

$$= \frac{3}{2+\ell_s^3} \left[s_\perp^{(B)} + s_\perp^{(GR)} + s_\perp^{(R)} \right]$$

$$= \left[\frac{\omega_D}{\omega_{D\perp}} \right]^3 , \qquad\qquad (5.44)$$

where $\ell_s = c_t/c_\ell$ and the weights $S_\sigma{}^Z(0)$ are given in (2.165). Note that $t_d(\text{bulk})$ is independent of ℓ_s in this approximation. Equation (5.44) allows one to correct very easily the desorption times $t_d(\text{bulk})$ as calculated in a bulk model, to take proper account of the surface modes. Table 2.7 presents the various cutoff frequencies $\omega_D^{(\sigma)}$ in the different phonon modes and the weights $s_\perp^{(\sigma)}$. Note that the shear horizontal (SH) mode does not contribute to the desorption rate for mobile adsorption. Indeed, for typical systems the dominant contribution of some 64% at $\ell_s = 0.5$ comes from the Rayleigh mode with the generalized Rayleigh mode adding 29% and the bulk modes making up the remaining 7%. In comparison, in the bulk Debye model, the transverse modes account for 94% of the desorption rate with the longitudinal modes adding only 6%. However, one sees from Table 2.7 that the cutoff frequency $\omega_D^{(R)}$ for the Rayleigh mode is the smallest and decreases for increasing ℓ_s, so that for systems in which an effective transition in the surface potential requires energy of the order $\omega_D^{(R)}$, the Rayleigh mode contributes little or nothing, if $|E_o| > \hbar\omega_D^{(R)}$. For most systems, (5.44) holds very well, so that the desorption time calculated in the bulk model is typically too long by a factor of two to three.

Relation (5.44) can be given an intuitive interpretation by using the concept of an effective surface Debye temperature for desorption as introduced in section 2.3 via the phonon density of states. We recall [see, e.g., Maradudin et al. (1971) for a critical discussion] that atoms in the free surface of a solid typically undergo mean square displacements that are up to about twice as large as those of atoms deep inside the bulk. The solid thus appears softer in the surface region, implying a smaller effective Debye temperature for effects to which mostly surface atoms contribute, such as scattering, or adsorption and desorption phenomena. One can propose the following definition of an effective surface Debye temperature for desorption,

$$\frac{T_{D,mobile}(surface)}{T_D} = \frac{\omega_{D\perp}}{\omega_D} \quad . \tag{5.45}$$

This ratio is also given in Table 2.7. Note that $\omega_{D\perp}$ is close to $\omega_D^{(R)}$ and also about $(1/\sqrt{2})\omega_D$, a value sometimes used in literature. Also note from (5.45) and Table 2.7 that the ratio $\omega_{D\perp}/\omega_D$ depends only very weakly on ℓ_s. It is interesting that the definition of the surface Debye temperature (5.45) as it emerges in desorption kinetics is identical to that in the general theory of the dynamics of surface atoms by Iosilevskii (1971). This is not surprising because the arguments leading to his definition can be essentially reduced to statements about the density of states, and thus are equivalent to the arguments used when (2.170,171) were proposed. In terms of our s-weights, his Z-quantities are given by

$$Z = \left[6\pi^2 \frac{3}{2 + \ell_s^3}\right]^{1/3}$$

$$Z_\perp = [6\pi^2/(s_\perp^{(B)} + s_\perp^{(GR)} + s_\perp^{(R)})]^{1/3}$$

$$Z_\| = [6\pi^2/(1+s_\|^{(B)} + s_\|^{(GR)} + s_\|^{(R)})]^{1/3} \quad . \tag{5.46}$$

We should, however, point out that the calculation of desorption times in strongly coupled gas-solid systems at $k_B T \gtrsim \hbar\omega_D$, requires the full theory, and not just the simple recipe to modify a calculation with bulk phonons by using an effective surface Debye temperature.

5.5   Inclusion of Parallel Phonon Momentum for Mobile Desorption

The microscopic expression for the one-phonon transition probabilities (4.1) for a mobile adsorbate, contains a delta function to conserve parallel momentum $K+P=K'$ where K and K' are the initial and final particle momenta parallel to the surface, and P is the phonon momentum parallel to the surface. With (4.3), one argued that P could be neglected from this conservation law, so that an essentially one-dimensional theory based on (4.4) emerges, as used in the preceeding sections. In analysing Taborek's (1982) critical cone experiment, Gortel and Kreuzer (1985) developed a fully three-dimensional theory of mobile desorption in which full momentum conservation is kept; surface corrugation, however, is not included. To calculate desorption times, they start from (3.53) (where the readsorption term is absent)

$$\frac{dn_i(\mathbf{K},t)}{dt} = \sum_{i',\mathbf{K'}} W(i,\mathbf{K};i',\mathbf{K'})\, n_{i'}(\mathbf{K'},t) - \sum_{i',\mathbf{K'}} W(i',\mathbf{K'};i,\mathbf{K})n_i(\mathbf{K},t)$$

$$- \sum_{k_z',\mathbf{K'}} W(k_z',\mathbf{K'};i,\mathbf{K})n_i(\mathbf{K},t) \quad, \tag{5.47}$$

where i and k_z denote the bound and continuum states in $V_s(z)$, respectively, and $\mathbf{K}$ is the particle momentum parallel to the surface. The summation over $\mathbf{K'}$, including the momentum-conserving δ-function contained in W, amounts to

$$\sum_{\mathbf{K'}} \delta_{\mathbf{K'},\mathbf{K+P}} \cdots = \int d^2\mathbf{K'}\, \delta_{(2)}(\mathbf{K'}-\mathbf{K}-\mathbf{P})\cdots$$

$$= \int_0^{2\pi} d\phi'\delta(\phi'-\phi_{K+P}) \int_0^\infty dE'\delta\left(E'-\frac{\hbar^2}{2m}(\mathbf{K+P})^2\right)\cdots \quad, \tag{5.48}$$

where $E' = \hbar^2 K'^2/2m$ and ϕ' is the aziumthal angle of $\mathbf{K'}$ with respect to some axis in the surface plane. Because initial equilibrium occupation probabilities depend only on E', and not on the direction of $\mathbf{K'}$, and because on an isotropic surface the W's do not favor any given direction of parallel phonon momentum, we can assume that the nonequilibrium occupation functions $n_i(\mathbf{K},t)=n_i(\hbar^2K^2/2m,t)=n_i(E,t)$ remain isotropic. Thus the angular integration in (5.47) can be performed to yield

$$\frac{dn_i(E,t)}{dt} = \sum_{i'} \int dE' W_{ii'}(E,E')n_{i'}(E't) - \sum_{i'} \int dE' W_{i'i}(E',E)n_i(E,t)$$

$$- \sum_{k_z} \int dE' W_{k_z i}(E',E)n_i(E,t) \quad, \tag{5.49}$$

where

$$W_{ii'}(E,E') = \frac{1}{2\pi^2\hbar^2\rho_0}\left|\langle i|\left(\frac{dV_s}{dz}\right)|i'\rangle\right|^2 \sum_\sigma \theta\left(\omega_D{}^{(\sigma)} - \left|E_i - E_{i'} + E - E'\right|\right)$$

$$*(E_i - E_{i'} + E-E')\,\left[\exp[\beta(E_i-E_{i'}+E-E')]-1\right]^{-1} \int_0^\infty dc\, A_\sigma(c)\, \frac{\theta(B)}{\sqrt{B}} \quad,$$

with
$$\tag{5.50}$$

$$B = \frac{2E}{mc^2}(E_i-E_{i'}+E-E')^2 - \left[E' - E - \frac{1}{2mc^2}(E_i-E_{i'}+E-E')^2\right]^2 \quad. \tag{5.51}$$

Here, c is the apparent sound velocity along the surface, and the $A_\sigma(c)$ are proportional to the squared phonon amplitudes for the vibrations perpendicular to the surface. For the bulk Debye model they are

$$A_L(c) = \frac{\sqrt{c^2-c_\ell{}^2}}{c_\ell c^4}\,\theta(c-c_\ell) \quad, \qquad A_T(c) = \frac{c_t}{c^4\sqrt{c^2-c_t{}^2}}\,\theta(c-c_t) \tag{5.52}$$

for the longitudinal and transverse branches, respectively, whereas for the surface Debye model we have

$$A_{SH}(c) = 0$$

$$A_+(c) = A_-(c) = \frac{1}{c^4} \, \theta(c-c_\ell) \, \frac{\alpha(\beta^2+1)^2}{(\beta^2-1)^2+4\alpha\beta}$$

$$A_{GR}(c) = \frac{1}{c^4} \, \theta(c_\ell-c) \, \theta(c-c_t) \, \frac{8\alpha^2\beta(\beta^2+1)^2}{(\beta^2-1)^4+16\alpha^2\beta^2}$$

$$A_R(c) = \frac{1}{c_R^3} \, \delta(c-c_R) \, \frac{2\alpha^2\beta(1-\sqrt{\alpha\beta})^2}{(\alpha-\beta)(\alpha-\beta+2\alpha\beta^2)} \quad , \quad \text{where} \tag{5.53}$$

$$\alpha^2 = |(c/c_\ell)^2-1|$$

$$\beta^2 = |(c/c_t)^2-1| \quad . \tag{5.54}$$

These weights, multiplied by c_t^3 and integrated over c, yield $s^{(L)}$ and $s^{(T)}$ in (2.166) for the bulk modes, and $s_l^{(\sigma)}$ in (2.165) for the surface modes. The matrix elements for a Morse surface potential are listed in (4.19-21).

To solve (5.49) by matrix diagonalisation, one discretizes the lateral energy for $0<E<E_{max}$ into a sufficiently large number N of discrete points. Typically, one can take E_{max} of the order of several k_BT and N less than 100.

Gortel and Kreuzer (1985) discussed helium desorption from a nichrome substrate that, in Taborek's experiment, was instantly flash heated from T_i = 2 K to T_f = 8 K. Thus the initial bound state occupation probabilities

$$n_i(E,t=0) = \frac{N_g}{V} \, (\frac{2\pi\hbar^2}{mk_BT})^{3/2} \, \exp[(E_i+E)/k_BT_i] \tag{5.55}$$

are taken at the initial temperature T_i whereas the transition probabilities W are evaluated at T_f.

One notes first the $\ln(n_i(E,t))$ versus t, obtained from (5.49), is very linear with a slope corresponding to a desorption time t_d = 6.25 x 10^{-7} s; the linearity pertaining for $0 \leq t \leq 5t_d$. This implies that out of the many eigenvalues λ_K of the transition probability matrix only the lowest one λ_0 contributes significantly to the time evolution. We note in passing that the one-dimensional version of the one-phonon theory based on (5.1), yields $(\lambda_0(\text{1-dim}))^{-1}$ = 6 x 10^{-7} s. To see how much the individual occupation functions $n_i(E,t)$ change, we have plotted in Fig. 5.8

$$n_j(E,t)e^{\lambda_0 t} \, \exp\left[(\frac{E_j}{T_f} + \frac{E}{T_i})/k_B\right] \tag{5.56}$$

versus E for various $\lambda_0 t$. We can infer the following:

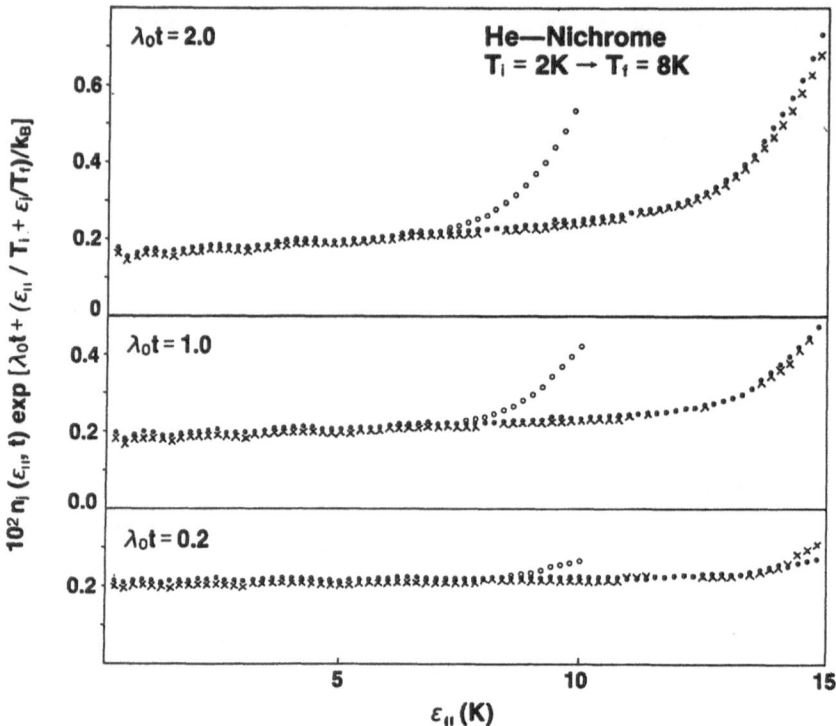

Fig.5.8. The nonequilibrium distribution functions (5.56) for a desorption flash from $T_i=2$ K to $T_f=8$ K as a function of lateral energy E for Helium on a Nichrome surface. Points are for the ground state j=0; crosses for j=1, both for a cut-off energy $E_{max}/k_B=15$ K. Circles are for j=0 and $E_{max}/k_B=10$ K. The wiggles are indicative of the numerical errors due to the coarse-graining of E to 75 and 50 points, respectively. $\lambda_0^{-1}=6.25\times10^{-7}$ s is the desorption time. Parameters: $\sigma_0=2.73$, r=83.0, $\omega_D/\gamma c_t=1.728$, $T_D=500$ K, i.e., $\gamma^{-1}=1$ Å and $V_0/k_B=45$ K. (Gortel and Kreuzer 1985)

(i)   from the fact that the curves for j = 0,1,2 are more or less identical, we can conclude that the motion in $V_s(z)$ perpendicular to the surface thermalizes to T_f as early as $0.2/\lambda_0$, i.e., with more than 80 % of the adsorbate still present;

(ii)   from the fact that the curves are flat, we see that the lateral motion remains characterized by the initial temperature T_i, i.e. does not thermalize even after 2 desorption times;

(iii) the rise of all curves near the upper cutoff E_{max} is numerical; increasing E_{max} also shifts the region of upward bending to higher lateral energy E.

(iv)   the fact that, for later times the slope of the curves slightly increases, indicates a slow increase in temperature. This is more apparent if we plot $\ln[n_0(E,t)\exp(\lambda_0 t)]$ versus E for various $\lambda_0 t$ in Fig. 5.9. A decreasing slope signals an increasing temperature. Also note the effect of 2 different cutoffs, E_{max}. From the change in slope for different λ_0, we can estimate the rate of lateral thermalization as 5×10^4 Ks^{-1}. Performing a temperature flash from T_i = 2 K to T_f = 3 K desorbs a helium film in t_d = 1.1 x 10^{-3} s and results in a lateral heating rate of 500 Ks^{-1}. Similarly, for a flash from T_i = 1 K to T_f = 2 K, we get t_d = 0.15 s, and a lateral heating rate of about 1 Ks^{-1}. Only if lateral thermalization is faster than desorption is one allowed to use thermodynamic arguments of detailed balance (Goodstein and Weimer 1983) to make statements about the desorption process. To avoid misunderstanding, we should point out that the transition probabilities in the master equation always satisfy detailed balance.

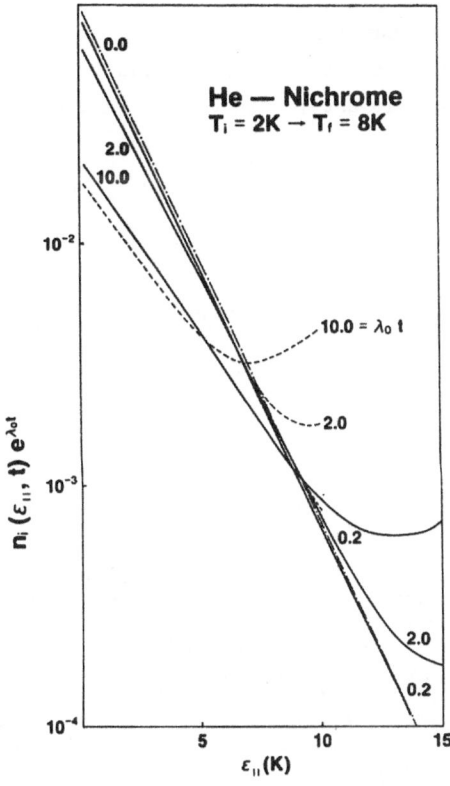

He — Nichrome
$T_i = 2K \rightarrow T_f = 8K$

$10.0 = \lambda_0 t$

Fig.5.9. The occupation functions as a function of lateral energy E for various times $\lambda_0 t$. The decrease in slope reflects the heating of the lateral degrees of freedom from the initial temperature T_i=2 K (dash-dotted line). The deviations from straight lines close to the upper cutoff energy E_{max} (=10 K for the dotted line) reflect (i) numerical errors and (ii) nonequilibrium effects. (Gortel and Kreuzer 1985)

5.6 Desorption from a Localized Adsorbate

We outlined in Sect. 4.2 the correlation function approach by Bendow and
Ying (1973a,b) to calculate the transition probabilities $W(\mathbf{k},i)$ for a particle
to go from a localized initial state i into a free continuum final state $\mathbf{k}$.
They calculate the rate to go from all initial states i into a given final
state $\mathbf{k}$ as a thermal average

$$R(\mathbf{k}) = \sum_{i} W(\mathbf{k},i) e^{-\beta E_i} \Big/ \sum_{i'} e^{-\beta E_{i'}} \quad . \tag{5.57}$$

Within the master equation approach, this is nothing but the lowest order
perturbation result (5.41) whose appropriateness has been discussed in Sect.
5.3. Summing (5.57) over the final states, we get the desorption rate and
various differential rates

$$R = \int d^3\mathbf{k} \; R(\mathbf{k}) = \int k^2 dk \; \sin\theta \; d\theta \; R(k,\theta) = \int \sin\theta \; d\theta \; R(\theta) \quad . \tag{5.58}$$

Bendow and Ying (1973b) calculate the desorption rate for Ne desorbing from
a Xe overlayer on graphite. The static surface potential resulting from
summing Gaussian two-body potentials between the Ne and Xe atoms (4.62) is
schematically depicted in Fig.5.10. The relevant parameters are reproduced
in Table 5.4.

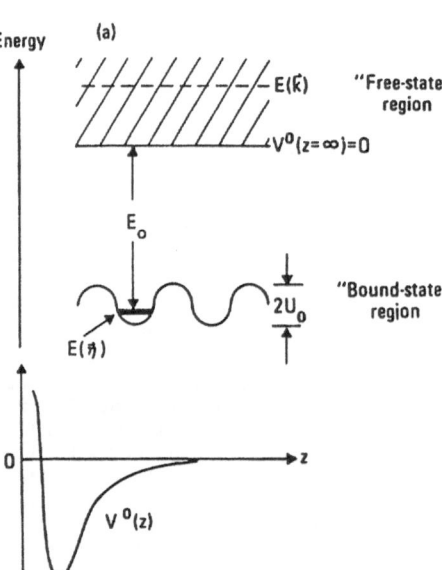

Fig.5.10. a Schematic diagram indicating energy levels of an adatom
while bound and free. The periodic
variation of the potential at $z=z_0$ in
an arbitrary direction along the surface plane is indicated along the horizontal. b Schematic surface potential
$V_0(z)$, averaged laterally along the
surface. (Bendow and Ying 1973)

Table 5.4. Ne-Xe-C parameters. (Bendow and Ying 1973b)

Xe substrate structure	simple cubic.
Xe lattice constant	$a_0 = 4.31$ Å.
Adatom vibrational energy	$\nu = 56$ K.
Vibrational range in plane	$r_2/a_0 = 6.2 \times 10^{-2}$.
Vibrational range in z	$r_3/a_0 = 2.8 \times 10^{-2}$.
Adsorption distance	$z_0/a_0 = 1$.
Bulk-mode Debye temperature	$T_D = 343$ K.
Phonon expansion parameter	$C_0 = 0.85 \times 10^{-4}$.
Well depth energy	$E_0 = 231$ K.
Potential variation in plane	$2U_0 = 100$ K.

Because the corrugation along the surface ($2U_0$) is assumed to be quite substantial, the lowest bound states form a narrow band of energies $E_0(K)=E_0+\hbar^2 K^2/2m$. Bendow and Ying (1973) calculate the desorption rate (5.57,58) from only these states.

Looking at the laterally averaged potential $V_0(z)$, we note that with an approximate depth V_0 and a range of typically a few angstroms, it will definitely develop more than one bound state, most likely about ten or so. They have not been included by Bendow and Ying (1973b) in the sum over i in (5.57). In calculating the transition probabilities (4.53), they include some 20 distinct lattice sites in the surface plane. For the phonons, they employ an averaged bulk Debye model.

Figure 5.11 illustrates the rate $R(k,\theta=0)$ (which is the desorption rate per unit solid angle in the forward direction, integrated over all angles in the plane) as a function of the adatom's final energy $E=\hbar^2 k^2/2m$, for Ne-Xe-C, calculated within the one-phonon approximation. Due to the neglect of multiphonon cascades, the curves are truncated at the single-phonon cutoff appropriate to the two values of E_0 employed. A general feature of the results is the strong peak near $E=0$, corresponding to an extremely narrow energy distribution of desorbed adatoms. For energies above the peak, R is seen to be very nearly an exponentially decreasing function of E. The fact that R peaks at a value of E somewhat above $E = 0$ is a consequence of the WKB wave function, which is proportional to $E^{1/2}$; for a free final state, for example, the peak would occur exactly at $E=0$.

Figure 5.12 illustrates the rate for desorption in the forward direction, $R(\theta=0)$, as a function of temperature. R is seen to follow a nearly exponential law, $R \approx \nu \exp(-\beta E_d)$, at low temperatures, departing somewhat at higher temperatures, where Bose-Einstein statistics become important. Note that the effective adsorption energy E_d exceeds E_0, being close in value to the energy for which the product $E(k)R(k)$ is maximal. For Ne-Xe-C, the prefac-

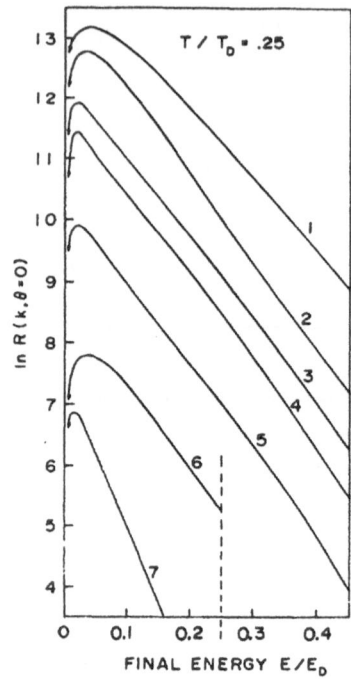

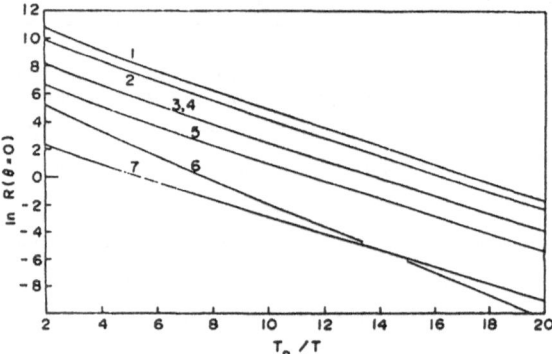

◄ **Fig. 5.11.** Logarithm of desorption rate in normal direction $R(k, \theta=0)$ vs final energy $E/\hbar\omega_D$. All curves are for parameters listed in Table 5.4 with the following variations: (1) $r_1=0.25a_0$, (2) $r_1=0.5a_0$, (3) no variation (4) $r_0=0.75a_0$, $z_0=0.75a_0$, (5) $r_2^2 \to 2r_2^2$, (6) $E_0=0.75\hbar\omega_D$, (7) $m \to 2m$. Calculations are carried out for $T/T_D = 0.25$. (Bendow and Ying 1973b)

Fig. 5.12. Logarithm of total rate in forward direction, $R(\theta=0)$ vs inverse temperature T_D/T. Labels correspond to those in Fig.5.11. (Bendow and Ying 1973b)

Table 5.5. Desorption rate prefactor ν and desorption time t_d at $T=0.25T_D$. (Bendow and Ying 1973b)

Case	ν $[s^{-1}]$	t_d^0 $[s]$	$t_d[T/T_D = 0.25]$ $[s]$
$r_1=0.25$	8.1×10^5	1.2×10^{-6}	7.2×10^{-4}
$r_1=0.50$	3.3×10^5	3.0×10^{-6}	1.8×10^{-3}
$r_1=1.0$	7.3×10^4	1.4×10^{-5}	8.4×10^{-3}
$r_1^2=2r_2^2$	1.6×10^4	8.6×10^{-5}	5.2×10^{-2}
bcc(100)	6.0×10^4	1.7×10^{-5}	1.0×10^{-3}

tor is $\nu \approx 10^5$ s^{-1}. Values of ν and $t_d^0 \approx \nu^{-1}$ corresponding to various values of the parameters are listed in Table 5.5. Figure 5.13 illustrates the angular properties of desorbed adatoms. Figure 5.13a shows that rate $R(\theta)$ peaks strongly in the normal direction compared to the $\cos\theta$ distribution. Increasing θ is also associated with a sharper peak in the energy distribution, as illustrated in Fig. 5.13b. Temperature dependencies are illustrated in Fig. 5.13c. As the peaks in Fig. 5.13b occur at nearly the same E, the R versus

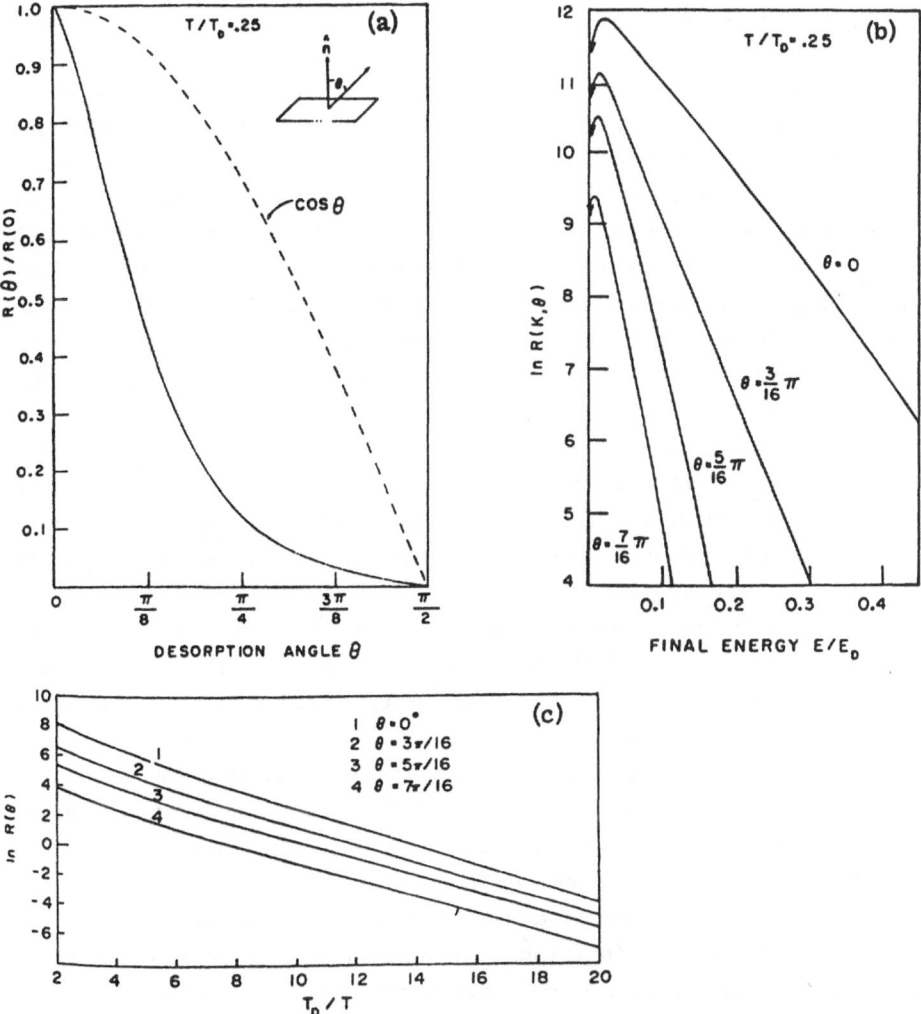

Fig.5.13. Angular properties of the desorption rate. **a** R(θ)/R(0) vs. desorption angle θ, for the parameters listed in Table 5.4, at T/T_D = 0.25. **b** energy distributions, in R(k,θ) vs E(k)/E_0, at different desorption angles **c** Total rate vs. T_D/T at different angles. (Bendow and Ying 1973b)

T^{-1} curves are nearly parallel. We note that the angular dependence, and also the magnitude of the prefactor in the desorption rate, depend strongly on the potential parameters, in particular on r_2 and r_3 affecting the localization of the adatom wave function. Bendow and Ying also studied the dependence of the desorption rate on the substrate structure. They find that the rate is about a factor of three larger on a simple cubic (100) face than it is on a bcc(100) face (Table 5.5).

One of the puzzling features of the theory by Bendow and Ying (1973a,b) is the fact that it produces prefactors ν in the desorption rate many orders of magnitude smaller than either the classical result (which identifies ν with the oscillation period of the adatom in the surface potential well), or with the results of the cascade model reported in the previous sections. The prefactor ν is found to be typically $\approx 10^5$ s^{-1} for the Ne-Xe-C system. To explain their results, Bendow and Ying (1973b) repeat an argument by Goodman (1971) that one-dimensional treatments of desorption yield abnormally large rates because all energy transfers necessarily lead to motion normal to the surface, while in the three-dimensional case, motion in the surface plane supplies an alternative sink to the energy transfer. Thus, they argue, smaller rates come about because

> (i) the three-dimensional density of adatom states leads to $\int dk k^2$ integrals in place of one-dimensional $\int dk$. Since the differential rate is strongly peaked about $k \approx 0$, the total rate is substantially decreased through the density of states in the three-dimensional case.

> (ii) The surface potential well depth V_0 is much larger in the three-dimensional case than in the one-dimensional one, generally by about a factor of 3-10, depending on the geometrical configuration.

The transition rate, however, is still proportional to only the depth $\tilde{V}_0$ of the two-body potential (4.62), as evidenced by the oscillation in sign of the contributions to the rate in the sum over sites. Since the rate decreases rapidly with increasing E_0, the decrease in the ratio of $\tilde{V}_0/V_0$ for the three-dimensional case thus leads to a sharply diminished desorption rate.

These arguments must be taken with a grain of salt because Bendow and Ying (1973b), in calculating the desorption rate in perturbation theory according to (5.57), take only the lowest localized bound states around ($-E_0$) into account. However, a surface potential of depth of the order of $E_0/k_B \approx 231$ K and a reasonable range of a few angstroms will develop a few, possibly as many as 10 or 20 bound states or more precisely bands of bound states

$$E_{\bar{1}} = E_{\bar{1}} + \hbar^2 K^2/2m \quad . \tag{5.59}$$

In particular, the ones above the corrugation threshold U_0 will become highly mobile so that their respective transition probabilities $W(\mathbf{k},\bar{1}) = W(k_z,\mathbf{K};K,\bar{1})$ can, e.g., be calculated from (4.22-26). Now we know from Sects. 5.1,2 that such higher bound states contribute significantly to the desorption rate. They thus ought to be included in the calculation. To

Table 5.6. Range of surface potential γ^{-1}, number of bound states N, ground-state energy E_0, desorption time t_d, and the transition probability R_{c0} from the ground state to all continuum states. Parameters: V^0/k_B=171.34 K; T_D=185 K; m/M_s=0.333 corresponding to helium on graphite. T=15 K.

γ^{-1} [Å]	N	$-E^0/k_B T_D$	t_d^{-1} [s^{-1}]	R_{c0} [s^{-1}]
0.48	3	0.595	4.22×10^8	3.88×10^8
0.67	4	0.685	1.05×10^8	6.55×10^7
0.95	5	0.752	2.85×10^7	6.63×10^6
1.34	7	0.801	9.36×10^6	3.65×10^5
1.90	10	0.837	3.16×10^6	6.45×10^3

show that small rates are not necessarily the result of three-dimensional localization, we show below the results of a model by Goldys et al.(1982) which produce large rates in such a situation. At this stage, we want to demonstrate (i) the strong effect of the potential range on the desorption rate and (ii) how important the contributions from higher bound states are. We have therefore done some further model calculations for a highly mobile adsorbate starting from the master equation (5.1). In Table 5.6 we list the exact desorption rate constant t_d^{-1} obtained by diagonalization, and also list the transition rate R_{c0} calculated from (4.29,25) from the lowest bound state to all continuum states; this would be the equivalent to the rate calculated by Bendow and Ying (1973). Varying the range of the surface potential, we see that the exact desorption rate varies by only two orders of magnitude whereas R_{c0} varies by five orders of magnitude. This indicates that for a surface potential of a rather large range as used by Bendow and Ying (1973b), the major contributions to desorption arise from transitions into the continuum originating about halfway up the potential well. Such cascades are important, not only for strongly bound systems with the ground state more than $\hbar\omega_D$ below the continuum, but also for shallow systems with only a few bound states as discussed at length in Sects. 5.1,2. The small rates calculated by Bendow and Ying (1973b) are therefore most likely the result of (i) a failure to include higher bound states in the surface potential and (ii) a choice of rather wide surface potentials for which desorption from these higher bound states becomes increasingly more likely.

To show that desorption from a highly localized state can yield reasonably large desorption rates, we now report on a model calculation by Goldys et al. (1982). They neglect all band structure effects, assuming a local surface potential

$$V_S(\mathbf{r}) = V_M(r) = V_0[e^{-2\gamma(r-r_0)} - 2e^{-\gamma(r-r_0)}] \quad , \qquad \text{for} \quad z > 0$$

$$= \infty \quad , \qquad\qquad\qquad\qquad\qquad\qquad \text{for} \quad z \leq 0 \quad . \qquad (5.60)$$

It will generally develop several bound states into which gas particles can be trapped to form the adsorbate. These states can be classified by specifying the angular momentum ℓ, its projection m_ℓ onto the z-axis, and a principal quantum number i.

The presence of the infinite wall $V = \infty$ for $z \leq 0$ implies that $\ell + m_\ell$ must be odd. The time-dependent occupation $n_{i\ell m_\ell}(t)$ of a state (i,ℓ,m_ℓ) is assumed to change in an isothermal desorption experiment according to the master equations

$$\frac{d}{dt} n_{i\ell m_\ell}(t) = \sum_{i',\ell',m_\ell'} W(i\ell m_\ell; i'\ell'm_\ell') n_{i'\ell'm_\ell'}(t)$$

$$- \left[R(\text{cont}; i\ell m_\ell) + \sum_{i'\ell'm_\ell'} W(i'\ell'm_\ell'; i\ell m_\ell) n_{i\ell m_\ell}(t) \right] , \quad (5.61)$$

Note that although the energies $E_{i\ell}$ do not depend on m_ℓ, the transition probabilities, and thus the time-dependent occupations in general, will depend on it. The transition probabilities for one-phonon processes are again calculated in second-order time-dependant perturbation theory from the dynamic part of the Hamiltonian

$$H_{dyn} = L^{-3} \sum_{\mathbf{p}\sigma} \frac{1}{\sqrt{\omega_{\mathbf{p}\sigma}}} \, \mathbf{e}_{\mathbf{p}\sigma} \cdot$$

$$* \sum_{i\ell m_\ell, i'\ell'm_\ell'} X(i\ell m_\ell; i'\ell'm_\ell') \, a_{i\ell m_\ell}^\dagger \, (b_{\mathbf{p}\sigma}^\dagger + b_{\mathbf{p}\sigma}) \, a_{i'\ell'm_\ell'} \quad . \qquad (5.62)$$

The sums over ι and ι' run over bound states and continuum states. In this calculation, we choose the bulk Debye model for the phonons. The extension to surface phonons is straightforward, with the results obtained in Chap. 4. The matrix element in (5.62) is

$$X(i\ell m_\ell; i'\ell'm_\ell') = L^3 \sqrt{\frac{\hbar}{2M_S N_S}} \int d^3\mathbf{r} \; \psi_{i\ell m_\ell}^*(\mathbf{r}) \, \frac{dV_0}{dr} \, \frac{\mathbf{r}}{r} \, \psi_{i'\ell'm_\ell'}(\mathbf{r}) \quad . \quad (5.63)$$

Next we factorize the wave functions

$$\psi_{i\ell m_\ell}(\mathbf{r}) = \frac{1}{r} R_{i\ell}(r) Y_{\ell m_\ell}(\theta, \phi) \qquad\qquad\qquad\qquad\qquad (5.64)$$

and find for the bound state - continuum transitions

170

$$R(cont;i\ell m_\ell) = \frac{2\pi}{\hbar} \ L^{-6} \sum_{p\sigma} \frac{1}{\omega_{p\sigma}} \sum_{k\ell'm_\ell^j} \left| e_{p\sigma} \cdot X(k\ell'm_\ell^j;i\ell m_\ell) \right|^2$$

$$* (e^{\beta\hbar\omega_{p\sigma}} - 1)^{-1} \ \delta \ (E_{k\ell'} - E_{i\ell} - \hbar\omega_{p\sigma}) \quad , \tag{5.65}$$

and for the bound state - bound state transitions with $E_{i'\ell'} > E_{i\ell}$

$$W(i'\ell'm_\ell^j;i\ell m_\ell) = \frac{2\pi}{\hbar} \ L^{-6} \sum_{p\sigma} \frac{1}{\omega_{p\sigma}} \left| e_{p\sigma} \cdot X(i'\ell'm_\ell^j;i\ell m_\ell) \right|^2$$

$$* (e^{\beta\hbar\omega_{p\sigma}} - 1)^{-1} \ \delta(E_{i'\ell'} - E_{i\ell} - \hbar\omega_{p\sigma}) \quad . \tag{5.66}$$

Next, we use the δ-function to perform the integration over the phonon modes, and introduce the dimensionless quantities, in addition to r and σ_0 in (5.20),

$$s_{i\ell}^2 = -2mE_{i\ell}/\hbar^2\gamma^2$$

$$n_\ell^2 = 2mE_{k\ell}/\hbar^2\gamma^2 \tag{5.67}$$

for bound and continuum states, respectively. We get for $s_{i\ell}^2 > s_{i'\ell'}^2$ (phonon absorption) and for the transitions into the continuum

$$W(i'\ell'm_\ell^j;i\ell m_\ell) = \omega_D \frac{24\sigma_0^4}{r^4} \frac{m}{M_s} \frac{s_{i\ell}^2 - s_{i'\ell'}^2}{\exp[\delta(s_{i\ell}^2 - s_{i'\ell'}^2)/r]-1}$$

$$* C^*(i'\ell'm_\ell^j;i\ell m_\ell) \cdot C(i'\ell'm_\ell^j;i\ell m_\ell) \ \theta \ (r-s_{i\ell}^2+s_{i'\ell'}^2) \ , \tag{5.68}$$

$$R(cont;i\ell m_\ell) = \omega_D \frac{24\sigma_0^4}{r^4} \frac{m}{M_s} \sum_{n\ell\ell'm_\ell^j} \frac{n_\ell^2 + s_{i\ell}^2}{\exp[\delta(n_\ell^2 + s_{i\ell}^2)/r]-1}$$

$$* C^*(n_\ell\ell'm_\ell^j;i\ell m_\ell) \cdot C(n_\ell\ell'm_\ell^j;i\ell m_\ell) \ \theta \ (r-n_\ell^2-s_{i\ell}^2) \ , \tag{5.69}$$

where we used the same Debye cutoff frequency ω_D for both longitudinal and transverse phonons, and defined a vector

$$C(i'\ell'm_\ell';i\ell m_\ell) = I(\ell'm_\ell';\ell m_\ell) \int_0^\infty d\xi \ f_{i'\ell'}^*(\xi) \frac{dV_s}{d\xi} \ f_{i\ell}(\xi) \quad , \tag{5.70}$$

with

$$I(\ell'm_\ell';\ell m_\ell) = \int_0^{2\pi} d\phi_\xi \int_0^{\pi/2} d\theta_\xi \ \sin\theta_\xi \ Y^*_{\ell'm_\ell^j}(\theta_\xi,\phi_\xi) \ Y_{\ell m_\ell}(\theta_\xi,\phi_\xi) \frac{\xi}{\xi} \quad . \tag{5.71}$$

The integrals (5.71) have been calculated analytically by Goldys et al. (1982). The radial wavefunction, e.g. for bound states, $f_{i\ell}(\xi) = (2\gamma)^{-1/2}R_{i\ell}(r)$ satisfies the Schrödinger equation

$$-\frac{d^2f_{i\ell}}{d\xi^2} + \left[\sigma_0^2 v(\xi) + \frac{\ell(\ell+1)}{\xi^2}\right] f_{i\ell} = -s_{i\ell}^2 \ f_{i\ell} \quad , \tag{5.72}$$

171

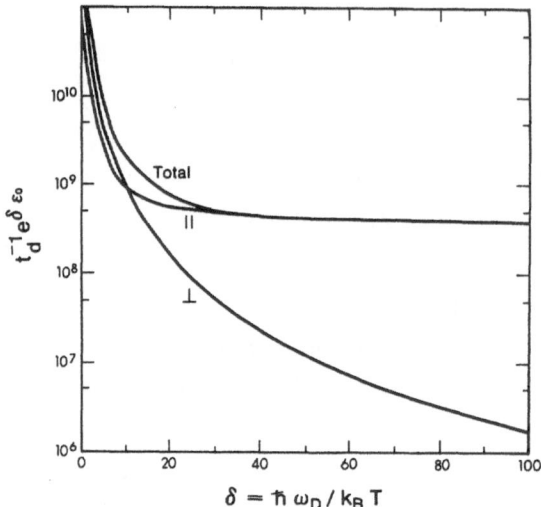

Fig. 5.14. Desorption rate prefactors for the three-dimensional system, defined by (5.60). (Goldys et al. 1982)

where we defined $v(\xi)=V_S(r)/V_0$, $\xi=\gamma r$ and $\xi=\gamma r$. Goldys et al. (1982) calculated the desorption rates by diagonalizing (5.61) after solving the Schrödinger equation (5.72) and calculating the matrix elements (5.68,69) numerically. Figure 5.14 presents the results for a fictitious gas-solid system with one bound state with $\ell = 1$ in a Morse potential (5.60) with $\gamma r_0=2$, $r=7.314$, and $\sigma_0=1.3965$ to fix the bound state at $\epsilon_0=0.065$. We note first that the prefactor is typically 10^9 s^{-1}, increasing by two orders of magnitude at higher temperatures. Reducing the range to $\gamma r_0=1$, we increase the prefactor by about an order of magnitude yet. Prefactors as low as 10^5 s^{-1}, as obtained by Bendow and Ying (1973b), can only be obtained for unreasonably large ranges of the surface potential. Figure 5.14 also demonstrates that for this model phonons polarized parallel to the surface are more effective in mediating desorption. This is in contrast to mobile adsorbates where desorption is only mediated by phonons polarized perpendicular to the surface.

We close the discussion of desorption times for localized adsorbates by introducing an effective surface Debye temperature as it was done in (5.44-45) for mobile adsorption (Goldys et al. 1982). Again we note that, for low temperatures where the upper limits of all integrals over phonon frequencies can be extended to infinity, and where the phonons polarized parallel to the surface are most effective, the desorption times calculated with bulk Debye phonons and surface Debye phonons, respectively are related by

$$\frac{t_d(bulk)}{t_d(surface)}\Big|_{localized} = \frac{3}{2+\ell_s{}^3} (s_\parallel^{(H)} + s_\parallel^{(B)} + s_\parallel^{(GR)} + s_\parallel^{(R)})$$

$$= (\frac{\omega_D}{\omega_{D\parallel}})^3 = (\frac{T_D}{T_{D,localized}(surface)})^3 \qquad (5.73)$$

where the weights $s_\parallel^{(\sigma)}$ are given in (2.165). This ratio, also given in Table 2.7, is of order 2 to 3 and quite close to the ratio for mobile adsorption in (5.44).

5.7 Cole-Toigo Corrections

Gortel, Kreuzer and Sommer (1983) have calculated the additional contributions to the desorption time as they arise from the Cole-Toigo corrections to the dynamical gas-solid coupling beyond the first derivative coupling, see (2.71). Using perturbation theory according to (5.41) and employing a bulk Debye model for the phonons, they get, for a mobile adsorbate,

$$t_d^{-1} = \frac{L^4}{(2\pi)^3} \frac{1}{M_S N_S} \sum_{n=0}^{N} \sum_{\sigma=L,T} \int d^2K \int_{-\infty}^{\infty} dk \int d^2Q$$

$$* \int d^2P \int_0^{(q_D{}^2-P^2)^{1/2}} dp_z \, \omega_{p\sigma}^{-1} (e^{\beta\hbar\omega_{p\sigma}} -1)^{-1}$$

$$*\exp\left[-\beta(E_n + \frac{\hbar^2 Q^2}{2m})\right](u_\sigma^{(0)} + p^2 u_\sigma^{(1)} + p^4 u_\sigma^{(2)})\delta(K-Q-P)$$

$$*\delta(E_n - \frac{\hbar^2(k_z^2+K^2-Q^2)}{2m} + \hbar\omega_{p\sigma})\left[\sum_{n=0}^{N} \int d^2Q \, \exp\left[-\beta(E_n + \frac{\hbar^2 Q^2}{2m})\right]\right]^{-1} \qquad (5.74)$$

where q_D is the Debye wave vector cutoff and

$$u_L^{(0)} = (\frac{p_z}{\gamma_p})^2 \left|\langle k_z| \frac{dV_S}{dz} |n\rangle\right|^2$$

$$u_T^{(0)} = (\frac{P}{\gamma_p})^2 \left|\langle k_z| \frac{dV_S}{dz} |n\rangle\right|^2$$

$$u_L^{(1)} = |\langle k_z| V_S |n\rangle|^2 -$$

$$- V_0(\frac{p_z}{p})^2\left[\langle k_z| \gamma(z)|n\rangle\langle n| \frac{dV_S}{dz} |k_z\rangle - \langle k_z| \beta(z)|n\rangle\langle n| \frac{dV_S}{dz} |k_z\rangle + c.c.\right]$$

$$u_T^{(1)} = V_0(\frac{P}{p})^2 \left[\langle k_z| \gamma(z)|n\rangle\langle n| \frac{dV_S}{dz} |k_z\rangle + c.c.\right]$$

$$u_L^{(2)} = V_0^2 \gamma^{-2} \left(\frac{p_z}{p}\right)^2 \left|\langle n|(\gamma(z) + \beta(z))|k_z\rangle\right|^2$$

$$u_T^{(2)} = V_0^2 \gamma^{-2} \left(\frac{P}{p}\right)^2 \left|\langle n|\gamma(z)|k_z\rangle\right|^2 \quad . \tag{5.75}$$

Here, $k = (K, k_z)$ is the momentum of the desorbed particle; K its momentum along the surface in the adsorbate; $p = (P, p_z)$ is the phonon wave number. The δ-functions take care of parallel momentum and energy conservation. The functions $\beta(z)$ and $\gamma(z)$ for a Morse potential are given in (2.75,76), and for the 9-3 potential in (2.72,73). The Cole-Toigo correction terms are quadratic and quartic in the phonon wave number p. In (5.74) E_n are the bound state energies in the surface potential with wave functions $\langle z|n\rangle$, there are (N+1) of them, and $\langle z|k_z\rangle$ is a continuum state in $V_s(z)$. Using a Morse surface potential, all matrix elements (5.75) can be evaluated analytically. Of the eight integrals in (5.74), three can be done with the help of the δ-functions, and two more are analytically possible, so that we get

$$t_d^{-1} = \frac{1}{2} \left(\frac{a}{2\pi}\right)^3 \frac{m}{M_s} \frac{\sigma_0^4}{r^2} \sqrt{\frac{\pi\delta}{r}} \; \omega_D \left[\sum_{n=0}^{N} e^{-\delta\epsilon_n}\right]^{-1}$$

$$* \sum_{i=0}^{2} \sum_{\sigma=L,T} \sum_{n=0}^{N} \int_0^{\infty} d\eta \exp\left(-\left(\frac{\delta}{r}\right)\eta^2\right) \int_0^{q_D} dP \int_0^{q_D-P} dx$$

$$* g_\sigma^{(i)}(x, P, n, \eta) \exp\left[-\frac{\delta r}{4P^2 q_D^2}(x-x_0)^2\right] , \tag{5.76}$$

where we introduced the Morse parameters r and σ_0 and the inverse temperature δ from (5.20,21), the Morse bound states $\epsilon_n = E_n/\hbar\omega_D$ from (2.38), phonon wave number cutoff q_d, and the lattice parameter a measured in γ^{-1}. The $g_\sigma^{(i)}$ are rather complicated functions related to the $u_\sigma^{(i)}$; also note that $x_0 = x_0(\sigma, n, P, \eta)$ is still a function of all the variables indicated. To get further, Gortel, Kreuzer, and Sommer (1983) used the method of steepest descent in the x-integral in (5.76), replacing x by x_0 in $g_\sigma^{(i)}$. This is justified for He on the alkali halides where the P integration is restricted to small values ($\approx 0.1q_D$) for $\eta^2 \leq r/\delta$ implying, with $\delta r \approx 10^4$, a very sharp peak from the Gaussian at small x. For the He-graphite system, δr is an order of magnitude smaller and the P range extends to at least $0.2q_D$, widening the Gaussian, now at large x, thus making corrections to the steepest descent method necessary. After the steepest descent method has been employed, the P-integral can then be done analytically, and only the remaining η integral has to be done numerically.

174

Table 5.7. Debye energy $\hbar\omega_D$, Debye cutoff q_D, depth V_0, range γ^{-1} and mini‐ mum position z_0 of the Morse potential; σ_0 and r from (5.20). (Gortel, Kreuzer, and Sommer 1983)

System	T_D [K]	q_D/γ	V_0 [K]	γ [Å^{-1}]	γz_0	σ_0	r
He-LiF	730	1.677	89.0	0.917	2.4	4.198	145.78
He-NaF	450	1.232	77.8	1.02	2.5	3.491	70.496
He-C	185	1.858	171.3	1.05	2.9	5.04	27.436

In Table 5.7, we list the pertinent parameters for the gas-solid systems for which desorption times have been calculated from (5.76), after invoking the steepest descent method. To present the Cole-Toigo corrections, we have written (5.76) as $r_d^{-1} = \lambda^0 + \lambda^1 + \lambda^2$, with $\lambda^{(i)}$ involving the $g_0^{(i)}$ contribu‐ tions which in turn are proportional to the phonon wave number to the power $2i$, see (5.75). Results are listed in Table 5.8, together with desorption rates calculated from the one-dimensional version of the cascade model (5.1). We note that the Cole-Toigo corrections never amount to more than a few percent for He desorbing from an alkali-halide surface, and they decre‐ ase in relative terms as the temperature is lowered. It is also very grati‐ fying to see that the one-dimensional version of the cascade model is again in excellent agreement with the full three-dimensional one. Also note that the Cole-Toigo corrections are rather sensitive to the position z_0 of the

Table 5.8. Contributions to the desorption rates in (5.76), $t_d^{-1} = \lambda^0 + \lambda^1 + \lambda^2$. The values of t_d^D are from the one-dimensional model (5.1) by diagonali‐ zation; parameters as in Table 5.7. The entries in brackets for ^{4}He-NaF are for $\gamma z_0 = 3.0$. (Gortel, Kreuzer, and Sommer 1983)

	T [K]	λ^0 [s^{-1}]	λ^1 [s^{-1}]	λ^2 [s^{-1}]	t_d [s]	t_d^D [s]
^{4}He-LiF	12.2	1.12x10^6	1.00x10^4	7.61x10^1	8.82x10^{-7}	8.77x10^{-7}
	7.3	1.57x10^4	1.18x10^2	7.26x10^{-1}	6.32x10^{-5}	6.30x10^{-5}
	4.0	4.196	2.92x10^{-2}	1.62x10^{-4}	0.22	0.22
^{4}He-NaF	22.5	9.66x10^7	1.73x10^6	2.53x10^4	1.01x10^{-8}	9.70x10^{-9}
			(2.1x10^6)	(3.8x10^4)		
	7.2	2.36x10^5	1.52x10^3	1.12x10^1	4.21x10^{-6}	4.20x10^{-6}
	4.5	8.95x19^2	4.26	2.71x10^{-2}	1.10x10^{-3}	1.10x10^{-3}

surface potential minimum; see e.g. the entries in brackets for He-NaF for which γz_0 was increased from 2.5 to 3.0. Indeed, for unrealistically large (γz_0), the $\lambda^{(2)}$ term eventually dominates, growing linearly with (γz_0). Gortel, Kreuzer, and Sommer (1983) have fixed the position of the minimum of the Morse potential to coincide with that of a surface potential obtained by summing Lennard-Jones potentials, with parameters in turn adjusted to fit surface bound states.

Gortel, Kreuzer, and Sommer (1983) have also calculated desorption times for He desorbing from graphite. We first note that the contribution $\lambda^{(0)}$ to the rate (i.e., without the Cole-Toigo corrections) is smaller than the rate in the one-dimensional version by some 10%. It is argued that this discrepancy is due to the steepest descent approximation. For the Cole-Toigo corrections, the error is expected to be larger. The contributions $\lambda^{(1)} + \lambda^{(2)}$ add about 35% to $\lambda^{(0)}$ at $T = 6$ K. However, looking at the integrands in (5.76), one should not be surprised if this contribution is reduced in a proper calculation that does not invoke the steepest descent approximation.

5.8 Multiphonon Contributions

The desorption times reported so far have all been calculated in the one-phonon approximation. In Sect. 4.2, we outlined various theories that included multiphonon processes in the calculation of the transition probabilities in the master equation. We now present the related results for desorption times.

We begin with the fourth order calculation by Gortel, Kreuzer, and Teshima (1980b) in which all one-phonon and two-phonon processes are included as they arise from first-, second- and third-derivative couplings, see (4.30). To make the calculation tractable, they restrict themselves to a gas-solid system in which the surface potential is approximated by a separable potential that develops just one bound state. For details, refer to Sect. 4.2. The parameters of the model are, accordingly, the bound state energy E_0, the range $\lambda = \gamma^{-1}$ of the surface potential [note that bound state energy and range determine the strength g in (4.38) of the potential uniquely],the Debye energy $\hbar\omega_D$ of the solid, the ratio m/M_S of the masses of a gas particle m, and a solid particle M_S, and the temperature T of the system. For a given bound state energy, the coupling between adsorbate and phonons increases as the range γ is decreased.

In terms of the transition probabilities $W^{(i)}(k,n=0)$ from the bound state n=0 to a continuum state k, the isothermal desorption time is given by

$$t_d^{-1} = \sum_{i=1}^{10} \sum_k W^{(i)}(k,n=0) \quad . \tag{5.77}$$

The last, and simplest, term in this sum was given as an example in (4.45). For the separable potential model chosen by Gortel, Kreuzer, and Teshima (1980b), the ten contributions can be reduced to double integrals, most of which have principle value singularities somewhere in the plane of integration, making numerical evaluation rather tricky.

In Fig. 5.15 we reproduce the isothermal desorption rate t_d^{-1} from (5.77) as a function of the bound state energy $\bar{\varepsilon}_0 = |\bar{E}_0|/\hbar\omega_D$ for a temperature such that $\delta = \hbar\omega_D/k_B T = 1$ for a rather weakly coupled gas-solid system with the range of the surface potential $\lambda = 2.5$ Å. The curves are labeled as follows: $(H^{(1)})^2$ is the contribution in second-order perturbation theory arising from the first-derivative coupling (dV_s/dz). $(H^{(1)})^4$ is its fourth-order contribu-

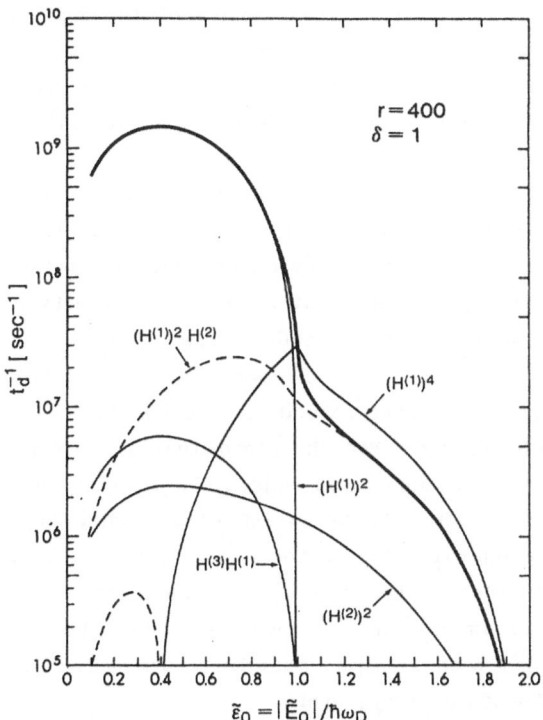

Fig. 5.15. Isothermal desorption rate t_d^{-1} as a function of the renormalized boundstate energy. The curve labeled $(H^{(i)})^n (H^{(j)})^m$ gives the contribution to the total rate (heavy line) arising from the n-th order in the Hamiltonian term H_{gs}^i and the m-th order in H_{gs}^j. The dashed portions of these curves are negative. From Gortel, Kreuzer, and Teshima (1980b)

tion. $(H^{(1)})^2 H^{(2)}$ is a contribution in which $H^{(1)}$ appears twice and the second-derivative coupling $H^{(2)} \propto d^2V/dz^2$ appears once, etc. Note that the cross term $(H^{(1)})^2 H^{(2)}$ is negative for all bound state energies $0 < \bar{\varepsilon}_0 < 2$. As $\bar{\varepsilon}_0$ approaches one, the second-order contribution $(H^{(1)})^2$ goes to zero because a gas particle trapped in a bound state with $\bar{\varepsilon}_0 > 1$ cannot be freed by absorbing a single phonon. For desorption from bound states with $\bar{\varepsilon}_0 > 1$, two-phonon processes are necessary. Note that in the latter region, the negative contribution from $(H^{(1)})^2 H^{(2)}$ is about half of that from $(H^{(2)})^4$. Of course, all contributions conspire at $\bar{\varepsilon}_0 = 1$ to make the total rate change continuously across this point at which all one-phonon processes vanish.

Because in this example the rate is dominated by the second-order contribution for $0 < \bar{\varepsilon}_0 \lesssim 0.9$ and for $k_D T \ll \hbar\omega_D$, we can expect that the rate drops appreciably as one-phonon processes become inoperative for $\bar{\varepsilon}_0 > 1$, although one can anticipate that for $1 \leq \bar{\varepsilon}_0 \leq 2$, fourth-order perturbation theory is sufficient for the calculation of desorption times, as it is the lowest order, giving nonvanishing contributions to the rate in this regime of bound state energies. Gortel, Kreuzer, and Teshima (1980b) also list the individual rate contributions for selected values of $\bar{\varepsilon}_0$ from which one can assess their relative importance. We note that the second-order rate arising from $(H^{(1)})^2$ goes to zero as the bound state energy $\bar{\varepsilon}_0$ approaches zero, but that the higher-order rates approach a nonzero, though very small, value in this limit. Although in a gas-solid system with $\bar{\varepsilon}_0 = 0$ there is no bound state, and there should therefore be no time evolution, the nonzero limits indicate a breakdown of the relaxation time approach inherent in the master equation. For the latter to be valid, one must have two distinct time scales in the system, namely, the microscopic time t_{micr} and the macroscopic relaxation time t_d, and one must require that the latter exceeds the former by a substantial margin as discussed in Sect. 3.2. For the microscopic time scale, we could e.g. take the oscillation period of an adparticle in the bound state $\bar{E}_0$, which, for $\bar{E}_0$ approaching zero, tends to infinity, leading to a breakdown of the relaxation time approximation on the way. Another way of looking at this result is to interpret the rate t_d^{-1} as the decay width $\Gamma = h t_d^{-1}$ of the bound state $\bar{E}_0$, in which case one must demand that Heisenberg's uncertainty relation $|\bar{E}_0| t_d \gg h$ or $\Gamma/|\bar{E}_0| \ll 1$ must hold to meaningfully talk about the time evolution of a given energy state.

Gortel, Kreuzer, and Teshima (1980b) present a number of model calculations at various temperatures and for several surface potential parameters, showing that two-phonon contributions can, indeed, become important for physisorption kinetics, particularly at high temperature. Several things can be learned from these model calculations:

(i) for consistency, multiphonon calculations must be carried out with the full Hamiltonian, i.e., in a perturbation approach by including higher derivative couplings.

(ii) A straightforward fourth-order calculation in $H^{(1)} \approx dV_s/dz$ overestimates two-phonon contributions typically by a factor of two, which in these model calculations is compensated by the negative contribution from the third-order terms $(H^{(1)})^2 H^{(2)}$.

(iii) The second order contribution from the higher derivatives like $(H^{(2)})^2$ are always negligible in these model calculations. This suggests that the second-order calculations by Allen and Feuer (1967), Bendow and Ying (1972), Efrima et al. (1980), Jedrzejek et al. (1981b), and Sedlmeir and Brenig (1981) underestimate multiphonon contributions.

With this caveat in mind, we next review the work by Bendow and Ying (1972,1973a,b) on multiphonon contributions to the desorption rate (again from a single bound state) as they arise by expanding the exponential involving the correlation function in their second-order expression for the rate (4.53) in a multiphonon series. Their results for the rate of desorption in the forward direction $R(\theta=0)$, see (5.58), is reproduced in Fig. 5.16. They attribute the discontinuities to the Debye cutoff at ω_D. As the calculations by Gortel, Kreuzer, and Teshima (1980b) indicate, they may also be due to the neglect of higher order terms in the perturbation theory. In any case, Bendow and Ying (1973b) conclude that the desorption rate is most likely dominated by the lowest order phonon processes allowed by energy conservation.

Efrima et al. (1981) and Jedrzejek et al. (1981b) have also reported multiphonon contributions to the desorption rate as they arise by expanding their correlation function expression for the transition probabilities, see (4.69,70). Within their one-dimensional model with a surface Morse potential, they consider a physisorption system, namely Ar on W, and a chemisorption system, CO on Cu. For the latter, the spacing between the lowest two energy states is larger than $\hbar\omega_D$ so that one-phonon processes cannot induce desorption, necessitating at least two-phonon processes. However, for a binding energy of about 0.6 eV one expects that the adsorption-desorption process is mediated to a large extend by electronic processes, rather than direct phonon processes, so that the applicability of the present model of phonon-mediated desorption kinetics to CO on Cu warrants further scrutiny. Restricting ourselves to the case of physisorption, we reproduce in

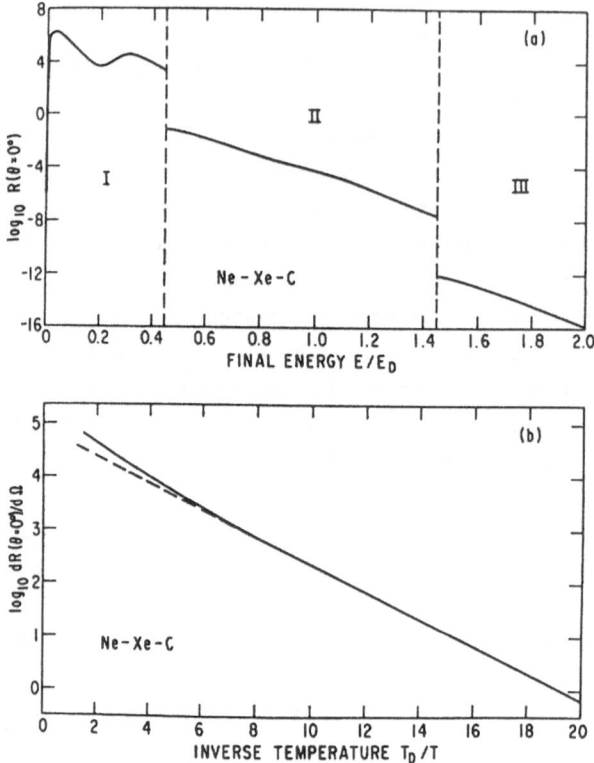

Fig.5.16. a Logarithm of the desorption rate in the normal direction, $R_k(\theta=0)$, v.s final adatom energy E/E_D at the fixed temperature $T/T_D=0.25$; $E_D = \hbar\omega_D$. Region I (bound state energy $E_o<$energy transfer$<$Debye energy E_D) includes one-, two-, and three-; region II ($E_D<$energy transfer$<2E_D$) just two- and three-; and region III ($2E_D<$energy transfer$<3E_D$) just three-phonon contributions. b Logarithm of the total desorption rate per unit solid angle in the normal direction vs. inverse temperature T_D/T. (Bendow and Ying 1972)

Table 5.9. The rate of Ar desorption from the W surface as a function of temperature in the one- (R_I), two- (R_{II}), three- (R_{III}) and multiphonon (R_{IV}) approximations. Powers of ten in parentheses. $T_D=312$ K; Morse potential parameters: $V_o=956$ K and $\gamma=1.44$ Å^{-1} so that the $0\to1$ transition is 51.1 cm^{-1}. The phonon width is 100 cm^{-1}. (Jedrzejek et al. 1981b)

T	R_I	R_{II}	R_{III}	R_{IV}
[K]	[s^{-1}]	[s^{-1}]	[s^{-1}]	[s^{-1}]
50	1.05(6)	1.11(6)	1.12(6)	1.12(6)
150	3.63(10)	3.99(10)	4.01(10)	4.01(10)
300	5.49(11)	6.53(11)	6.65(11)	6.65(11)
600	3.91(12)	5.72(12)	5.55(12)	5.88(12)

Table 5.9 their result for Ar desorption from W. At temperatures below the Debye temperature T_D, one-phonon processes are sufficient, whereas for $T >$ T_D many phonon processes start to contribute. We repeat again that their results are based on second-order perturbation theory which according to Gortel, Kreuzer, and Teshima (1980b) seriously underestimates multiphonon contributions so that (i) their exact result cannot be accepted without reservations and (ii) their claim that at $T = 600$ K more than three-phonon processes are necessary has to be verified as well. Also the use of a phonon width Δ to ensure numerical convergence warrants further study. They chose $\Delta = 100$ cm^{-1} which is about twice the energy difference between the lowest two energy levels and about a third of the Debye energy. If Δ is a measure of the phonon linewidth as Jedrzejek et al. (1981b) claim, then this number seems quite large for temperatures below T_D.

We should emphasize again that in gas-solid systems where the energy of the lowest bound state is in magnitude of the order of or larger than the Debye energy of the solid, the surface potential will develop more, and most often, many more than one bound state so that desorption does not only take place via simultaneous absorption of two or more phonons but via a cascade of one-phonon processes. Only if (i) any two neighbouring energy levels in the surface potential - most likely the lowest two - are separated by more than the Debye energy of the solid or (ii) if the temperature is of the order or even larger than the Debye temperature are two or more phonon processes necessary. Either situation is not very likely in physisorption because to make (i) happen, the surface potential must be deep and narrow which signals the onset of chemisorption, and in the high temperature situation (ii) weak physisorption is not possible to start with.

5.9 Multilayer Desorption

In Sect. 2.2, we outlined the mean field theory of multilayer physisorption as developed by Summerside et al. (1982) and Sommer and Kreuzer (1982a-d). In Sect. 4.4, we briefly commented on the calculation of the transition probabilities in the master equation (3.109). Sommer and Kreuzer (1982a-d) have calculated the isothermal desorption time as a function of coverage and temperature for ^{3}He desorbing from various substrates. They adopted the following strategy:

> (1) one starts with a gas-solid system in equilibrium at a pressure P and a temperature T, with the chemical potential determined by the (ideal) gas and the occupation functions given

by Fermi-Dirac statistics. Solving the Hartree-Fock equations (2.50) self-consistently, one gets the gas particle wave functions and the single particle energies. The resulting coverage can then be calculated from (2.54).

(2) With the (coverage-dependent) wave functions one next integrates the transition probabilities (4.7).

(3) With the continuum-bound state transitions dropped in (3.109), one then solves these equations by matrix diagonalization for a small time increment Δt with the right-hand side determined by the initial conditions.

(4) With the new occupation functions $n_i(\Delta t)$ (all continuum states are empty) corresponding to the reduced coverage $\theta(\Delta t)$ one reenters the Hartree-Fock equations (2.50) and recalculates wave functions and energies self-consistently as in step (1). In this way one generates the time evolution $\theta(t)$ from which the coverage-dependent desorption time can be extracted.

The implicit assumption in the above procedure is, of course, that the internal readjustment of the adsorbate during the desorption process is much faster than the desorption process itself. Note also that in addition to the explicit nonlinearity in the master equation (3.109), there is a much stronger implicit one through the dependence of the initial and final states in the transition probabilities (4.7) on the occupation functions in the Hartree-Fock equations (2.50).

Recalling that, at least at small coverage, the isothermal desorption time for systems like He-graphite can be calculated using perturbation theory as outlined in Sect. 5.3, one gets

$$t_d^{-1} = \frac{\pi}{M_s N_s} \sum_i \sum_p \sum_k \frac{1}{\omega_p} \left| \int d^3r \, \psi_i^*(r) \, \frac{\partial V_s(z)}{\partial z} \, \psi_k(r) \right|^2$$

$$*\delta(E_i - E_k + \hbar\omega_p) n_i n_p^{(ph)} / \sum_j n_j \quad . \tag{5.78}$$

We begin the discussion of physisorption kinetics for ^{3}He desorbing from the basal plane of graphite. The relevant potential parameters have been discussed in Sect. 2.2. For a graphical representation, we use the Frenkel-Arrhenius parametrization (1.3) with $E_d=Q$ calculated as the isosteric heat of adsorption. In Fig. 5.17, we plot the logarithm of the preexponential factor ν and the heat of adsorption Q versus coverage θ. We note that Q drops from a value $Q\approx-E_0+1.5k_BT$ at $\theta=0$, where $E_0=-135$ K is the lowest-energy bound state in the bare surface potential, to $Q\approx E_g + 1.5k_BT$ at $\theta\approx1.5$, where E_g is

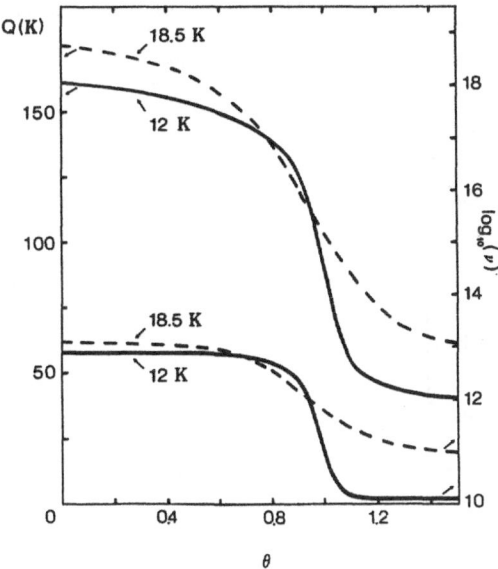

Fig. 5.17. Heat of adsorption Q and logarithm of the prefactor ν vs. coverage θ for ³He-graphite at two temperatures. (Sommer and Kreuzer 1982b)

the depth of the He-He interaction potential. This trend simply reflects the fact that a He atom is more tightly bound to the graphite surface than to a monolayer of helium on graphite. At the same time, the prefactor drops from $\nu \approx 10^{13}$ s^{-1} at θ=0 to $\nu \approx 10^{10}$ s^{-1} at T=12 K and $\nu \approx 10^{11}$ s^{-1} at T=18.5 K for θ > 1. The rapid decrease in ν as a monolayer gets completed can be readily understood from the expression (4.7) for the transition probabilities and the fact that, according to the lower panel of Fig. 2.22, the second bound state is moving quite far out for θ > 1 into the region of the second adlayer where the coupling to the phonons, quantified by the gradient of the bare surface potential (∇V_S) in (4.7), gets very weak. The fact that the prefactor ν changes in the same direction as the heat of adsorption, in particular that

$$\ln\nu = bQ + C \quad , \tag{5.79}$$

is referred to generally for thermally activated processes as a compensation effect, and has been observed in many chemisorbed systems and "families" of similar catalysts. For an introductory discussion we refer to Clark's book (1970).

Sommer and Kreuzer (1982b) also considered physisorption of ³He on solid argon. The calculated isotherms and the heat of adsorption agree well with experimental data by Wallace and Goodstein (1970). The desorption kinetics again shows a pronounced compensation effect, see Fig. 5.18. With Q and ν

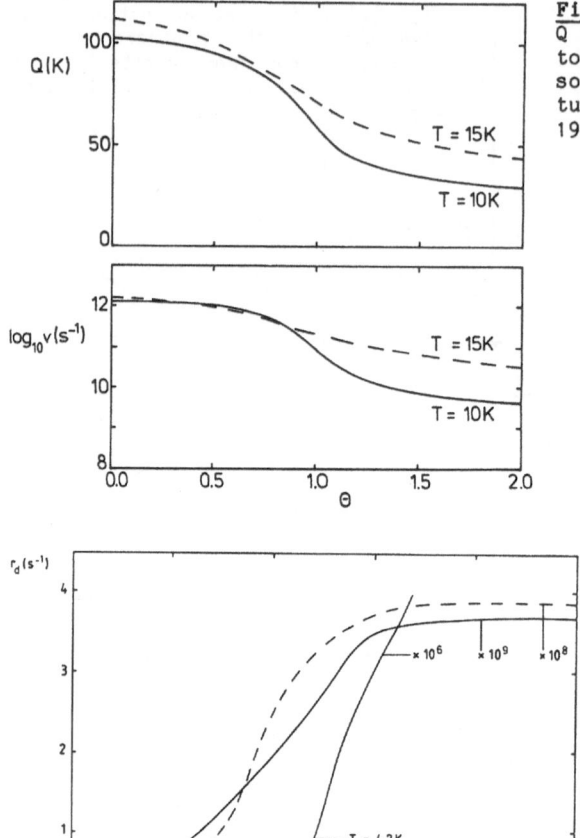

Fig. 5.18. Heat of adsorption Q and logarithm of the prefactor ν vs. coverage θ for ³He⁻ solid Ar(100) at two temperatures. (Sommer and Kreuzer 1982b)

Fig. 5.19. Desorption rate $r_d = \theta / t_d$ for ³He⁻Ar(100) for three different temperatures. (Sommer and Kreuzer 1982b)

varying so strongly as a function of coverage θ, the question arises whether it is at all meaningful to write the overall desorption rate r_d as a first-order reaction according to (1.2). Sommer and Kreuzer (1982b) have therefore plotted, in Fig.5.19, r_d versus θ for several temperatures. At T=10 K and 15 K, it is obvious that for θ≤0.5 desorption is of first order. However, at θ>1.5, the rate r_d becomes zero order for T>10 K; i.e., desorption at low coverage first order and thus proportional to coverage, goes over at high coverage into evaporation which is independent of coverage. That evaporation sets in so early, i.e., around θ≥1.5, is explained as arising from the rather small binding energy of ³He on the substrate argon. Indeed, for ³He desorbing

184

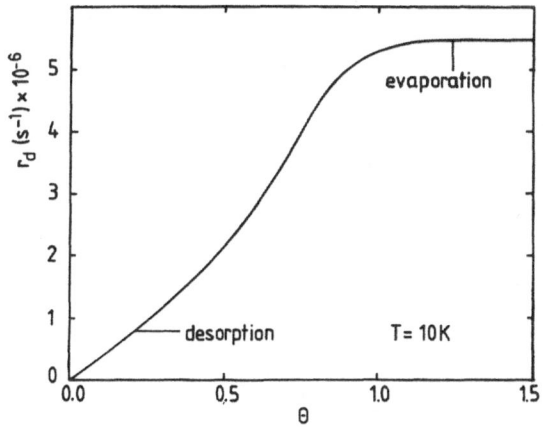

Fig. 5.20. Desorption rate $r_d = \theta / t_d$ for ^{3}He-constantan. (Sommer and Kreuzer 1982b)

from graphite, no evaporation is predicted up to $\theta \gtrsim 2.0$. On the other hand, for ^{3}He desorbing from constantan and Nichrome, two Cu-Ni alloys, a transition from first-order desorption to zero-order evaporation is predicted at coverages slightly larger than a monolayer, see Fig. 5.20.

The mean field approach to the phonon-mediated desorption kinetics of multilayer physisorption, as developed by Sommer and Kreuzer (1982a-d), seems to be the first microscopic calculation of the coverage- and temperature-dependent desorption times. So far, the computations have had to be restricted to the He isotopes as adsorbates. The main results are the following:

(1) The desorption rates increase with coverage, in some systems changing from first-order desorption at submonolayer coverage to zero-order evaporation above a monolayer.

(2) Using the Frenkel-Arrhenius parametrization for the desorption time, one finds a pronounced compensation effect, i.e., a concurrent variation of Q and $\ln \nu$.

(3) A typical reduction in ν around monolayer coverage by two to three orders of magnitude is observed in qualitative agreement with experimental data by Sinvani et al. (1982) for ^{4}He desorbing from Nichrome and constantan heaters at and below monolayer coverage. Their data is summarized in Table 5.10. We note that these desorption times are orders of magnitude shorter than those reported previously by Cohen and King (1973).

Sommer and Kreuzer (1982b) have suggested a number of improvements in their theory, both at the mean field and at the kinetics levels. It is worth noting that a quite different approach to the compensation effect in chemisorption has recently been taken by Leuthäuser (1980) and Zhdanov

Table 5.10. Summary of data for ⁴He desorbing from Nichrome and Constantan (*) heaters. First two columns give the temperature and chemical potential of the film before the heat pulse. $Q=E_d$ is the heat of adsorption and t_d^0 is the prefactor in (1.3). (Sinvani et al. 1982)

T	$-\mu$	Q	$10^{10}t_d^0$
[K]	[K]	[K]	[s]
3.75	94	60±4	2
3.50	94	65±2	1
3.48	94	62±2	1.3
3.48	84	56±2	1.6
3.48	80	54±2	2.7
3.48	72	45±2	5
3.27	64	33±2	13
2.71	61	32±2	13
1.50	30	20±2	15
3.49*	94	59±2	4
3.80*	56	43±4	2

(1981a,b) who start from a lattice gas model and calculate desorption times within the framework of absolute rate theory.

5.10 Adsorbent Cooling in Thermal Desorption

A particle adsorbed on a surface of a solid will desorb if it is supplied with enough energy to break its surface bond and, possibly, to overcome an activation barrier. In thermal desorption, this energy is drawn from the solid itself, either directly from the lattice degrees of freedom in phonon-mediated desorption or from the electronic degrees of freedom in chemisorbed systems. For laser- or particle induced desorption, the energy is supplied by the laser or the particle beam. In the latter cases, it is obvious that the desorption rate will be a function of the beam intensities, i.e., of the rate at which the energy is supplied. It is also easily seen that the rate of thermal desorption does not only depend on the coupling of the adsorbate to the internal degrees of freedom of the solid (vibrational or electronic), but also on the capacity of the latter to supply the energy during the desorption time at the surface.

A simple estimate, given in Sect. 1.1, shows that in typical systems no significant energy depletion, i.e., cooling, will occur in the surface region as a result of the desorption process. Gortel and Kreuzer (1983a) have sub-

stantiated this estimate by a detailed calculation in which they couple the
master equation (5.1) to Fourier's heat equation. The tacit assumption here
is that, during desorption, the solid can be described in the local equilibrium approximation, implying that the phonon system is controlled by a
local, time-dependent temperature field $T(r,t)$ inside the solid, which is
subject to Fourier's equation of heat conduction

$$\partial T/\partial t - \chi \frac{\partial^2 T}{\partial z^2} = 0$$

$$\frac{C_V \chi}{V_m} \left[\frac{\partial T}{\partial z} \right]_{z=0} = \dot{E}(t) \quad . \tag{5.80}$$

Here V_m, C_V and $\chi = \lambda V_m / C_V$ are the molar volume, molar heat capacity, and the
thermal diffusivity of the solid, $\dot{E}(t)$ is the energy current density through
the surface accounting for the energy taken up by the desorbing particles

$$\dot{E} = \theta_0 N_S \sum_i n_i(t) \left[\sum_{i'} (E_i - E_{i'}) W(i',i) + \sum_k (E_i - E_k) W(k,i) \right] / \sum_{i''} n_{i''}(0) \quad , \tag{5.81}$$

θ_0 is the initial coverage and N_S is the maximum number of particles per
unit area that can be adsorbed in a monolayer. Equations (5.1,80,81)
describe the dynamics of the coupled adsorbate-phonon system. Here the
sums over i and j exhaust all the bound states and the sum over k only
encompasses the continuum. We note that the bound state-bound state transitions $W(i,i')$ and the bound state-continuum transitions $W(k,i)$ depend explicitly on temperature via the phonon occupation functions. The energy current density $\dot{E}$ is obviously only appreciable if the number of particles $\theta_0 N_S$
initially adsorbed per unit surface area is a reasonable fraction of a monolayer. On the other hand, we must be aware that the kinetic theory contained in (5.1) is only valid at vanishingly small coverage. However, the mean
field theory of multilayer physisorption has shown that the desorption kinetics in the helium-graphite system does not change significantly up to
about half a monolayer; this justifies the ansatz (5.81).

It is useful to solve (5.80) using the appropriate Green function (Tikhonov and Samarskii 1963)

$$T(z,t) = \frac{1}{2\sqrt{\pi\chi}} \int_{-\infty}^{0} ds \left[\exp\left[-\frac{(z-s)^2}{4\chi(t-t_0)} \right] + \exp\left[-\frac{(z+s)^2}{4\chi(t-t_0)} \right] \right] \frac{T(s,t_0)}{\sqrt{t-t_0}}$$

$$+ \frac{V_m}{C_V} \frac{1}{\sqrt{\pi\chi}} \int_{t_0}^{t} dt' \exp\left[-\frac{z^2}{4\chi(t-t')} \right] \frac{\dot{E}(t')}{\sqrt{t-t'}} \quad . \tag{5.82}$$

187

To solve (5.1) and (5.82) simultaneously, one discretizes time into steps $t_0=0, t_1, t_2, \ldots$ and assumes that the temperature $T(z,t)=T_n(z)$ remains constant in each time interval $t_n < t < t_{n+1}$. Starting at $t_0 = 0$ with a uniform temperature T_0 throughout the solid, one calculates the transition probabilities and solves (5.1) by diagonalization. The resultant time-dependent occupation functions (5.4) are inserted in (5.81) to yield the energy current density for times $t_0 < t < t_1$, which inserted in (5.82) yields the temperature field $T(z,t_1)$. Its value at the surface $z = 0$ (which can actually be calculated algebraically if we put $\dot{E}$ constant in each time interval) is used to calculate $W(i,i')$ and $W(k,i)$ for the next time interval for which the above procedure is repeated. The overall solution is found to be stable for time intervals that are smaller than about one fifth of the desorption time at the particular time.

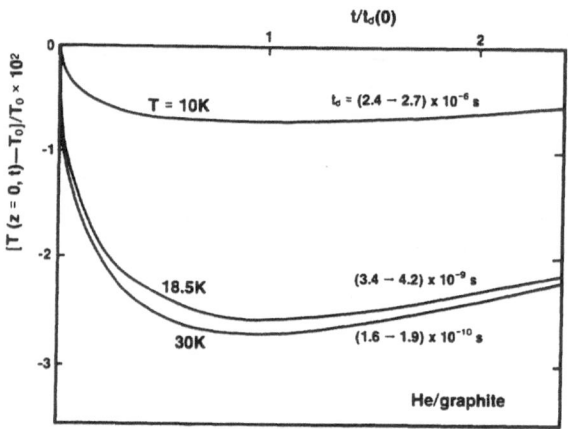

Fig. 5.21. Relative surface temperature drop for the desorption of half a monolayer of helium from graphite at three temperatures as a function of time (in units of desorption time $t_d(0)$ at time $t = 0$). (Gortel and Kreuzer 1983a)

In Fig. 5.21 we present the drop in surface temperature during the desorption process. The system is helium desorbing from graphite for different initial temperatures. For T>10 K, we have approximated the specific heat of graphite by (Keesom and Pearlman 1955) $C_V=(0.208 \, T^2-6.8)$ mJK^{-1}mol^{-1}. Its thermal conductivity we took as (Touloukian et al. 1970) $\lambda=AT^2$ with $A \approx 10^{-2}$ Jm^{-1}s^{-1}K^{-2} being the one for pyrolitic graphite perpendicular to the crystal planes. For all initial temperatures, there is a fast drop in surface temperature within one desorption time. The maximum relative temperature drop does not exceed a few percent. It will be larger, the smaller

the thermal conductivity and specific heat are. The former controls the
rate of energy replenishment from the interior of the solid, whereas the
latter is by definition the temperature change accompanying the energy loss
due to desorption. From these facts alone, one would expect that the absolute temperature drop at T_1 = 10 K is about four times larger than at T_2 =
18.5 K, which is obviously not the case. What is more important is the fact
that at lower temperature the desorption time grows exponentially, making $\dot{E}$
much smaller as well. The rise of temperature after about one desorption
time is controlled by thermal diffusion alone. We also note that the overall temperature drop is roughly proportional to the initial adsorbate coverage.

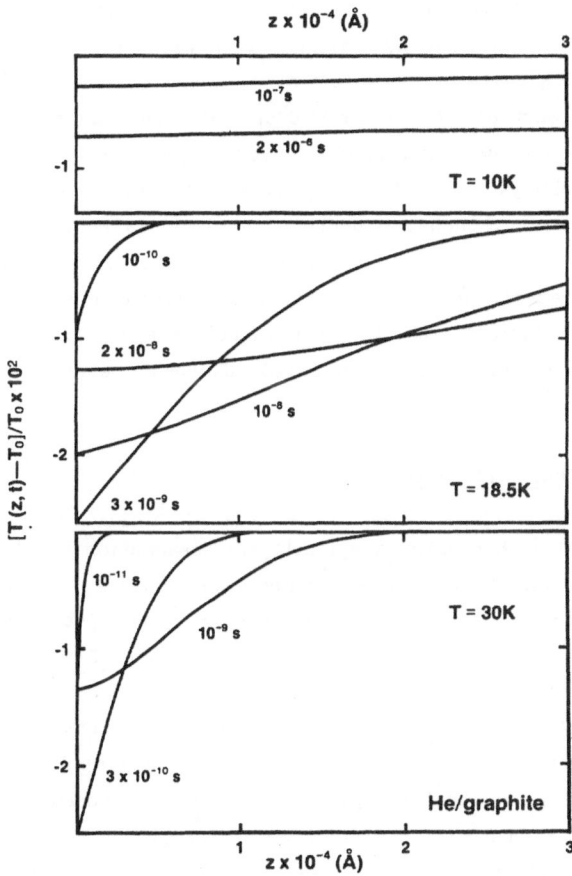

Fig. 5.22. Temperature depth profile during the desorption of half a monolayer of helium from graphite at three temperatures at the times indicated
after the desorption process has been started. (Gortel and Kreuzer 1983a)

To study the temperature profile $T(z,t)$ as a function of distance z into the solid we turn to Fig. 5.22. In the upper panel we plot $T(z,t)$ for $T = 10$ K at some very early time $t_1 = 0.042t_d$, around the maximum temperature drop at $t_2 = 0.83t_d$, and at some late time $t_3 = 4.2t_d$. In a time t after desorption starts, the temperature inhomogeneity has spread a distance $L = (\chi t)^{1/2}$ into the solid which for t_1 is already some 6.4×10^4 Å and grows to 6.4×10^5 Å at t_3. Desorption at 10 K is so slow that the thermal response of the solid is fast enough to replenish readily the phonons used up in the process. This is no longer the case at $T = 18.5$ K, where desorption is faster by 3 orders of magnitude. Thus after 10^{-10} s, when about as much has desorbed as after 10^{-7} s at $T = 10$ K, the temperature inhomogeneity has extended some 2000 Å into the solid and about 10^4 Å after 3×10^{-9} s. At $T = 30$ K, desorption is so fast that all the necessary energy is taken out of a thin layer less than 10^4 Å, which then thermalizes again with the rest of the solid on a much longer time scale of some 10^{-7} s.

Comparing the calculated surface temperature drop with the estimates according to (1.14) in Table 1.4, we find good agreement, keeping in mind that a factor e^{-1} or so can always be justified in (1.14). If the adsorbent is a very thin film of thickness d, then the depth L in (1.14) is not given by (1.15) but is equal to d. Performing a desorption experiment at high temperature from a very thin adsorbent could lead to a significant temperature drop. However, for a "realistic" situation, the latter is always negligible, allowing the theorist to assume that the solid adsorbent remains in thermal equilibrium.

5.11 Flash Desorption and Thermalization

We have so far reviewed theories that calculate isothermal desorption times. We recall that in an isothermal desorption experiment, one starts from a gas-solid system in equilibrium at some initial temperature T_i and pressure P_i, and reduces the latter suddenly to a much smaller final value $P_f \ll P_i$ that is maintained subsequently by continuous pumping. In the theory, this experiment is modeled by putting the occupation functions of the continuum states equal to zero. In contrast, in a flash desorption experiment one creates the nonequilibrium situation, that triggers the desorption process by a sudden increase in the temperature of the solid from T_i to T_f.

To describe the flash desorption experiment, readsorption out of the gas phase must be included in the theory which has been done by Gortel, Kreuzer, and Spaner (1980) by imposing the appropriate final state reached by the gas-solid system after a long time without specifying the physical mechanism responsible for thermalization. Indeed, there are two processes which can equilibrate the gas phase: (i) collisions between particles in the gas phase and (ii) collisions of gas particles with the walls. If the mean free path ℓ for gas particle collisions is large compared to the size L of the experimental chamber, then only collisions with the walls provide an effective means of changing the energy distribution of the gas phase to a thermal one at the temperature of the walls. If $\ell \leq L$, thermalization is speeded up by collisions between gas particles. For temperatures below room temperatures and pressure lower than 10^{-2} Torr and $L \approx 10^{-1}$ m, one always gets $\ell \gg L$.

In this section, we will set up the appropriate boundary conditions for the master equation (5.1) to describe a flash desorption experiment at low coverage and the subsequent thermalization due to wall collisions (Gortel and Kreuzer 1983b). We assume that at times $t < 0$ the gas-solid system is in equilibrium at T_i and P_i, so that the gas particle occupation probabilities are given by (5.2). Having raised the temperature of the solid to T_f for times $t \geq 0$, we calculate the phonon-mediated transition rates $W(\iota,\iota')$ at T_f. The experimental situation we thus describe is one in which all walls surrounding the gas phase are at T_f. From (5.1), we can then calculate the time evolution of the relative bound state occupation (5.5). Likewise, we get the time-dependent pressure

$$P(t)/P(0) = \sum_k E_k n_k(t) / \sum_k E_k n_k(0) \quad , \tag{5.83}$$

where both sums go over all continuum states of energy $E_k = \hbar^2 k^2/2m$ with m the mass of a gas particle.

To calculate the time-dependent occupation functions $n_k(t)$ of the continuum states, Gortel and Kreuzer (1983b) considered a one-dimensional model appropriate for a highly mobile adsorbate; they also employed the one-phonon transition probabilities (4.7). To solve (5.1) numerically by matrix diagonalization, one must first truncate the wave number k at some upper limit k_u to get a finite number of equations. To find an estimate for k_u, we note that one-phonon processes can connect energy levels that are at most one Debye energy apart. Restricting ourselves for physisorption to temperatures $k_B T_f < \hbar\omega_D$ (otherwise multiphonon processes become important), we know that in the final equilibrium only continuum states with $E_k \approx k_B T_f < \hbar\omega_D$ are appreciably occupied. This suggests that k_u can be chosen such that $\varepsilon_k = E_k/\hbar\omega_D$ is at most 2 or 3 to allow for the possibility of some transient occupation of higher energy states due to continuum-continuum transitions during the desorption process and the thermalization towards the final equilibrium. From

$$\varepsilon_k = \frac{1}{r} \left(\frac{\pi}{\gamma L}\right)^2 k^2 \leq \varepsilon_u \quad , \quad k=1,2,3,\ldots.k_u \tag{5.84}$$

we get

$$k_u = \sqrt{r\varepsilon_u} \, \frac{\gamma L}{\pi} \quad . \tag{5.85}$$

Note that in (5.84) and in the remainder of this section k and k_u are no longer wave numbers but just integers. To get a feeling for numbers, we look at the helium-graphite system for which $\gamma^{-1} \approx 1$ Å, $r \approx 25$, so that for a typical experimental chamber of, say, 0.30 m diameter, k_u is of the order of 10^{10}. Obviously, one cannot diagonalize a matrix of that size, so that a coarse-graining of the energy spectrum is mandatory. We therefore group the numbers k into N_{cg} blocks, each containing n_{cg} and label these blocks by $R = 1,\ldots,N_{cg}$, where N_{cg} can be chosen of the order of 100 or 200 for a reasonably sized computer. We next define average occupations

$$\bar{n}_R = \sum_{k=k_1}^{k_2} n_k \quad , \qquad k_i = R \pm \tfrac{1}{2} n_{cg} \tag{5.86}$$

which are subject to the rate equations

$$\dot{n}_i = \sum_{i'=0}^{N} W(i,i')n_{i'} - \sum_{i'=0}^{N} W(i',i)n_i + \sum_{R=1}^{N_{cg}} W(i,k)\bar{n}_R - \sum_{R=1}^{N_{cg}} n_{cg} W(k,i)\, n_i$$

$$\dot{\bar{n}}_R = \sum_{R'=1}^{N_{cg}} n_{cg}W(k,k')\bar{n}_{R'} - \sum_{R'=1}^{N_{cg}} n_{cg}W(k',k)\bar{n}_R + \sum_i n_{cg}W(k,i)n_i$$

$$- \sum_{i=0}^{N} W(i,k)\, \bar{n}_{R'} \quad , \tag{5.87}$$

where k and k' are representative numbers within the blocks R and R', respectively; e.g., $k=n_{cg} (R+\tfrac{1}{2})$ and $k' = n_{cg} (R'+\tfrac{1}{2})$. These choices and also the values for n_{cg}, N_{cg}, and ε_u must, of course, be made so that the features of the solution of (5.87) that are important for the kinetics, are independent of them. Equations (5.87) can now be solved by matrix diagonalization to yield (5.4). Likewise, we find for the pressure

$$\frac{P(t)}{P(0)} = \sum_{\kappa} T_\kappa e^{-\lambda_\kappa t} \quad ,$$

$$T_\kappa = \sum_{R} R^2 A_{\kappa\kappa} / \sum_{R',\kappa'} R'^2 A_{\kappa'\kappa'} \quad . \tag{5.88}$$

We will now present numerical results, given by Gortel and Kreuzer (1983), on the time evolution of the relative bound state occupation (5.5) and the pressure (5.88) for flash desorption and subsequent thermalization of helium desorbing from a graphite surface. Results for desorption from a LiF surface are given elsewhere. In a first series of calculations, we start from an initial temperature of T_i = 6.17 K (δ_i = 30) and flash the solid at time t=0 to T_f = 7.4 K (δ_f = 25). The resulting time evolution of pressure and coverage is plotted in Fig. 5.23 against log t to highlight the short time transient behaviour. For the top panel, we have chosen a small experimental chamber of only 3 cm linear dimension. Note first that in-

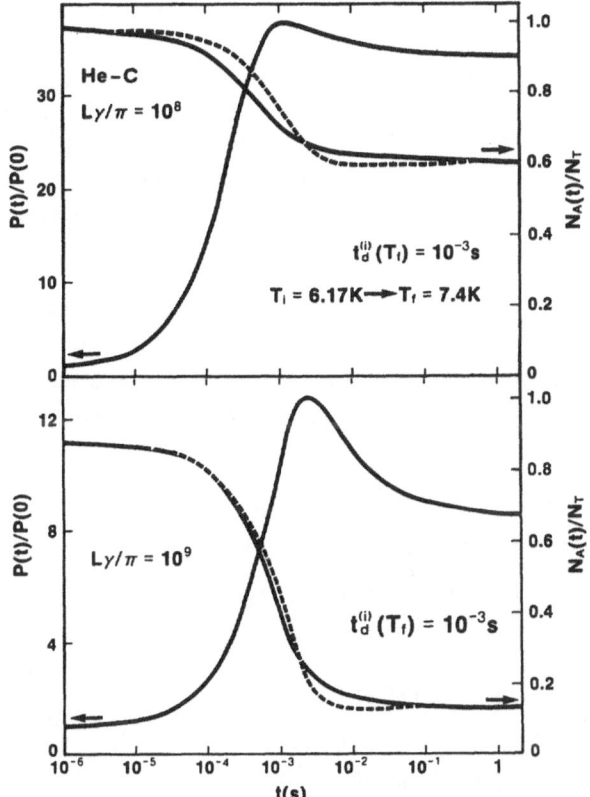

Fig. 5.23. Flash desorption in the He-graphite system of size $L\gamma/\pi$ after a flash of the temperature of the solid from T_i to T_f. The isothermal desorption time at T_f is $t_d^{(i)}(T_f)$. The time evolution of pressure $P(t)/P(0)$ and relative bound state occupation $N(t)/N_g$ are calculated from (5.88) and (5.5), respectively, where N_g is the total number of particles in the gas phase and in the adsorbate. The dashed line is calculated from (5.90). Debye temperature of graphite T_D = 185 K. Morse potential parameters V_0 = 171.34 K and γ^{-1} = 0.95 Å. (Gortel and Kreuzer 1983b)

itially at $T_i = 6.17$ K more or less all particles are actually trapped in the bound states so that $N(t=0)/N_g \approx 1.0$, where N_g is the total number of particles in the gas phase and in the adsorbate. Within a time of the order of 10^{-3} s, it decreases to $N(t=\infty)/N_g \approx 0.605$, however, not in an exponential fashion because the transition matrix $\mathbf{W}$ has about 30 eigenvalues of the order 10^3 s^{-1}, which contribute with comparable weights to the sum (5.5). The time evolution for isothermal desorption at $T_f = 7.4$ K,

$$N(t)/N_g = (N(0)/N_g)\, \exp(-t/t_d^{(i)}) \quad , \tag{5.89}$$

describes complete desorption to $N(t = \infty) = 0$ with $t_d^{(i)} = 10^{-3}$ s. Accounting for the residual gas phase at $t = \infty$, a phenomenological ansatz, employed by Gortel, Kreuzer, and Spaner (1980),

$$N(t) = N(0) + [N(\infty) - N(0)]\exp(-t/t_d^{(i)}) \tag{5.90}$$

produces a fair fit to the flash desorption "data" (dashed lines). To calculate the equilibrium values of pressure and relative bound state occupation for $t=0$ and $t\to\infty$, we first determine the chemical potential through normalization

$$1 = \sum_\iota n_\iota^{eq} = \frac{1}{N_g}\, e^{\beta\mu} (L/\lambda_{th})^2 \left[\sum_j e^{\beta E_j} + L/\lambda_{th} \right] \quad , \tag{5.91}$$

where ι also includes the summation over the particle momentum $\mathbf{K}$ parallel to the surface and $\lambda_{th} = h/(2\pi m k_B T)^{1/2}$ is the thermal wavelength. We then get

$$\frac{N(t=\infty)}{N(t=0)} = \left[\frac{\displaystyle\sum_{j=0}^{N} e^{-\delta_f \epsilon_j}}{\displaystyle\sum_{j=0}^{N} e^{-\delta_i \epsilon_j}} \right] \frac{L/\lambda_{th}(T_i) + \displaystyle\sum_{j=0}^{N} e^{-\delta_i \epsilon_j}}{L/\lambda_{th}(T_f) + \displaystyle\sum_{j=0}^{N} e^{-\delta_f \epsilon_j}} \quad , \tag{5.92}$$

$$\frac{P_f}{P_i} = \frac{P(t=\infty)}{P(t=0)} = \left(\frac{\delta_i}{\delta_f}\right)^{3/2} \frac{\displaystyle\sum_{j=0}^{N} e^{-\delta_i \epsilon_j} + L/\lambda_{th}(T_i)}{\displaystyle\sum_{j=0}^{N} e^{-\delta_f \epsilon_j} + L/\lambda_{th}(T_f)} \quad . \tag{5.93}$$

We note that only in the large volume limit $L \to \infty$ will the pressures be in the ratio of the temperatures. For the present model system, all terms in (5.92,93) contribute, resulting in a pressure ratio $P_f/P_i = 34.4$ in the upper panel of Fig. 5.23, whereas in an infinite box it would be only $P_f/P_i = T_f/T_i = 1.2$; the difference being due to a substantial shift in the chemical poten-

194

tial. Note from the top panel of Fig. 5.23 that the final pressure is not reached monotonically, but that a slight overshoot occurs during the major part of the desorption process, followed by a much slower thermalization process during which wall collisions slowly adjust the occupation functions to the final equilibrium one. The desorption process thus puts the particles into continuum states that have too high an energy for T_f. This implies that in a very large box, the flux of desorbing particles, e.g., detected in a mass spectrometer, will not have a Maxwell velocity distribution, but rather one that is determined by the bound state-continuum transition rates in the master equation. We will comment on such deviations from a Maxwell velocity distribution in Chap. 6 on time of flight spectra. For the lower panel of Fig. 5.23, we have enlarged the experimental chamber to L = 30 cm. Fewer of the total number N_g of particles are initially in the adsorbate [$N(t=0)/N_g$ = 0.877] and more of them desorb [$N(t=\infty)/N_g$ = 0.133], the flash desorption process being closer to an isothermal one. Also note that the pressure overshoot is relatively larger, although the final pressure rise is lower by about a factor of 3 compared to the smaller box in the top panel.

In Figure 5.24 we look at the same system as in Fig. 5.23, but enclosed in a yet larger box of length about 1 m. In the left panel, we first follow a flash adsorption experiment decreasing the solid temperature at time t=0 from T_i = 7.4 K to T_f = 6.17 K. The adsorption and thermalization process being over after a few seconds, we then increase, in the right panel, the temperature to 7.4 K again to monitor a flash desorption experiment. As the

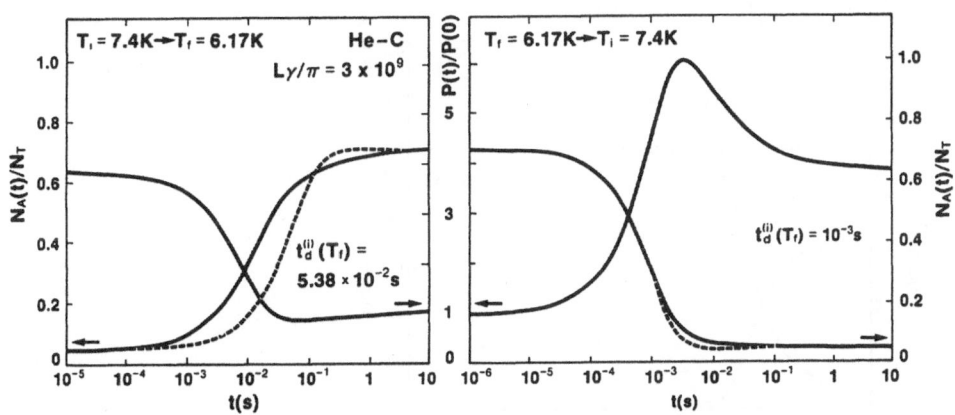

Fig. 5.24. Flash adsorption after a sudden temperature drop and flash desorption after a sudden temperature rise, back to the initial temperature in the He-graphite system. For details see Fig. 5.23. (Gortel and Kreuzer 1983b)

kinetic processes at the surface are mainly determined by the temperature of the phonon bath of the solid, the flash adsorption from 7.4 to 6.17 K is orders of magnitude slower than the flash desorption from 6.17 to 7.4 K. Also note from the dashed curve in the left panel of Fig. 5.24 that the ansatz (5.90) with $t_d^{(i)}$ the isothermal desorption time at $T_f = 6.17$ K is a rather poor description of the flash adsorption process indicating the importance of continuum-continuum transitions to thermalize the gas phase during the course of the flash adsorption. Also note the slight undershoot of the pressure.

Fig. 5.25 demonstrates the rather dramatic nonequilibrium effects in the He-graphite system after a substantial temperature flash of the solid from $T_i = 6.17$ K to 13.3 K. In this situation, the adsorbate evolution is very well described by the ansatz (5.89). Interesting is the fact that, in the

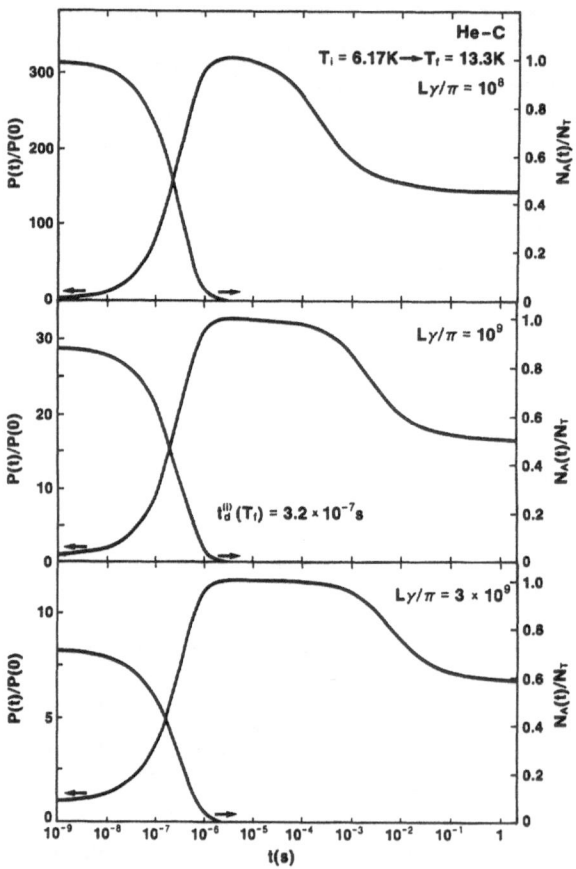

Fig. 5.25. Flash desorption in the He-graphite system as in Figs. 5.23,24, but for a larger temperature rise. (Gortel and Kreuzer 1983b)

bigger boxes, the pressure, after rising substantially during a time of the order of the isothermal desorption time $t_d^{(1)}$ (T = 13.3 K) = 3.2×10^{-7} s, remains constant for about a millisecond before it gradually decreases over about one second to its final equilibrium value.

In this section, we have reviewed a theory to study transient effects in a flash desorption experiment. In situations where the mean free path for gas particle collisions is much larger than the size of the experimental chamber, particles desorbing after the adsorbent temperature has been raised to its final value will thermalize due to wall collisions. Whereas the number of adsorbed particles decreases in a more or less exponential fashion within a typical flash desorption time, the gas pressure will rise with the same time scale, but overshoot its final value by as much as a factor of two depending on the temperature difference in the flash and the size of the vacuum chamber.

The theoretical model assumes that all walls of the vacuum chamber are covered by the adsorbent (as is the case, e.g., with films deposited on the inside of glass systems) and are subject to the temperature flash. If only a small single crystal surface or a filament is the adsorbent, then the desorbing particles will most likely thermalize with the other walls in the vacuum chamber. If the latter adsorb the gas readily, then a pressure device may be arranged to record the transient pressure only. It is important to note that it would be wrong to measure the number of desorbed particles and calculate a pressure using the ideal gas law. Such a procedure is only admissible for the final, not the transient, pressure.

The model, studied by Gortel and Kreuzer (1983b), is one-dimensional. This is appropriate for highly mobile adsorbates. It would appear that in a one-dimensional model, only the momentum component perpendicular to the surface from which the particle desorbed will be thermalized. That this is not the case, can best be seen in an idealized experiment for which we choose a perfectly cubic vacuum chamber with all surfaces covered by the adsorbent and uniformly flashed to a higher temperature. A particle leaving surface A with momentum k_1 perpendicular to A and momentum k_2 parallel to it, will either hit the surface B opposite to A thus thermalizing k_1, or it will hit a surface C perpendicular to A thermalizing the component k_2 perpendicular to C. In a one-dimensional model, it is assumed that the thermalization of k_1 proceeds at the same rate as that of k_2. It is feasible to set up a three-dimensional model in which the angular distribution of the desorbing particles is taken into account as well.

6. Time of Flight Spectra

6.1 Experiments

Isothermal desorption and temperature programmed desorption experiments are typically designed to measure a single time scale, namely the desorption time. Its temperature dependence can, via the Frenkel-Arrhenius parametrization (1.3), yield information on the heat of adsorption $Q=E_d$ or the depth V_0 of the surface potential well and on the prefactor ν, the former being an equilibrium property and the latter being a measure of the dynamic gas-solid coupling. One hopes that more can be learned about the dynamics of desorption if the velocity and angular distributions of the desorbed particles are measured in a time of flight spectrum.

The first experiment on the angular distribution of desorbed particles must be Knudsen's (1915) who evaporated mercury atoms from a surface element inside a glass sphere and observed optically that these atoms formed a uniform deposit all over the sphere. This demonstrates that the angular distribution of desorbed, or better evaporated, atoms follow Knudsen's cosine law, i.e., that the number of particles evaporating under an angle θ to the surface normal is proportional to $\cos\theta$. Other experiments of a similar kind confirmed these findings. With the improvements in molecular beam techniques and the advent of fast time of flight detectors the subject has attracted renewed interest (Palmer et al. 1970; Smith and Palmer 1972). A number of authors (van Willigen 1968; Dabiri et al. 1971; Bradley et al. 1972; Cardillo et al. 1975; Comsa et al. 1977, 1979, 1980 and 1982) reported strong forward peaking of the desorption flux in permeation experiments which could be fitted to a $\cos^d\theta$ law with d anywhere between 3 and 6. This phenomenon was associated with activation barriers connected with dissociation and recombination of chemisorbed molecules. Although no comprehensive review of time of flight measurements in chemisorbed systems exists, we must restrict the following discussion to physisorbed systems. Here an early rapid flash desorption experiment was performed by Cohen and King (1973) who adsorbed about 1.5 monolayers of helium on a constantan (a Cu-Ni alloy) wire nominally 2.54×10^{-3} cm thick and at a temperature below 2 K. A current pulse of 50 ns minimum duration leads to a temperature rise $\Delta T/\Delta t \geq 10^7$ Ks^{-1} causing complete desorption within 10^{-8} s. Their analysis based on time of flight distributions for final temperatures between 4 K and 18 K suggests that the second layer of helium is bound with about (31 ± 1) K and that the prefactor in the Frenkel-Arrhenius parametrization (1.3) is $\nu^{-1} = t_d^0 = (2 \pm 0.5) \times 10^{-7}$ s. Recently, Sinvani et al. (1982) have remeasured this system and found a prefactor $t_d^0 \approx (2-4) \times 10^{-10}$ s. They assert that the discrepancy can be traced to Cohen and King keeping the constantan wire hot for times long compared to t_d so that their deconvolution analysis of the time of flight signals could not have detected the short times measured by Sinvani et al. (1982). As the latter's measurement of the desorption kinetics is a direct one we discuss their procedure at some length. They achieve heating and cooling rates of 10^9K/s by using evaporated thin film (≈ 600 Å) Nichrome or constantan heaters and thin (≈ 2000 Å) Sn superconducting transition bolometer detectors on sapphire substrates. The fast thermal response arises because the thin film devices are in intimate contact with a phonon-transparent substrate.

A schematic diagram of the experimental geometry is shown in the inset of Fig. 6.1. When a voltage pulse is applied to the heater, the bolometer detects a signal due to the helium atoms desorbed by the heat pulse. As the heater pulse width, t_p, is increased from its minimum value (≈ 30 ns)

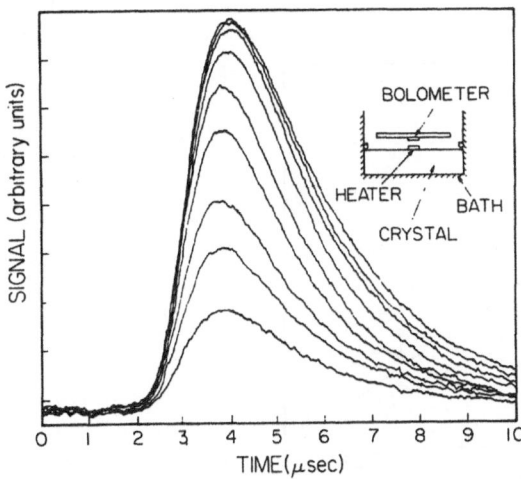

Fig.6.1. Bolometer signal vs. time for pulse width, t_p, in ascending order: 0.03, 0.06, 0.08, 0.15, 0.22, 0.5, 1, 1.5 and 2.5 μs. In all cases, the heater temperature was 8.2 K, the bath temperature 3.48 K, and the chemical potential 72 K. Inset shows experimental arrangements. (Sinvani et al. 1982)

more and more atoms are desorbed and the bolometer signal increases until $t_p \approx t_d$. For sufficiently long pulses, however, the signal saturates and becomes independent of the pulse width, as seen in Fig. 6.1, because the helium film reaches a steady state before the pulse is turned off.

Characterizing each spectrum by its maximum, Sinvani et al. (1982) find empirically that data of the kind shown in Fig. 6.1. can be fitted to the form

$$S_m(t_p) = S_m(\infty)(1-e^{-t_p/t_d}) \quad , \tag{6.1}$$

where $S_m(t_p)$ is the height of the maximum in the bolometer signal as a function of t_p, the pulse width, and $S_m(\infty)$ is the corresponding value for very long pulses. This is shown in Fig. 6.2. where $\log[1-S_m(t_p)/S_m(\infty)]$ is plotted versus t_p for the data of Fig. 6.1. The desorption time t_d is obtained from the slope of the resulting straight line.

In order to use the values of t_d to extract the parameters $\nu^{-1} = t_d^0$ and E_d in the Frenkel-Arrhenius parametrization (1.3) it is necessary to determine the heater temperature, T_s. This is done by making use of acoustic mismatch theory (Anderson 1981) which is applicable to phonon transport at a metal-insulator interface. Since the pulses used are always long compared to the thermal relaxation time of the heater, the heater temperature T_s may be regarded as constant during the pulse. According to acoustic mismatch theory, T_s is given by

$$T_S = (\frac{CW}{A} + T^4)^{1/4} \quad , \tag{6.2}$$

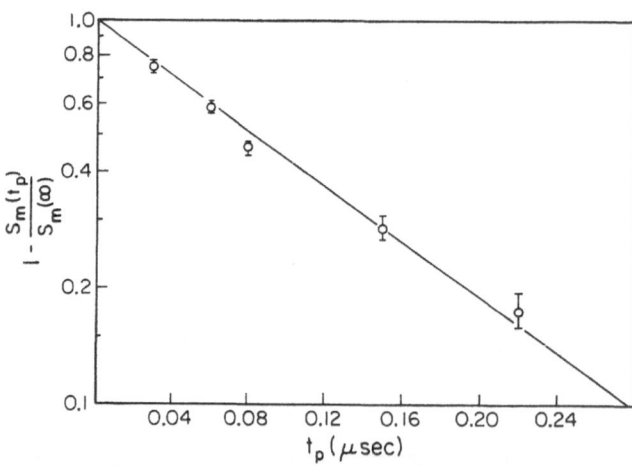

Fig. 6.2. $\log[1 - S_m(t_p)/S_m(\infty)]$ versus t_p for data in Fig. 6.1. Straight line gives $\tau = 0.11$ µs. (Sinvani et al. 1982)

where A is the heater area, W is the power dissipated in the heater film, T is the ambient temperature, and C depends on the heater and substrate elastic properties.

An independent estimate of the heater temperature may be obtained from an analysis of the measured times of flight. Typical models of desorption suggest that the desorbed atoms should have an approximately Maxwellian velocity distribution characterized by a temperature T_a which should be the same as the substrate temperature T_s, as we will see in detail in Sect. 6.2. The bolometer signal may thus be analyzed to find T_s. The precise result depends somewhat on the model used, but the values of T_s that emerge from the different models vary by only about 20%. Within these limits, the value of T_s deduced from the bolometer signal agrees with that given by (6.2) for each heater power.

Using the values of t_d and T_s determined by the methods described above, the Frenkel–Arrhenius parametrization (1.3) is tested by plotting log t_d versus T_s^{-1}. The results are very good straight lines; two examples are shown in Fig. 6.3. The slope and intercept of the line give the energy parameter E_d and the prefactor $\nu^{-1} = t_d^0$, respectively.

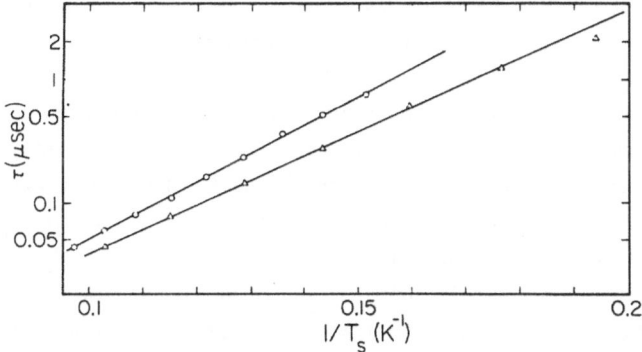

Fig. 6.3. Plots of τ versus $1/T_s$, verifying (1.3). Upper and lower curves respectively are 5th and 6th rows of Table 5.10. (Sinvani et al. 1982)

In order to complete the description of the experimental conditions, the pressure in the gas at equilibrium must be specified. Since the pressure in the gas governs the time for the film to return to equilibrium after each heat pulse, it was measured by Sinvani et al. in situ even at the lowest pressures ($\approx 10^{-9}$ Torr) by observing t_{rc}, the shortest interval between pulses such that the signal was not affected. Since the film is far from equilibrium during most of its recovery, the outgoing flux is negligible compared to the incoming flux. Then, kinetic theory suggests that

$$t_{rc} \approx \eta \, \frac{\sqrt{2\pi m k_B T}}{P} \, , \tag{6.3}$$

where η is to be determined. This was done by measuring t_{rc} as a function of the pressure, P, in a regime in which P was high enough to be measured directly by conventional means. It is found that P, deduced from (6.3) with $\eta = 10^{15}$ atom/cm^2, agrees with the directly measured pressure to within 30% for pulse powers between 0.2 and 2 W. Because of the weak logarithmic dependence of the chemical potential μ on pressure P, a 30% uncertainty in the pressure yields an uncertainty in μ of only 1%. Thus, μ may be regarded as well known in these experiments.

The amount of helium adsorbed per unit area in the film is a definite, but unknown function of T and μ. One can put an upper limit on it, however, by observing that for any T and μ, less helium will be adsorbed on the metallic heater than on the more strongly binding surface of graphite. The thickest film had $-\mu \approx 30$ K, corresponding to 1.2 layers on graphite (Elgin and Goodstein 1974).

The results of a series of experiments on Nichrome heaters have already been presented in Table 5.10. The most important features evident from this table are that the prefactor $\nu^{-1} = t_0$ is of the order of 10^{-10} to 10^{-9} s, and that the energy parameter E_d is strongly correlated with $-\mu$, being roughly equal to 2/3 of its magnitude.

Sinvani, Taborek and Goodstein (1983) have used a similar experimental set-up to study desorption from a ^{4}He film due to a hot but low-intensity beam of phonons from a sapphire substrate. The apparatus is shown schematically in Fig. 6.4. A sapphire crystal (9.5 mm thick and 57 mm in diameter) is sealed between two cells immersed in a He bath at 1.5 K. On the bottom of the crystal in the evacuated cell there is an evaporated thin film (≈ 600 Å) constantan heater. A controlled quantity of He gas is admitted to the upper cell to form ≈ 2 layers of He film on all surfaces including the sapphire crystal. A superconducting bolometer (≈ 2000 Å, Sn) is evaporated on a separate sapphire substrate placed 1.25 mm from the upper face of the crystal, directly above the heater on the lower face. A mask with a 1 mm hole is placed between the crystal and the bolometer to limit the desorbing area. Heater and bolometer areas are each ≈ 0.1 mm^2.

When a current pulse is supplied to the heater its temperature rises rapidly (≈ 10 ns) to a constant value (≈ 10 K), giving rise to a burst of phonons. These propagate ballistically in the sapphire, some of them reaching the opposite surface where they desorb atoms. Phonon intensity at the desorbing surface is lower than that at the heater by a factor of $\approx 6 \times 10^3$.

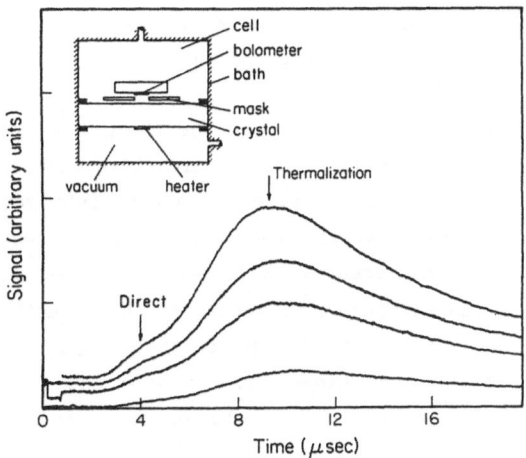

Fig. 6.4. Bolometer signals versus time for different heater powers 0.23, 0.184, 0.146, and 0.073 W (pulse width 1 µs). The time scale does not include the phonon time of flight (1.72 µs) in the crystal. Inset shows experimental arrangement. (Sinvani, Taborek, and Goodstein 1983)

The desorbed atoms, in turn, propagate ballistically since the mean free path in the gas is long compared to 1 mm. Those that arrive at the bolometer cause a signal S(t) which is amplified and recorded as a function of time t after the arrival of the heater pulse at the crystal surface.

Typical bolometer signals for different heater temperatures, T_h, are shown in Fig. 6.4. Each signal is a sum of two Maxwellian peaks: a small one at short times (appearing as shoulders in Fig. 6.4.) due to direct phonon processes, and a large one at later times due to thermalization. In each case, within experimental uncertainty, the small, hot peak has a characteristic temperature T_h, and the cooler peak has a temperature just above ambient.

To analyze their data quantitatively, Sinvani et al(1983) have assumed that for narrow pulses the time of flight signal is given by

$$S(t) = Av^n(\tfrac{1}{2} mv^2 + E_b) \exp(-mv^2/2k_BT_a) \quad .$$ (6.4)

Here $mv^2/2$ is the kinetic energy of atoms arriving at time $t=\ell/v$, where ℓ is the distance to the bolometer, and T_a is their temperature. In (6.4) E_b is the atom-surface binding energy, A is a scale factor and n depends on geometry, with n=5 for a point source and n=3 for an extended source, with a point detector in both cases. Using (6.4), the temperature of each peak can be found from

$$T_a = \frac{2E_k}{nk_B} \frac{1 + E_b/E_k}{(n+2)/n+E_b/E_k} \quad ,$$ (6.5)

where $E_k = \frac{1}{2} m(\ell/t_m)^2$ is the kinetic energy of an atom arriving at the time t_m of the peak maximum. For E_b they use the chemical potential of the gas, which was measured independently, $|\mu| = 22k_B$.

Analysis of spectra like those in Fig. 6.4. shows that the thermal peaks fit (6.4) with n = 3, possibly because of the 1mm diameter effective source. Fitting t_m and the scale factor A for each spectrum, one can then subtract the thermal peak from the data, isolating the peak due to direct processes. This procedure is shown in Fig. 6.5. The direct process peak may then be analyzed to find t_m and hence T_a which is found to be equal to the acoustic mismatch value of T_h for the heater. It should be emphasized that while this excellent quantitative agreement depends on theoretical analysis and supporting experimental evidence, the basic qualitative conclusions are clearly evident in the raw data of Fig. 6.4.

In Fig. 6.6 typical data are shown from an experiment which differed from that in Fig. 6.4 in that the bolometer was closer to the desorbing surface

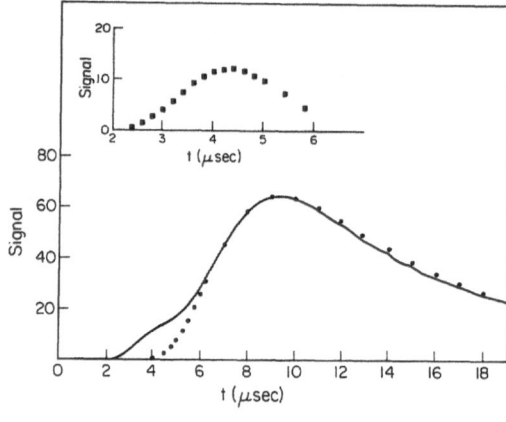

Fig. 6.5. The fit of the thermal peak to (6.4) with n = 3 (dots). The insert is the direct process peak after subtracting the thermal peak from the measured signal. (Sinvani, Taborek, and Goodstein 1983)

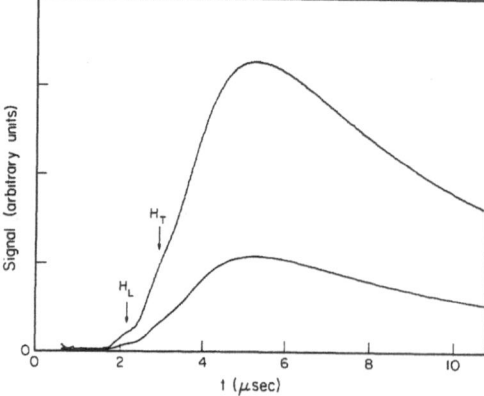

Fig. 6.6. The signal, at two different amplifications, when the bolometer is placed closer to the desorbing film, shows desorption of hot atoms by two phonon polarizations (T_h = 18 K and pulse width, 0.09 μs). The time scale includes the phonon time of flight through the crystal. (Sinvani, Taborek, and Goodstein 1983)

(0.3-0.4 mm) and a smaller mask was used (0.3 mm aperture). The result is similar to Fig. 6.4 except that an additional small peak (H_L) can be distinguished before the usual hot peak (H_T). The difference in arrival times and ratio of intensities between H_L and H_T agrees precisely with the difference in arrival times (0.72 µs) and intensities ratio (1:6) of the longitudinal and transverse phonon beams measured at the upper surface of the crystal. Sinvani, Taborek, and Goodstein (1983) thus conclude that H_L and H_T are single-phonon peaks due, respectively, to longitudinal and transverse phonons.

To understand the time of flight curves in Figs. (6.4,6) it is best to distinguish two time regimes. First, ballistically propagating hot longitudinal and transverse phonons arrive at the adsorbate causing desorption at $T_L \approx 18$ K. Because the surface potential well for ^{4}He on sapphire is much deeper than $k_B T_L$ only a few particles to be found in the upper bound states will desorb. Cascade processes as described in Chap. 5 will be limited due to the fact that the hot phonon cloud will only be in contact for the duration of reflection of the pulse of width 0.09 µs; in this sense the hot peaks may be attributed to single phonon processes. By that time thermalization of the hot phonons will be well in progress leading eventually to a rise in the ambient temperature from 2 K to 2.3 - 2.6 K causing further desorption. As we saw in Sect. 5.2, desorption of helium is always dominated by individual bound state-bound state and bound state-continuum transitions due to absorption or emission of one phonon at a time. However, the overall desorption processes will involve a cascade of many such one-phonon processes. As we saw above, the ballistic hot phonons are more likely to cause desorption from the upper energy levels of the surface potential from which atoms need to acquire the energy from a single phonon only whereas the main peak in Figs. (6.4,6) results from the slow desorption at the low temperature involving a cascade of one-phonon processes. This point will be substantiated in detail by the theoretical developments in Sect. 6.3.

Taborek (1982) has used the rapid flash desorption technique to study striking nonequilibrium effects in the desorption of ^{4}He from the surface of the metallic heater itself. Mounting a sapphire crystal with a Nichrome heater film on one surface onto a turntable he was able to measure the angular dependence of the time of flight spectrum; see the insert in Fig. 6.8. For a temperature flash from $T_i = 2$ K to $T_s = 8$ K within 10 ns Taborek (1982) observes very strong forward peaking which he claims is evidence for desorption mediated by a single phonon. We reproduce his argument: Assuming that the surface-atom interaction potential is translationally invariant in the plane of the surface, the kinematical constraints on the one-phonon desorption process are given by conservation of energy and parallel momentum

$$\hbar\omega_p = \hbar c|p| = \hbar^2 k_z^2/2m + E_b \quad , \qquad (6.6)$$

$$\mathbf{K} = \mathbf{P} \quad , \qquad (6.7)$$

where c is the phonon phase velocity, $p=(P,p_z)$ is the phonon wave vector, $k=(K,k_z)$ is the wave vector of the free atom, and E_b is the binding energy of the atom to the surface.

Figure 6.7 shows curves of constant energy in momentum space for atoms and phonons. For a given incident phonon, the wave vector $\mathbf{k}$ of the desorbed atom can be found by a simple geometric construction which shows that the atoms are emitted into a cone with a critical half-angle of approximately 10°. Because the atom energy is a quadratic function of momentum while the phonon energy is linear in momentum, the exact value of the critical angle is energy dependent. This results in a rainbowlike effect in which low-energy atoms can escape from the surface at larger angles than can high-energy atoms, which are emitted almost normal to the surface, as shown in Fig. 6.7b.

Because desorption spectra based on this kinematical model are qualitatively rather similar to the experimental results in Fig. 6.8, Taborek (1982) takes this as evidence that desorption proceeds via single-phonon absorption. Several cautionary points must be made: (i) the master equation approach with transition probabilities mediated by one-phonon processes yields clear statements on the importance of bound state – bound state cascades vis-a-vis the final bound state-continuum transition causing desorption. The phy-

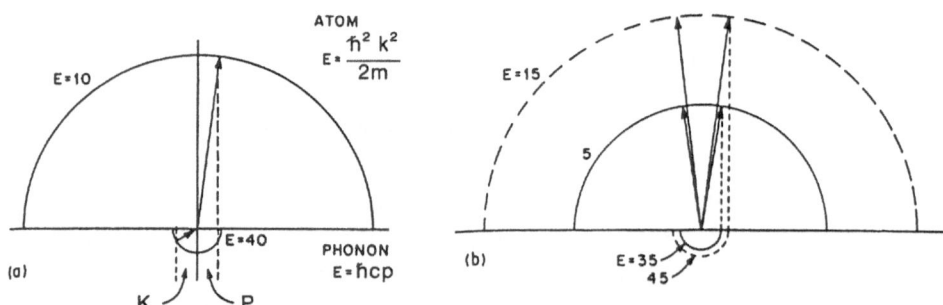

Fig. 6.7. Curves of constant energy in momentum space for atoms above the interface, and phonons below, drawn to scale with $c = 3\times10^5$ cms^{-1} and $E_b = 30$ K (all energies in kelvins). **a** Geometric construction which shows that parallel-momentum conservation leads to atoms emitted almost normal to the interface. **b** the angular width of the atomic critical cone depends on the energy, with high-energy atoms emitted into a narrower cone. This results in an atomic rainbow in the desorption spectrum. (Taborek 1982)

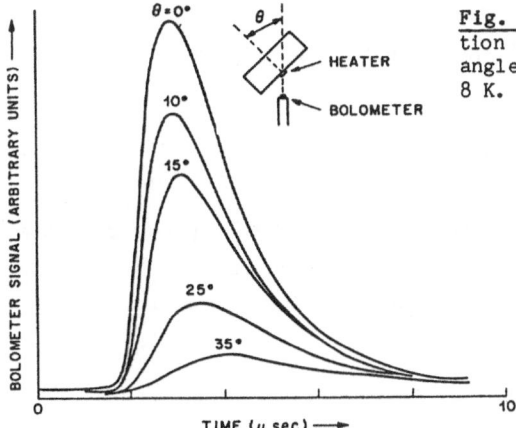

Fig. 6.8. Bolometer signal as a function of time for several values of the angle θ from the surface normal. T_S = 8 K. From Taborek (1982).

sically relevant question ought to be whether the elementary transitions are caused by one-phonon or multiphonon processes. Cascades of one-phonon processes will only be irrelevant in extremely shallow potentials. Indeed, we shall see in Sect. 6.3 that the fully three-dimensional cascade model leads, for Taborek's experimental conditions, to strong forward peaking and to the rainbow effect in the desorption flux as a result of a lack of thermalization of the lateral degrees of freedom in the adsorbate. (ii) Desorbing more than a monolayer of helium in nanoseconds produces a high-density cloud of desorbed particles in which scattering can induce further forward peaking as suggested by Cowin et al (1978). We will discuss in Sect. 6.3 how to lay out experiments to distinguish between these two sources of forward peaking.

6.2 Flux of Desorbed Particles

Experimentalists tend to present their data on time of flight spectra as detector signal versus time whereas most theoretical papers present the flux of desorbed particles as a function of velocity. Some care should be exercised in comparing the two.

Let us assume that every particle crossing the detector entrance, described as a surface S, will be registered. The number of particles registered per unit time is then

$$\frac{dN_{reg}}{dt} = \int_S \mathbf{J}_{reg}(\mathbf{r},t) \cdot d\mathbf{S} , \qquad (6.8)$$

where $d\mathbf{S}$ is a surface element of the detector entrance and $\mathbf{J}_{reg}(\mathbf{r},t)$ is the

current density of registered particles i.e., the number of particles crossing per unit time a unit area perpendicular to j_{reg} at position r and at
time t. The latter is given by

$$J_{reg}(r,t) = \frac{N_g}{V_g} \int d(\mathbf{v})\mathbf{v}f(r,\mathbf{v},t)d^3\mathbf{v} \qquad (6.9)$$

in terms of the particle distribution function $f(r,\mathbf{v},t)$ that gives the probability of finding, at time t, some of the N_g gas particles in a volume element d^3r around r with velocities in $d^3\mathbf{v}$ around $\mathbf{v}$. It is normalized to

$$\int f(r,\mathbf{v},t)\ d^3r\ d^3\mathbf{v} = \int n(\mathbf{v},t)d^3\mathbf{v} = V_g \qquad , \qquad (6.10)$$

where V_g is the volume of the gas phase. In (6.9) $d(\mathbf{v})$ is the response
function of the detector to a particle with velocity $\mathbf{v}$. If the detector is
a bolometer measuring the energy deposited, then $d(\mathbf{v})$ is proportional to v^2;
if the particles are detected after ionization as is the case in an ionization gauge or a mass spectrometer, then $d(\mathbf{v})$ is proportional to v^{-1}.

The distribution function f for particles in the gas phase is subject to a
kinetic equation

$$\left[\frac{\partial}{\partial t} + \mathbf{v}\cdot\frac{\partial}{\partial r} + \frac{\mathbf{F}}{m}\cdot\frac{\partial}{\partial \mathbf{v}}\right]f(r,\mathbf{v},t) = (\frac{\partial f}{\partial t})_{source} + (\frac{\partial f}{\partial t})_{coll} \quad . \qquad (6.11)$$

The first term on the right hand side is a source term for particles that
enter the gas phase either by desorbing from a surface or by effusing
through a hole in the container wall. The second term on the right-hand
side of (6.11) accounts for collisions between gas particles. For the pressures and temperatures at which desorption and effusion experiments are
carried out they are usually unimportant except possibly in the early stages
of desorption. To estimate their importance we observe that during a
desorption time t_d most of the N_A particles in the adsorbate will have
desorbed from the adsorbent surface S_{em} yielding an average particle density
per unit volume

$$n = \frac{N_A}{V} = \frac{N_A}{S_{em}\bar{v}t_d} \quad . \qquad (6.12)$$

Here $\bar{v}t_d$ is the distance away from the surface which a particle, desorbing
at t=0, will have traveled at a speed $\bar{v}$. The latter is taken to be the
thermal velocity of a particle at the temperature of the solid

$$\bar{v} = v_{th} = \sqrt{2k_BT/m} \quad . \qquad (6.13)$$

For times longer than t_d the cloud will spread and diminish in density. We
208

want to estimate (6.12) for a monolayer of helium desorbing from graphite or a metal: typical monolayer density is 10^{15} cm^{-2}. At T=10 K we have $\bar{v}$=2x10^4 cms^{-1} and t$_d$=10^{-8} s initially at monolayer coverage, shortening by about an order of magnitude at zero coverage. Thus $\bar{v}$t$_{d\approx}$2x10^{-4} cm and n=N$_A$/V$\approx$5x10^{18} cm^{-3}. We next recall that the mean free path of gas particles between collisions is given by elementary kinetic theory as

$$\ell = \frac{1}{\sqrt{2}\ \pi\sigma^2 n} \quad , \tag{6.14}$$

where σ is the hard sphere radius of the particle. For He, ℓ=1.5x10^{-4} cm at the above density. Note that $\ell \approx \bar{v}$t$_d$ so that collisions in the cloud of desorbed He may be important in the initial stages of desorption. Cowin et al. (1978) estimate for heavier chemisorbed molecules, like CO, desorbing from metals that complete desorption of a monolayer at elevated temperature should result in collision-induced forward peaking of the desorption flux.

For now, we neglect collisions and concentrate on the source term in (6.11), which we write as

$$\left(\frac{\partial f}{\partial t}\right)_{source} = \frac{\partial n(\mathbf{v},t)}{\partial t}\ g(\mathbf{r}) \quad , \tag{6.15}$$

where $g(\mathbf{r})$ is determined by the geometry of the source. We then note that

$$f(\mathbf{r},\mathbf{v},t) = \int_0^t d\tau\ \frac{\partial n(\mathbf{v},\tau)}{\partial \tau}\ g(\mathbf{r}-\mathbf{v}(t-\tau)) \tag{6.16}$$

is the solution of (6.11) so that we can calculate the particle current density from (6.9) and ultimately the rate (6.8) at which particles are registered by a detector

$$\frac{dN_{reg}}{dt} = \frac{N_g}{V_g} \int_0^t d\tau \int d^3\mathbf{v}\ d(\mathbf{v})\ \frac{\partial n(\mathbf{v},t-\tau)}{\partial t} \bigg|_S g(\mathbf{r}-\mathbf{v}\tau)\mathbf{v}\cdot d\mathbf{S} \quad . \tag{6.17}$$

This is as far as the general formulation can be carried. Let us now specify a situation where the lateral dimensions of both the source and the detector are small compared to their separation so that (6.17) becomes

$$\frac{dN_{reg}}{dt} = \frac{N_g}{V_g} S_{det} \int_0^t d\tau \int d^3\mathbf{v} d(\mathbf{v})\mathbf{v}\cdot \frac{\mathbf{L}}{L}\ \frac{\partial n(\mathbf{v},t-\tau)}{\partial t}\ g(\mathbf{L}-\mathbf{v}\tau) \quad , \tag{6.18}$$

where S$_{det}$ is the detector area, L is the radius vector from the center of the source to the center of the detector, and L=$|\mathbf{L}|$. For the source of area S$_{em}$ = b^2 we put

209

$$g(\mathbf{r}) = S_{em}^{-1} \, \delta(z)\,\theta(b-|x|)\,\theta(b-|y|) \quad , \tag{6.19}$$

where the factor S_{em}^{-1} assures that $g(\mathbf{r})$ is normalized to one. This allows us to do the integral over $\mathbf{v}$ so that we get for $b \ll L$

$$\frac{dN_{reg}}{dt} = \frac{Ng}{V_g} \, S_{det} \, \frac{L}{t^4} \int_0^t d\tau \; d(\mathbf{v}=L/\tau) \; \frac{\partial n(\mathbf{v}=L/\tau, t-\tau)}{\partial t} \quad . \tag{6.20}$$

Let us first look at a pulsed effusion experiment for which we have for the duration t_p of the opening of the shutter,

$$\frac{\partial n(\mathbf{v},t)}{\partial t} = S_{em} \; v_z \; \left(\frac{m}{2\pi k_B T}\right)^{3/2} \exp\left(-\frac{mv^2}{2k_B T}\right) \; \theta(t)\theta(t_p-t) \tag{6.21}$$

in terms of the Maxwell–Boltzmann distribution function. Thus for $t \gg t_p$ we get

$$\frac{dN_{reg}}{dt} = t_0^{-1} \, \frac{Ng}{V_g} \, \frac{S_{em} S_{det}}{L\pi^{3/2}} \, \frac{t_p}{t_0} \; d\left(\mathbf{v} = \frac{L}{t}\right) \left(\frac{t_0}{t}\right)^5 e^{-(t_0/t)^2} \cos\theta \quad , \tag{6.22}$$

where $t_0 = L(m/2k_B T)^{1/2}$ is the time a particle with thermal velocity needs to reach the detector a distance L away. If the detector is insensitive to the velocity of the particle we put the function $d(\mathbf{v})$ equal to a constant. For a mass spectrometer, in which the incoming particles are ionized by electron impact, $d(\mathbf{v}) \propto v^{-1}$, so that the signal registered reads

$$\frac{dN_{reg}}{dt} = \text{const} \; \frac{Ng}{V_g} \, S_{em} S_{det} \left(\frac{t_0}{t}\right)^4 e^{-(t_0/t)^2} \cos\theta \quad . \tag{6.23}$$

For a bolometer with $d \propto v^2$ we get

$$\frac{dN_{reg}}{dt} = \text{const} \; \frac{Ng}{V_g} \, S_{em} S_{det} \left(\frac{t_0}{t}\right)^7 e^{-(t_0/t)^2} \cos\theta \quad . \tag{6.24}$$

From now on we assume that the experimental signals are corrected for the detector efficiency so that we put the function $d(\mathbf{v}) = 1$.

Equation (6.22) is Knudsen's famous cosine law for the effusion of an equilibrium gas through a hole. It is also the flux of gas particles arriving at a unit surface of the walls surrounding the gas.

To understand the scattering of a gas at a solid surface and its accommodation Knudsen (1909) – and after him Gaede (1913), Clausing (1930) and many others – have argued that the surface must emit as many particles per unit time into the direction θ, because otherwise density inhomogeneities would build up in contradiction to the assumptions of a uniform gas. Clausing (1930) goes further and argues verbally that otherwise the second law of thermodynamics would be violated because with such spontaneously created density inhomogeneities (increase in entropy) a perpetual motion machine could clearly be conceived. To further "prove" the generality of Knudsen's

210

cosine law, Clausing (1930) brings in detailed balance which, he argues, would allow angular dependent scattering at a real surface but that would again violate the assumptions of homogeneity and isotropy of the gas down to the surface. Knudsen's cosine law - like Lambert's law for blackbody radiation - is acceptable and is, no doubt, a very good description for the effusion of a gas in equilibrium through a hole into an evacuated chamber and even for evaporation.

The general validity of the cosine law has been questioned by Palmer et al. (1970) who also present evidence that the principle of detailed balance implies a relationship between the angular dependence of adsorption and desorption, respectively. With their desorption signal varying as $\cos^d \theta$ with $2.5 < d < 4.0$, they deduce that adsorption should show a behaviour like $\cos^a \theta$ with $a \approx d - 1$. This argument has been carefully analysed by Cardillo et al. (1975) for H_2 on various Cu surfaces. These authors, however, end somewhat pessimistically noting that the angular distribution of desorption may not be a unique or even sensitive reflection of the gas-surface interaction potential nor of the dynamical process of desorption, and suggesting that complementary measurements of velocity distributions are needed.

To develop the nonequilibrium theory of the time of flight spectrum of thermally desorbed particles, we will follow a master equation approach and briefly outline the assumptions and the calculational procedures (Gortel, Kreuzer, Schäff, and Wedler 1983). Let us follow the cloud of desorbing particles as it moves away from the surface. It first has to escape, in the desorption process proper, the region in which particles are in interaction with the surface. The appropriate distribution function in this region is controlled by the master equation so that desorbed particles emerge from it with an occupation function $n(\mathbf{k},t)$. We must, however, be aware that $n(\mathbf{k},t)$ is the occupation function of wave number $\mathbf{k}$ at time t anywhere in the gas phase. In reality, desorbed particles will within a desorption time t_d be confined to a (microscopic) depth of order $z_0 = \bar{v} t_d$ away from the surface where $\bar{v}$ is some average velocity.

We can therefore identify $\partial n / \partial t$ in (6.20), up to a normalization factor, with the left-hand side of the master equation (3.47) in which the readsorption and continuum-continuum scattering processes are neglected:

$$\partial n(\mathbf{k},t) / \partial t = \sum_i W(\mathbf{k},i) n(i,t) \quad . \tag{6.25}$$

Here $\hbar k = m v$ and the sum over i exhausts the N+1 bound states in the surface potential. With the help of the formal solution (5.4) we get

$$\partial n(\mathbf{k},t)/\partial t = \sum_{\kappa=0}^{N} \sum_{l} W(\mathbf{k},l) \, A_{l\kappa} \, e^{-\lambda_\kappa t} \quad , \tag{6.26}$$

so that (6.20) reads [for $d(\mathbf{v})=1$]

$$\frac{dN_{reg}}{dt} = \text{const } \frac{N_g}{V_g} \, S_{em} S_{det} \left(\frac{t_0}{t}\right)^4$$

$$* \sum_{\kappa=0}^{N} \sum_{l} A_{l\kappa} \int_{0}^{t} d\tau \exp[-\lambda_\kappa(t-\tau)] \, W(\mathbf{k},l) \quad , \tag{6.27}$$

where the components of $\mathbf{k}$ are set equal to

$$k_x = \left(\frac{m}{\hbar}\right)\left(\frac{L}{\tau}\right) \cos\phi \, \sin\theta$$

$$k_y = \left(\frac{m}{\hbar}\right)\left(\frac{L}{\tau}\right) \sin\phi \, \sin\theta$$

$$k_z = \left(\frac{m}{\hbar}\right)\left(\frac{L}{\tau}\right) \cos\theta \quad . \tag{6.28}$$

The coefficients $A_{l\kappa}$ can be calculated from (5.34). For most experimental conditions the time of flight is much larger than the desorption time t_d so that for $t \gg t_d \gtrsim \lambda_\kappa^{-1}$ we get

$$\frac{dN_{reg}}{dt} = \text{const } \frac{N_g}{V_g} \, S_{em} S_{det} \left(\frac{t_0}{t}\right)^4 \sum_{\kappa=0}^{N} \sum_{l} A_{l\kappa} \, \lambda_\kappa^{-1} \, W(\mathbf{k},l) \quad , \tag{6.29}$$

where the components of $\mathbf{k}$ are again given by (6.28) but for $\tau = t$.

We have seen in Sect. 5.3 that for the calculation of desorption times perturbation theory is usually quite adequate. We will see below that this is even more so for time of flight spectra. We therefore limit the sum over κ in (6.29) to $\kappa = 0$, set $\lambda_0^{-1} = t_d$ and equate, according to (5.35), A_{l0} to the initial Boltzmann occupation of the bound states. The further use of detailed balance allows us then to write

$$\frac{dN_{reg}}{dt} = \text{const } \frac{N_g}{V_g} \, S_{em} S_{det} \left(\frac{t_0}{t}\right)^4 e^{-(t_0/t)^2} \sum_{l} W(l,\mathbf{k}) \, t_d \quad . \tag{6.30}$$

Note that the dynamics of the microscopic processes, contained in $W(l,\mathbf{k})$, determines through (6.28) the time as well as the angular dependence of the time of flight spectra (6.30). In principle, both can differ significantly from the dependence given in (6.22) for a pulsed effusion experiment.

6.3 One-phonon Processes

Gortel, Kreuzer, Schäff, and Wedler (1983) have calculated time of flight spectra for thermal desorption of helium using the master equation approach with the transition probabilities calculated in the one-phonon approximation. For a flat surface they assume the surface potential to be of Morse type. Neglecting parallel phonon momentum the time of flight signal can be calculated with the transition probabilities (4.22-25) either from (6.29) or by employing the perturbation result (6.30). From the latter one gets explicitly

$$\frac{dN_{reg}^{(1)}}{dt} = const \; t^{-5} \cos\theta \; \exp\left[-\frac{m}{2k_BT} \left(\frac{L}{t}\right)^2 \sin^2\theta\right]$$

$$* \sum_{n=0}^{N} e^{-E_n/k_BT} \; f_n\left(\frac{\cos\theta}{t}\right) \left[\frac{m}{2} \left(\frac{L}{t}\right)^2 \cos^2\theta - E_n\right]^3$$

$$* \left[\exp\left[\left(\frac{m}{2} \left(\frac{L}{t}\right)^2 \cos^2\theta - E_n\right)/k_BT\right] - 1\right]^{-1}$$

$$* \frac{3}{2+\ell_s} \sum_{\sigma} s(\sigma) \; \theta\left[\hbar\omega_D(\sigma) + E_n - \frac{m}{2} \left(\frac{L}{t}\right)^2 \cos^2\theta\right]. \tag{6.31}$$

where

$$f\left(\frac{\cos\theta}{t}\right) = \frac{\sqrt{-E_n}}{n! \; \Gamma(2\sigma_0-n)} \; \frac{\sinh\left(\left(\frac{2\pi mL}{\hbar\gamma t}\right)\cos\theta\right) \left| \Gamma\left(\sigma_0+\frac{1}{2}+i \; \frac{mL}{\hbar\gamma t} \cos\theta\right)\right|^2}{\cos^2(\pi\sigma_0) + \sinh^2\left(\frac{\pi mL}{\hbar\gamma t} \cos\theta\right)} \; . \tag{6.32}$$

The weights $s(\sigma)$ are given in (4.10) for the bulk Debye model and are equal to $S_\sigma^z(0)$ in (2.165) for the surface Debye model. Including parallel phonon momentum the fully three-dimensional theory (4.1) yields from (6.30)

$$\frac{dN_{reg}^{(3)}}{dt} = const \; t^{-5} \cos\theta \; \exp\left[-\frac{m}{2k_BT_i} \left(\frac{L}{t}\right)^2\right]$$

$$* \sum_{n=0}^{N} \left(\frac{m}{2} \left(\frac{L}{t}\right)^2 \cos^2\theta - E_n\right)^2 f_n\left(\frac{\cos\theta}{t}\right)$$

$$* \sum_{\sigma} \int_0^\infty dy \; s(\sigma)(y) \int_0^{\omega_D(\sigma)/\omega_D} dw \; w \; \exp(w\hbar\omega_D/k_BT_i)$$

$$*[\exp(w\hbar\omega_D/k_BT_f)-1]^{-1} \; [a^2-b^2]^{-1/2}\theta(a^2-b^2) \quad, \quad \text{where} \tag{6.33}$$

$$a = \frac{2m\,\omega_D}{\hbar c_t \gamma^2} \; \frac{L}{t} \; \sin\theta \; w\sqrt{y}$$

$$b = \frac{2mE_n}{\hbar^2\gamma^2} + \left(\frac{\omega_D}{c_t\gamma}\right)^2 w^2 y + \frac{2m\,\omega_D}{\hbar\gamma^2} \; w - \left(\frac{m}{\hbar\gamma} \; \frac{L}{t} \cos\theta\right)^2 \quad. \tag{6.34}$$

The weights $s^{(\sigma)}(y)$ are given by

$$s^{(\sigma)}(y) = \tfrac{1}{2} c_t^2 y^{-3/2} A_\sigma(c_t \sqrt{y}) \quad , \tag{6.35}$$

where $A_\sigma(c)$ are given in (5.52) and (5.53) for bulk and surface phonons, respectively. For the latter they are simultaneously the integrands in the expressions (2.165) for $S_\sigma{}^z(0)$, connecting them with the phonon density of states at the surface. We note in (6.31,33) the appearance of an extra factor t^{-1} and the explicit $\cos\theta$ dependence. Both arise from the microscopic dynamics and are not due to geometric considerations as they were in (6.22).

In order to obtain (6.33) the general perturbation theory result (6.30) has been modified by assigning a temperature T_i to the motion of the adsorbate along the surface and a temperature T_f to the motion normal to it. The motivation for this ad hoc procedure can be found in Sect. 5.5 where it was shown that in a flash desorption experiment for mobile adsorbates in which the temperature of the solid is suddenly raised from T_i to T_f the motion of the adsorbed particles normal to the surface thermalizes before any significant desorption takes place but the motion along the surface remains characterized by the initial temperature T_i throughout most of the desorption process. For an isothermal experiment we set $T_i = T_f$ in (6.33).

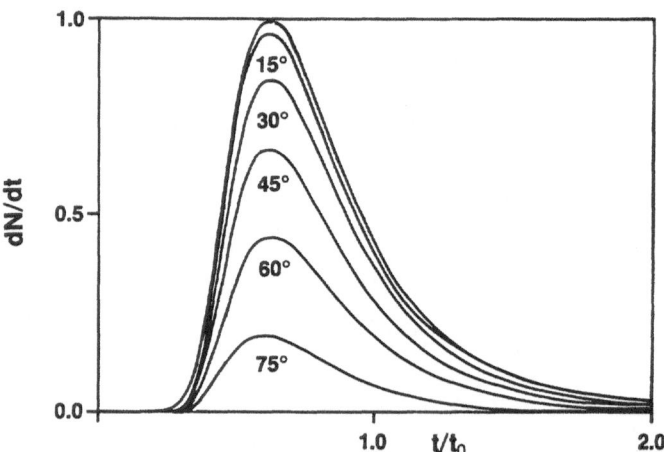

Fig. 6.9. The time of flight spectra for helium desorbing from graphite at $T = 4$ K (number of particles registered per unit time by a detector) normalized to one at its peak in the direction normal to the surface, as a function of time; t_0 is the time needed by a thermal atom to reach the detector, see after (6.22). Angles are measured with respect to the surface normal. Calculations based on (6.31) or (6.33) cannot be discriminated on the graph. (Gortel, Kreuzer, Schäff, and Wedler 1983)

In Fig. 6.9 we reproduce the time of flight spectra for helium desorbing from a graphite surface at T = 4 K with the detector at different angles θ with respect to the surface normal. Within the accuracy of the drawing, the curves are the same whether they are calculated from the one-dimensional version (6.31) or the full three-dimensional theory (6.33) with $T_i = T_f$. This implies that the phonon wave vector parallel to the surface is indeed of little consequence to the desorption process. Thus the numerical complications arising from the evaluation of the double integrals in (6.33) can be avoided and a theory based on the simple momentum conservation law (4.4) is totally adequate.

We note that the maxima in the time of flight spectra in Fig. 6.9 appear at essentially the same time; they decrease as a function of angle θ slightly stronger than cos θ, roughly like $\cos^{1.2}\theta$ as can be seen in Fig. 6.10 (curve 1). Experimentalists usually give the angular dependence of the total particle flux into a given direction, i.e., the area under any one of the curves in Fig. (6.9). The difference between the angular dependence of the peak heights and that of the total flux for the present system are imperceptible. It is also interesting to compare the time of flight spectra calculated from a microscopic theory to a Maxwellian spectra. For the latter Gortel et al (1983) took (6.23). It seems more appropriate to base a comparison on (6.22) in which case a Maxwellian spectrum at T=4 K differs only marginally from the dynamical time of flight spectrum calculated from either (6.31 or 33).

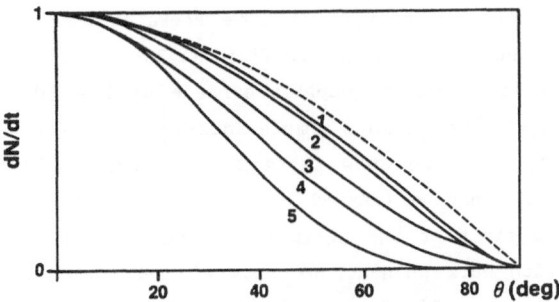

Fig. 6.10. Angular dependence of the peak in the time of flight spectra in Fig. 6.9 as a function of angle (curve 1). The same system at T = 20 K follows curve 2. Curve 3 is for the He-Nichrome system at T = 8 K as discussed in Sect. 5.5. Curve 4 is a fictitious system with all parameters as for Hegraphite but with the range of the surface potential decreased to 0.2 Å. For curve 5 we chose the right range but decreased the depth to a mere 6 K. Curves 4 and 5 seem to show the strongest deviations from Knudsen's cosine law (dashed curve) possible in the phonon cascade model. (Gortel, Kreuzer, Schäff, and Wedler 1983)

We have already noted that the spectra in Fig. 6.9 are the same within the accuracy of the plot whether they are calculated (i) with the one-dimensional theory based on (6.31), (ii) with the three-dimensional theory based on (6.33) ($T_i = T_f = T$) with bulk phonons (5.52), or (iii) with surface phonons (5.53). The relative contributions of the various phonon modes are as follows: within the bulk Debye model for the phonons, 94% of the total time of flight spectrum is generated by the transverse modes and 6% by the longitudinal phonons. Within the surface Debye model, the Rayleigh mode contributes about 64%, the generalized Rayleigh mode about 29%, and the ± modes add the remaining 7%. These are, indeed, the same ratios with which the various modes contribute to the desorption time according to (5.44) and Table 2.7 in situations where the upper frequency cutoffs can be allowed to be infinite. In this case the w-integration in (6.33) is always limited by the θ-function, which restricts it, for the present parameters, to a narrow range around $w_0 \approx m/(2\hbar\omega_D) (L/t)^2\cos^2\theta - E_n/\hbar\omega_D$ so that the y-integration decouples giving the weights $s_l^{(\sigma)}$ of Table 2.7 and reducing (6.33) to (6.31) apart from overall factors. Also note that the ratio of the sum of the weights $s_l^{(\sigma)}$ in the bulk and surface Debye phonon models, respectively, just give the factor by which the desorption time calculated in the surface phonon model is shorter than that obtained in the bulk phonon model; it is, e.g., 2.31 for $l = 0.6$. An extensive discussion can be found in Sects. 2.4 and 5.4.

To see the different effects of the bulk versus surface phonon model, one has to pick a gas-solid system where (i) several surface bound states are appreciably occupied, i.e.,, one has to work at some elevated temperature, and (ii) where the upper frequency cutoffs $\omega_D^{(\sigma)}$ in (6.33) are important. We look at helium desorption from a graphite surface at T = 30 K. The angular dependence is a bit stronger than a simple Knudsen's cosine law, something like $\cos^{1.8}\theta$. In Fig. 6.11 we show the time of flight spectra calculated from (6.33) with bulk phonons (dashed curve) and with surface phonons (solid curve). Looking first at the curve for 0° calculated from the three-dimensional theory with bulk phonons, we note a small secondary maximum just after the main peak. In the one-dimensional version this structure is followed by a discontinuous jump to the peak. Similar discontinuities develop at earlier times. To understand this feature we recall that the particles were, prior to desorption, in any one of the bound states (of energy $E_n < 0$) of the surface potential V_s. Because the theory assumes that one-phonon processes are dominant the maximum kinetic energy of a particle emerging from E_n is

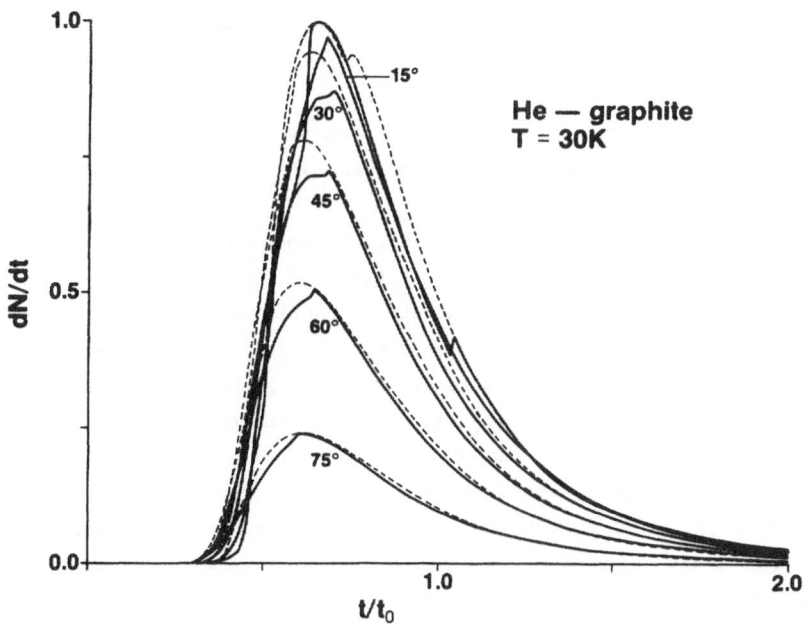

Fig.6.11. Time of flight spectra for helium desorbing from graphite at 30 K. Solid curves: surface phonons; dashed curves: bulk phonons. (Gortel and Kreuzer 1985)

$$\frac{m}{2} v^2 = E_n + \hbar\omega_D \ . \tag{6.36}$$

Thus particles from E_n arrive at times

$$t > L[2 (E_n + \hbar\omega_D)/m]^{-1/2} \ , \tag{6.37}$$

or with the thermal time of flight t_0, defined below (6.22)

$$\tau = \frac{t}{t_0} > \left[\frac{k_BT}{\hbar\omega_D + E_n} \right]^{1/2} \ . \tag{6.38}$$

With the potential parameters for the helium-graphite system σ_0 = 4.88, r= 27.46, we find that particles will emerge from the deepest bound state $E_0/\hbar\omega_D$ = $-$ 0.699 for τ > 0.576, from $E_1/\hbar\omega_D$ = $-$ 0.4164 for τ > 0.414, from $E_2/\hbar\omega_D$ = $-$ 0.2066 for τ > 0.355, from $E_3/\hbar\omega_D$ = $-$ 0.0695 for τ > 0.329, and from $E_4/\hbar\omega_D$ = $-$ 0.0053 for τ > 0.317. These jump discontinuities are washed out in the three-dimensional theory with bulk phonons as a result of the two integrals in (6.33) that average over the initial lateral momentum of adsorbed particles and phonons. If, however, we use proper surface phonons in the three-dimensional theory these discontinuities appear again whenever the Rayleigh phonons set in at times given by (6.38) with ω_D replaced by $\omega_D^{(R)}$, which for $\ell_S = c_T/c_L = 0.5$ is $\omega_D^{(R)}$ = 0.756 ω_D. On smooth surfaces

217

where surface modes can develop and can be described by elasticity theory of a semi-infinite continuum, such spectra should be observable. If the surface, however, is not smooth, one could expect a depression of surface phonons so that desorption is basically triggered by bulk-like phonons resulting in time of flight spectra as calculated with bulk phonons (dashed curves).

Gortel and Kreuzer (1985) have also analyzed the desorption of helium from a monolayer of helium adsorbed on Nichrome for a nonequilibrium desorption experiment in which the substrate is rapidly flashed from an initial temperature T_i = 2 K to a final T_f = 8 K to mimic Taborek's experiment discussed in Sect. 6.1. Their analysis of the desorption process for this system was already presented in Sect. 5.5, the main result being the lack of thermalization of the adsorbate motion along the surface over the time of the desorption process but a very fast thermalization of the motion normal to it. In order to calculate the time of flight spectra for Taborek's non-equilibrium desorption experiment, one can, indeed, use (6.33) with the distribution of lateral energies in the adsorbate kept at T_i = 2 K and the perpendicular motion of the surface potential thermalized to the temperature of the solid T_f = 8 K. The resultant Fig. 6.12 shows all the features of the experimental results: (i) a very strong angular dependence, and (ii) a substantial shift in the peak position to later times. For a temperature flash T_i = 1 K to T_f = 2 K less pronounced forward peaking is predicted, see Fig. 6.13.

One can conclude this discussion by saying that forward peaking in time of flight desorption spectra will occur for highly mobile adsorbates if the thermalization of the lateral degrees of freedom is slow compared to the phonon-mediated desorption process. For particles heavier than helium the surface potential is more or less corrugated, leading to lateral scattering and better thermalization within the adsorbate. Collisions between parti-cles in the adsorbate do not promote thermalization as no net energy transfer from the solid would be involved. We note, however, that scattering in the cloud of desorbed particles is argued to produce forward peaking as well (Cowin et al. 1978; Cowin 1985). Careful experiments ought to distinguish between these two sources of forward peaking. The one-phonon theory makes definite predictions on the degree of forward peaking and the peak shifts as a function of the temperature jump, compare Fig. 6.13. As Cowin et al. (1978) and Cowin (1985) have shown, forward peaking should be reduced and eventually eliminated as the initial coverage is reduced. The above theory, of course, is applicable for small coverages only. We would

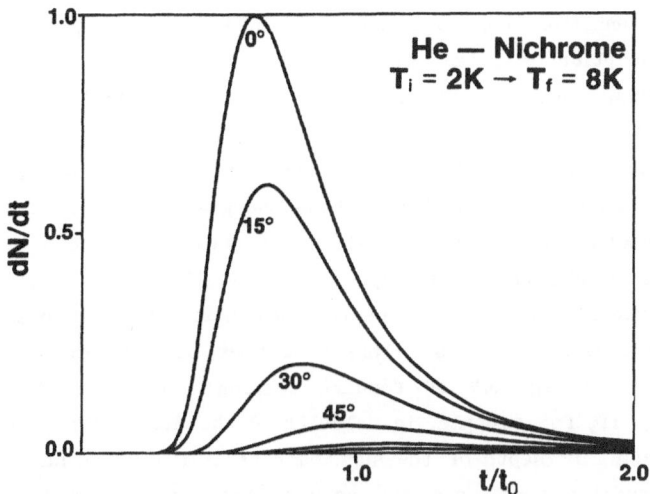

Fig.6.12. Flash desorption spectra for T_i = 2 K to T_f = 8 K, appropriate for Taborek's experiment reproduced in Fig. 6.8. (Gortel and Kreuzer 1985)

like to stress again that within the master equation approach based on (4.1) the theory considers elementary bound state-bound state and bound state-contnuum transitions as mediated by the absorption or emission of one phonon at a time. The overall desorption process will, except for extremely shallow surface potentials, always involve a cascade of such one-phonon

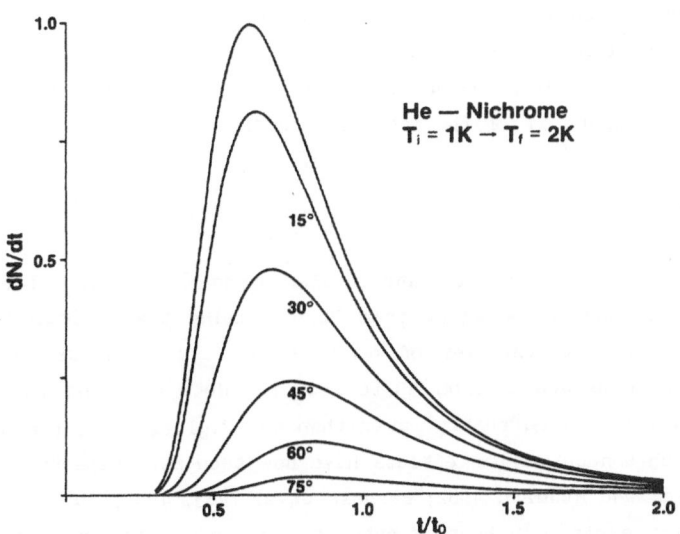

Fig.6.13. Flash desorption spectra as in Fig. 6.12 but for a smaller temperature jump. (Gortel and Kreuzer 1985)

processes. The final transition from some excited bound state into the continuum involves the absorption of a last single phonon. From which of the bound states that happens, depends on the particular gas-solid system. What can be said in general is that the bound state-continuum transitions have a maximum for bound state energies within a Debye energy $\hbar\omega_D$ from the top, decaying sharply as one moves to lower bound state energies. This implies that even in shallow surface potentials, in which the ground state can be emptied by absorbing a single phonon desorption will still predominantly proceed from higher bound states which are replenished from the lower ones by more one-phonon absorption processes. This fact must be kept in mind when explanations based on Fig. 6.7 of forward peaking and the rainbow effect are put forward. If, for example, in calculating the desorption time for helium desorbing from LiF (depth of the surface potential $\approx$ 80 K; $\hbar\omega_D$ = 730 K) one allows desorption from only the ground state, the desorption times are about a factor of four longer than those from a proper calculation in which desorption can take place from all four bound states, between which, in addition, fast bound state - bound state transitions maintain a quasiequilibrium. It thus seems somewhat artificial to try to discriminate single-phonon processes from one-phonon cascades as both are based on the same transition probabilities. In this sense we would interpret the experiment by Sinvani, Taborek, and Goodstein (1983) (see Figs. 6.4,5) as evidence that first, i.e., after the arrival of the hot ballistic phonons, some desorption takes place at a transient high temperature established at the adsorbate roughly for the duration of the phonon pulse. Because thermalization to a much lower temperature (slightly above ambient) takes place rather rapidly after the ballistic phonons are reflected, the hot desorption peaks are rather small (about 5% of the total desorbed).

6.4 Classical Models

Tully (1981) has calculated the velocity and angular dependence of the time of flight signals of Ar and Xe desorbing from Pt(111) using the stochastic trajectory method. The mean energies of desorbing Ar and Xe atoms are shown in Fig. 6.14 at temperatures from 50 to 2500 K. Note that the mean energies are in many cases considerably lower than the $2k_BT$ expected for a Maxwellian beam. Such nonequilibrium effects have now been seen repeatedly in experiments (Wedler and Ruhmann 1982; Burgess et al.1983, 1984; Hurst et al. 1983) that use extremely high heating rates as they occur in laser-induced thermal desorption. A possible explanation may be that particles

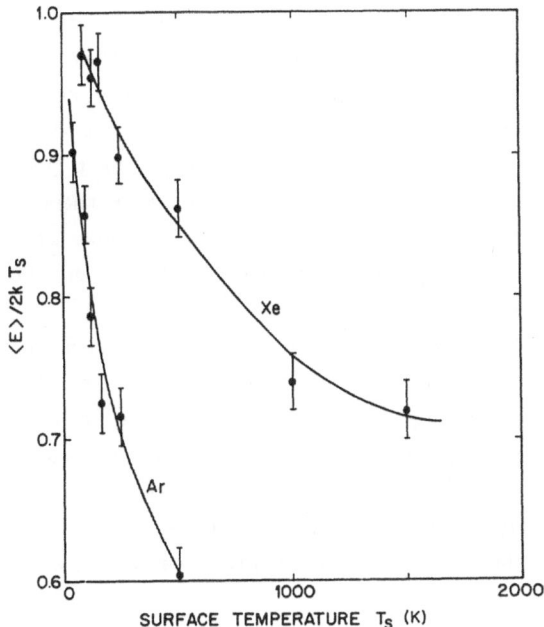

Fig. 6.14. Mean translational energy of thermally desorbed atoms, in units of $k_B T_s$ where T_s is the temperature of the surface. The points including the computational error bars have been calculated by the stochastic trajectory method. (Tully 1981)

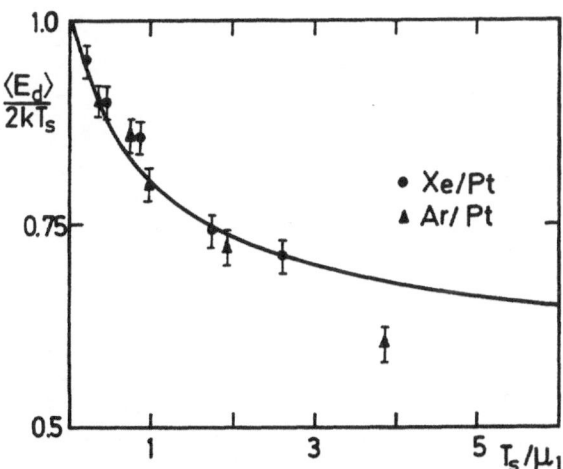

Fig.6.15. Mean desorption energies, as a function of the surface temperature T_s, with μ_1 as the first moment of the transition probability. The data points are Tully's computer simulations shown in Fig.6.14. (Leuthäuser 1983)

climbing out of a reasonably deep surface potential well will desorb as soon as they make it over the top. Because the probability that they will fall back down into the well is close to one for deep wells, they do not spend enough time within k_BT of the top to actually acquire the full thermal energy. If this argument is correct then one should expect that particles escaping from shallow wells should emerge with more or less thermal energy. This is in fact the case for helium desorption. That Ar shows less thermal behavior than Xe in Tully's computer experiments might be due to the fact that its coupling to the phonon bath of the solid is weaker.

Leuthäuser (1983) has used the Iche-Nozières-Brenig approach to calculate the energy and angular distributions of desorbing molecules using, in particular, Gaussian transition probabilities. He can reproduce Tully's results, as seen in Fig. 6.15. We refrain from giving details of this theory because it will be dealt with at length in Chap. 7. For completeness we should mention that Goodman and Garcia (1982) and Goodman (1983) have studied non-equlibrium effects in desorption but at a rather simple phenomenological level.

7. Sticking and Accommodation

The sticking coefficient S is, no doubt, an important, but unfortunately also a very controversial, parameter in the study of desorption and evaporation as well as atomic beam scattering. This is particularly true for physisorption where few experimental data are explained by too many, and often inconclusive, theories. The same can be said for accommodation. We will try to sketch the latest developments.

7.1 Experimental Results

7.1.1 Sticking Coefficient

A survey of experimental data on the kinetics of physisorbed gas-solid systems including sticking coefficients was presented in Table 1.2 where we also quoted experiments that give only estimates. In this section we report on those experiments that measure the coverage dependence of the sticking coefficients. Stoll et al. (1971) have studied the scattering of rare gases (He, Ne, Ar, Kr, and Xe) from Pt(111) surfaces for surface temperatures between 296 K and 663 K using effusive thermal energy beams of 296 K and 973 K. Approximate mass balances between the total scattered flux and the incident flux are used to estimate initial trapping probabilities which are interpreted as initial sticking coefficients. Their data, reproduced in Fig.7.1, have been analyzed by Weinberg and Merrill (1971) in terms of a simple classical model which uses thermal accommodation coefficient data as input; see also Modak and Pagni (1973). Wang and Gomer (1979) have measured the sticking coefficient of Xe on clean and oxygen-covered (110) and (100) planes of W, see Figs. 7.2,3. The initial increase with coverage is attributed to better energy accommodation on xenon-covered than on clean tungsten. Govers et al. (1980) have performed molecular beam experiments on the sticking and accommodation of physisorbed molecular hydrogen on a low-temperature substrate, namely a (doped Si slab) bolometer covered with a cryodeposit of background gas consisting mainly of frozen air, water

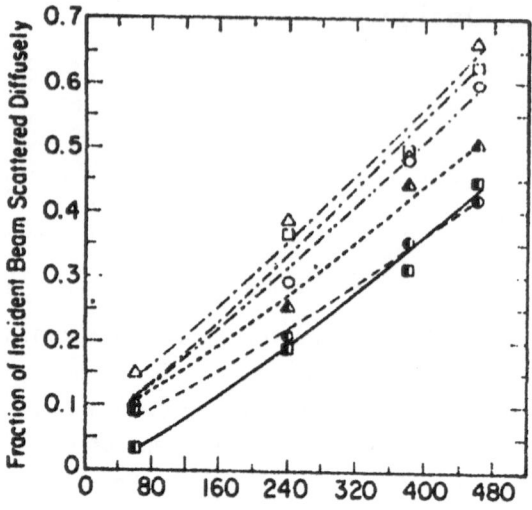

Fig. 7.1. Trapping of rare gases on Pt(111). The solid is at a temperature T_s, and the beam at T_g. Open triangles: T_s=100 °C, T_g=23 °C. Open squares: T_s=300 °C. Open circles: T_s=500 °C. For the half-filled symbols the beam temperature has been raised to T_g=700 °C. (Stoll et al. 1971)

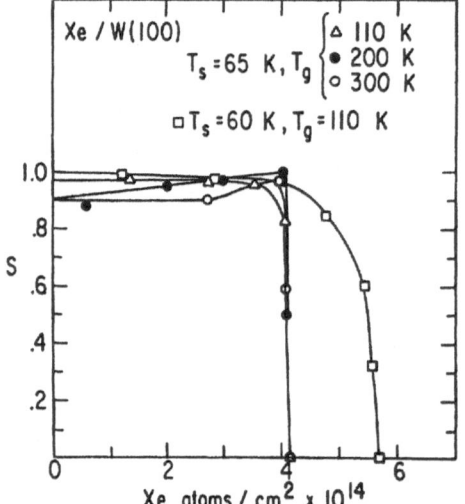

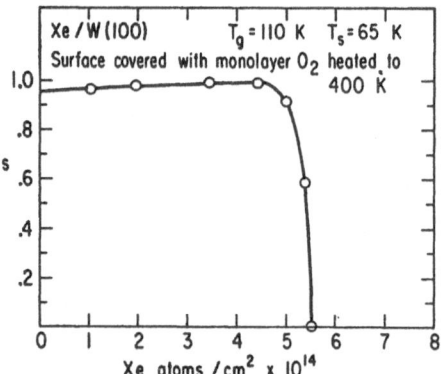

Fig. 7.2. S versus coverage for Xe on W(100): T_s and T_g as indicated. (Wang and Gomer 1979)

Fig. 7.3. S versus coverage for Xe on W(100) covered with a monolayer of O (O/W = 1.0) and heated to 400 K before Xe adsorption at T_s = 65 K. (Wang and Gomer 1979)

ice, and a small amount of hydrocarbons. They monitor the time dependence of the bolometer signal and record in a mass spectrometer those particles that are reflected from the bolometer during the beam deposition of molecular H_2 and D_2. Correlating time (lower abscissa in Figs. 7.4,5) and coverage

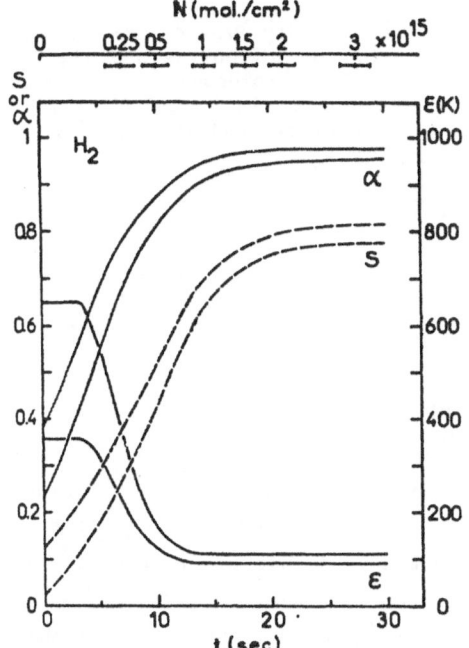

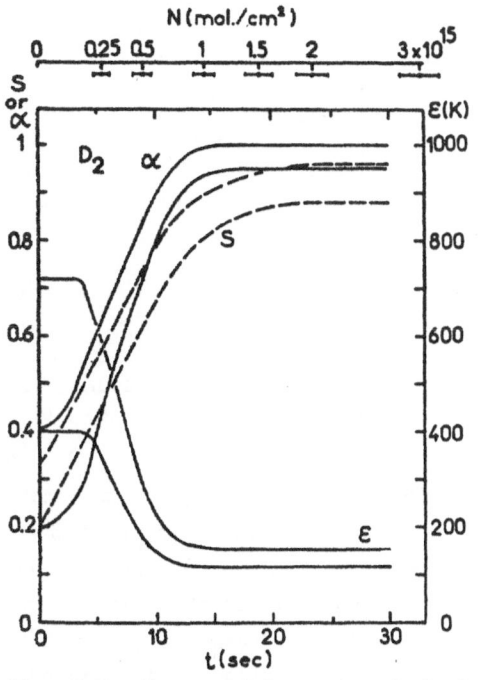

Fig. 7.4. Upper and lower bounds to S (- -) and α (-) for H_2 when the binding energy ε is allowed to vary between the limits shown ($360 \leq \varepsilon_0 \leq 650$ K and $95 \leq \varepsilon_\infty \leq 115$ K). These bounds include the uncertainty due to a possible 10% error in the measurements of the mass spectrometer signal. The upper abscissa shows the H_2 coverage. (Govers et al. 1980)

Fig. 7.5. Upper and lower bounds to S (- -) and α (-) for D_2. ε_0 was scaled to the limiting values estimated for H_2, while the limits ε_∞ were estimated by the requirement $\alpha_\infty(D_2) = \alpha_\infty(H_2)$: 400 K$\leq \varepsilon_0 \leq 720$ K, 116 K$\leq \varepsilon_\infty \leq 156$ K. Allowance was made for a possible 10% error in the measurements of the mass spectrometer signal. The upper abscisa shows the D_2 coverage. (Govers et al. 1980)

(upper abscissa) they can measure the sticking and accommodation coefficients as a function of coverage. Amazingly their data was reproducible over periods of more than a year; they are reproduced in Figs. 7.4,5. Christmann and Demuth (1982) have performed a careful study of Xe adsorption on a nickel (100) surface reporting the relative sticking coefficient as a function of Xe coverage, see Fig. 7.6. The fact that the sticking coefficient does not drop to zero as monolayer coverage is approached is attributed to efficient condensation of xenon films. Similar studies for other rare gases have been announced by these authors. The only measurement of the sticking coefficient of a ^{4}He atom on a solid surface, namely a Nichrome film of thickness about 600 Å, at low temperature (T<4 K) has been reported by Sinvani, Cole, and Goodstein (1983). For incident He atoms with beam temperatures of 10 to 20 K and for the surface at T=3.6 K they find $2/3 \leq S < 1$

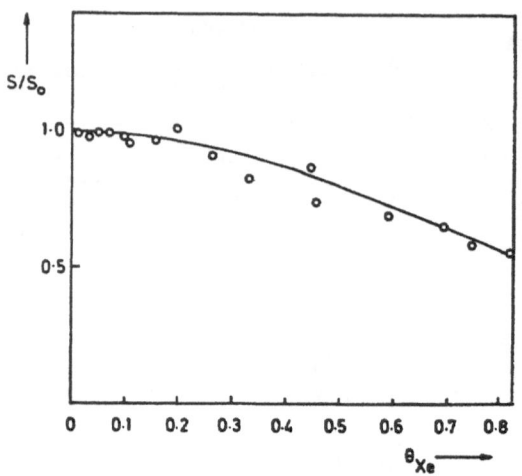

Fig. 7.6. Relative sticking probability S/S_0 for Xe adsorp‑ tion on Ni(100) versus the Xe surface coverage θ; maximum coverage is 5.65×10^{14} cm^{-2}. (Christmann and Demuth 1982)

for He coverages up to a monolayer, with $S \approx 0.8$ the most likely value. Grunze (1984) has interpreted the adsorption of N_2 on Ni(110) at $T_S = 87$ K as physisorption. Isosteric heat of adsorption and sticking coefficient as function of temperature are shown in Fig. 7.7. The constant value $S=1$ up to saturation coverage is explained by assuming a mobile precursor state that allows the molecule to accommodate its translational energy. King et al. (1968) have reported extensive data on N_2 adsorption on W, Mo, Ti, Pd and Ni films that were sintered at various temperatures. Effects of surface geo- metry and high index planes are examined and a variety of physisorbed and chemisorbed states is identified, all leading to rather complicated func- tional dependences of sticking coefficients on coverage.

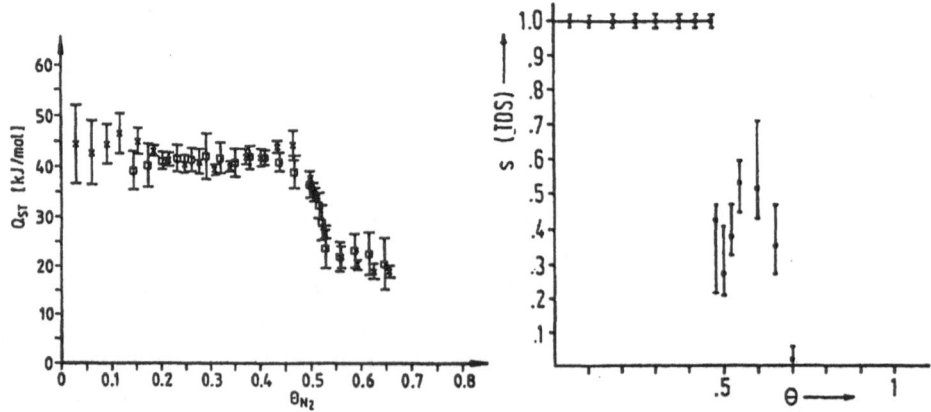

Fig. 7.7. Isosteric heat of adsorption a and sticking coefficient b for N_2 adsorption on Ni(110) as a function of coverage (T_S = 87 K). (Grunze 1984)

7.1.2 Accommodation Coefficient

Thermal accommodation and adsorption coefficients of gases have been reviewed comprehensively by Saxena and Joshi (1981). They outline the history
of the subject and survey the available experimental methods, in particular
the variants of the hot wire method; theories are sketched to the extend
that they are used to understand and evaluate the experiments. The experimental data on the thermal accommodation coefficient α consist of 589 data
sets on 159 gas-solid systems extracted from 206 references. It would be
futile to duplicate this effort here for physisorbed gas-solid systems.
Rather we restrict our attention to those data that have become standards
in physisorption. The early work of various gases on platinum is summarized
in Fig. 7.8. Data on the isothermal accommodation coefficient α for the
rare gases on tungsten as measured by Kouptsidis and Menzel (1967, 1970) and
Thomas (1967) are reproduced in Fig. 7.9. The isotopic dependence of α for

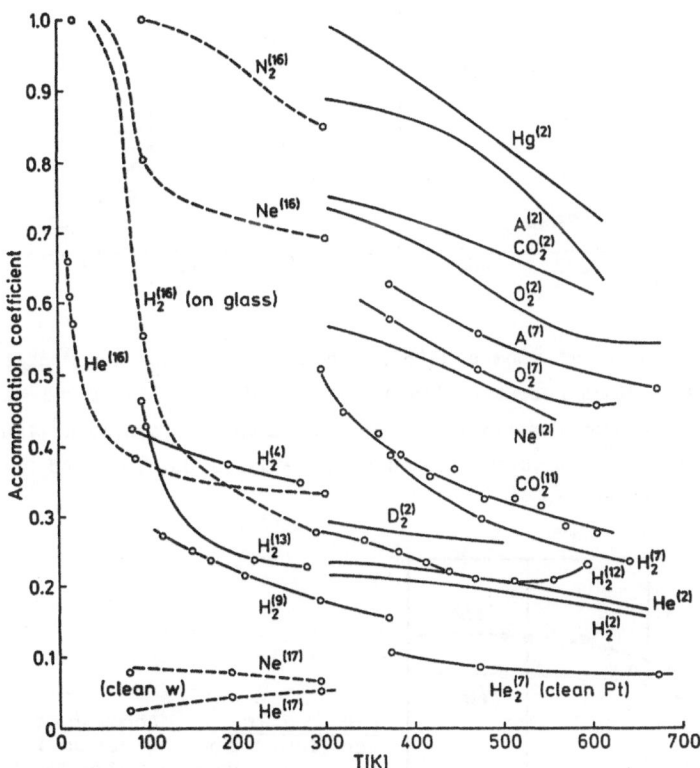

Fig. 7.8. The accommodation coefficients of various gases on platinum (except as noted) vs. absolute temperature. (Thomas 1967)

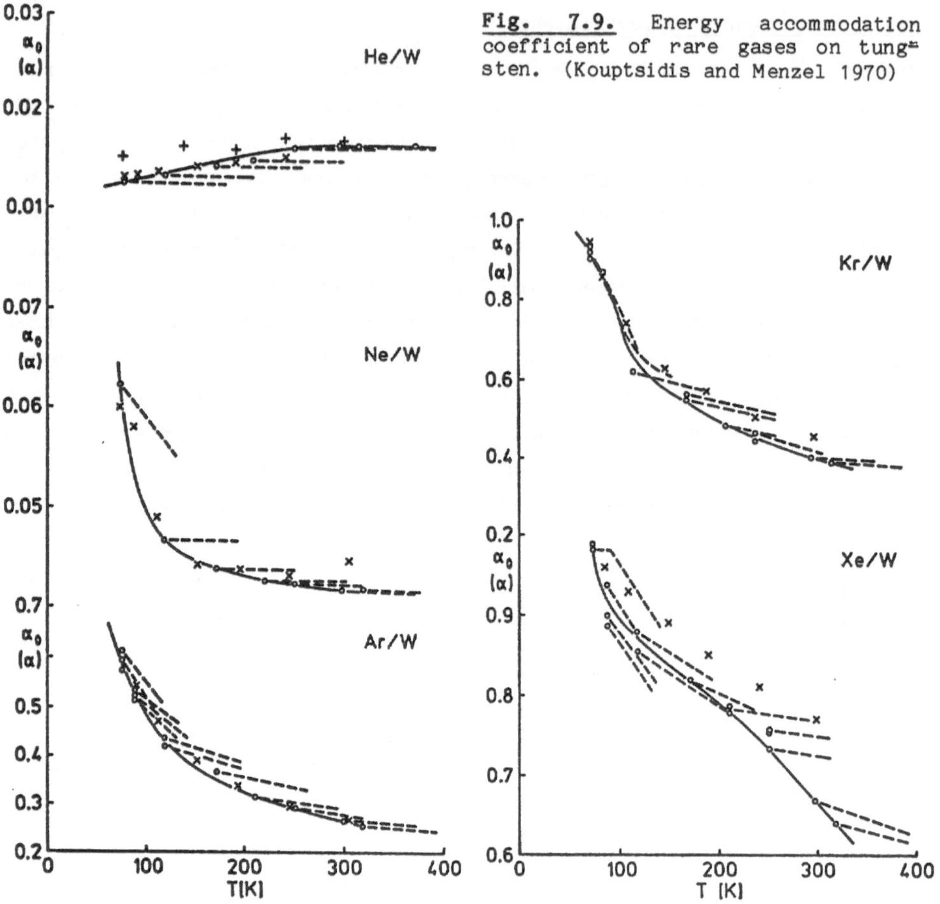

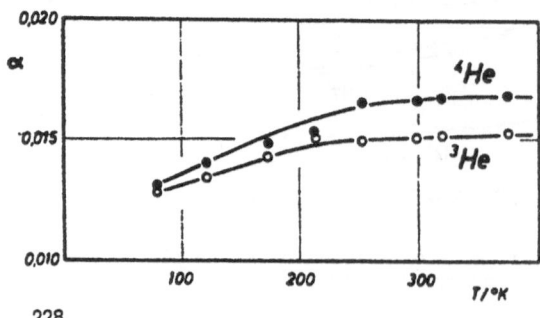

Fig. 7.9. Energy accommodation coefficient of rare gases on tungsten. (Kouptsidis and Menzel 1970)

helium has been measured by Thomas (1967) and by Kouptsidis and Menzel (1969), see Fig. 7.10. More recently Shields (1975, 1983) has developed an acoustical method to measure energy and tangential momentum accommodation coefficients. He has, in particular, looked at rare gases on clean and O_2- and CO-covered Pt and Ag surfaces.

Fig. 7.10. Accommodation coefficient for the helium isotopes on tungsten. (Kouptsidis and Menzel 1969)

7.2 Theory

7.2.1 Sticking

The theory of the sticking coefficient has been plagued for a long time by a controversy about the low temperature behavior. Classical theories obtain a sticking coefficient of unity as the temperature of the solid approaches zero, whereas quantum mechanical theories yield $S(T_S = 0) = 0$. A great many confusing papers have been written on this "puzzle". Its creators claim that experiments support the classical result and contradict the quantum mechanical behavior; unfortunately, they rarely try in their polemic of "author's opinions" to state explicitly below which temperature classical and quantum mechanical theories deviate. As recent analyses, in particular by Brenig and his co-workers, have shown, such deviations should only show up at temperatures well below those accessible by experiments to date.

7.2.1 (a) Quantum Theories of Sticking

We have seen in Chap. 3 how one can derive an expression for the sticking coefficient from the master equation, (3.68-71). At least two features of that derivation are unsatisfactory:

(i)   We recall that the sticking coefficient gives the fraction of particles trapped at the surface out of an incoming beam. Because the master equation deals with occupation functions one should first introduce a microscopic definition of a particle flux.

(ii)   Our starting expression (3.68) for the sticking coefficient has, on the right-hand side, a time dependent occupation function $n_k(t)$ which, in going to (3.69), is replaced by the equilibrium occupation.

This is, as we have seen in Sect. 5.3, a good approximation at best at low temperatures. Indeed, a chopped beam of particles hitting a clean surface will initially not encounter any adsorbate. Thus particles being trapped by undergoing a continuum – bound state transition will rapidly descend to the bottom of the surface potential well through bound state – bound state transitions keeping the uppermost energy levels empty rather than thermally occupied, Fig. 5.4. Thus it seems that bound state – bound state and, in the same spirit, bound state – continuum transitions ought to be included in the definition of the sticking coefficient. A promising theory was developed by Brenig (1982) who presented a very thorough discussion of the role of the master equation for the description of desorption, sticking, and energy accommodation. He considers a chopped beam of particles impinging on a clean surface. The initial evolution over times of the order of the transit time τ_0 that a particle takes to traverse the surface potential must be described by solutions of the equations of motion, i.e., either Newton's equations or preferably Schrödinger's equation. Typically τ_0 can be taken as the oscillation time of a particle at the bottom of the surface potential. If the lifetime of an excited bound state in the surface potential, given by some average of the bound state – bound state transitions $W(i,i')^{-1}$ or by a large eigenvalue of the matrix $\tilde{W}$ in (3.24), say λ_κ^{-1}, is large compared to τ_0, then a description with the master equation may be feasible. Three further, possibly overlapping, stages in the time evolution of the impinging beam can be identified. For times $\tau_0 \ll t \approx t_\kappa = \lambda_\kappa^{-1}$ the particle will undergo single transitions in the surface potential described by $W(\iota,\iota')$, which for continuum – continuum transitions lead to direct or, as Brenig calls it, prompt scattering. In times $t_\kappa \ll t < t_d = \lambda_0^{-1}$ a cascade of transitions will have taken place in the surface potential leading to a stationary distribution with particles also trapped in the surface region. In this time regime, sticking can be defined meaningfully. As time progresses to $t \gg t_d$ these particles will again desorb leading to a delayed particle pulse in a detector.

Experimental evidence for this picture is nicely exhibited in Fig. 7.11 (Hurst et al. 1979). It shows the time of flight spectrum of the Xe beam of mean energy $< E_{kin}>/k_B = 1615$ K incident at 75° onto a Pt(111) surface. Curve (d) gives the specularly reflected beam at 75° with the same mean energy as the incident beam represented by curve (a). Curve (b) at 0° to the surface normal has a much lower mean energy and corresponds to those particles that get trapped at the surface and desorb with mean energy corresponding closely to the surface temperature. At 45° to the surface normal (curve c) both the direct (prompt) and the delayed (trapped and then desorbed) particles will determine the energy accommodation at the surface.

To define the sticking coefficient following Brenig (1982) we start from (3.46) and again treat the continuum occupation $n_k(t)$ as a known quantity, e.g. controlled by a flux of incoming particles. We can then write the solution of the master equation (3.46) as

$$n_j(t) = \sum_\kappa e^{-\lambda_\kappa t} \sum_i \tilde{e}_i(\kappa) e_j(\kappa) n_i(0)$$

$$+ \sum_{i,k} \sum_\kappa e^{-\lambda_\kappa t} \int_0^t dt' e^{\lambda_\kappa t'} \tilde{e}_i(\kappa)\, e_j(\kappa)\, W(i,k) n_k(t') \quad , \qquad (7.1)$$

230

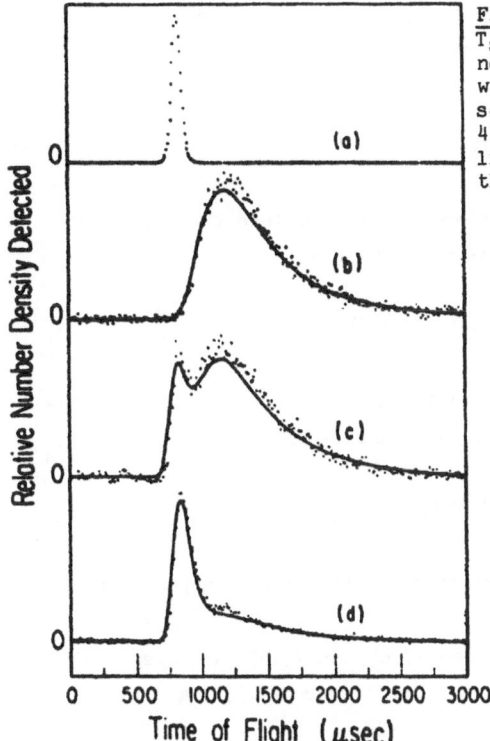

Fig. 7.11. Time of flight spectrum at $T_S = 185K$, angle of incidence 75° from normal. Curve (a): incident Xe beam with $\langle E_{kin}\rangle_i/k_B = 1615$ K; Curve (b): Xe scattered at 0° (normal); curve (c): 45°; curve (d): 75° (specular). Solid lines are from a model as described in the text. (Hurst et al. 1979)

Relative Number Density Detected

Time of Flight (μsec)

where $\bar{e}_i(\kappa)$ and $e_i(\kappa)$ are the ith components of the left and right eigenvectors of $W^{iso}(i,i')$ corresponding th the eigenvalue λ_κ, see (3.27,28). For the description of an adsorption experiment we can set the initial bound state occupation $n_i(0)$ equal to zero, so that the first term in (7.1) is absent. Next we recall that the eigenvalues λ_κ of W^{iso} in most situations group into a slow set for which we take λ_0 as a characteristic one, and all other, fast ones, $\lambda_\kappa \gg \lambda_0$. For times $\lambda_\kappa \ll t \ll \lambda_0$ we can then define a quasi-stationary distribution of temporarily adsorbed particles

$$n_j(a)(t) = e^{-\lambda_0 t} \sum_{k,i} \int_0^t dt' e^{\lambda_0 t'} \; \bar{e}_i(0)e_j(0)W(i,k)n_k(t') \quad , \tag{7.2}$$

so that (7.1) reads approximately,

$$n_j(t) = n_j(a)(t) + \sum_{i,k} \sum_{\kappa \neq 0} (1-e^{-\lambda_\kappa t})\lambda_\kappa^{-1} \; \bar{e}_j(\kappa)e_j(\kappa)W(i,k)n_k(t) \quad , \tag{7.3}$$

if we observe that the externally controlled gas phase occupation $n_k(t)$ varies slowly on the time scale $\lambda_\kappa^{-1}, \kappa \neq 0$, so that it can be taken outside the integral in (7.1). Brenig (1982) refers to $n_j^{(a)}(t)$ as the adsorbate

231

whereas $n_j(t)$ describes all the bound particles, the difference being the fact that $n_j^{(a)}(t)$ varies slowly on the time scale of desorption λ_0^{-1}. The time dependent coverage is then given by [see (3.48) for comparison]

$$\theta(t) = \frac{N_g}{N_s A} \, n^{(a)}(t) = \frac{N_g}{N_s A} \sum_j n_j^{(a)}(t) \quad . \tag{7.4}$$

It satisfies a differential equation which can be obtained by taking the time derivative of (7.2) and using the definition (7.4), as

$$\dot{\theta}(t) = \frac{N_g}{N_s A} \sum_{i,j,k} \bar{e}_i^{(0)} e_j^{(0)} W(1,k) n_k(t) - \lambda_0 \theta(t) \quad . \tag{7.5}$$

It is reminiscent of the phenomenological rate equation (1.2) and allows us to identify the sticking probability as [compare (3.68)]

$$S = \frac{(2\pi m k_B T)^{1/2} N_s}{P} \frac{N_g}{N_s A} \sum_{i,j,k} \bar{e}_i^{(0)} e_j^{(0)} W(1,k) n_k^{eq} \quad , \tag{7.6}$$

where we specified $n_k(t)$ to be the equilibrium distribution appropriate for a beam of thermal particles.

To understand the difference between (7.6) and (3.69) we use (7.2) with n_k^{eq} in it, integrate over time, use detailed balance (3.33), and insert into (7.4).

$$\theta(t) = \frac{N_g}{N_s A} \frac{1-e^{-\lambda_0 t}}{\lambda_0} \sum_{i,j} e_j^{(0)} \sum_k W(k,i) \, \bar{e}_i^{(0)} \, n_i^{eq} \quad . \tag{7.7}$$

Using now the definition (3.53) of W^{iso} we write the eigenvalue equation for the left eigenvector in the form

$$\sum_k W(k,i) \, \bar{e}_i^{(0)} = \lambda_0 \bar{e}_i^{(0)} + \sum_{i'} \bar{e}_{i'}^{(0)} W(i',i) - \sum_{i'} W(i',i) \, \bar{e}_i^{(0)} \quad , \tag{7.8}$$

which now is inserted into (7.7). The terms generated by the last two terms in (7.8) cancel after detailed balance is invoked, so that we get

$$\theta(t) = \frac{N_g}{N_s A} (1-e^{-\lambda_0 t}) \sum_i S_i n_i^{eq} \quad , \qquad \text{where} \tag{7.9}$$

$$S_i = \bar{e}_i^{(0)} \sum_j e_j^{(0)} \tag{7.10}$$

are the weights that define the quasi-stationary distribution over times $\lambda_\kappa^{-1} \ll t$, provided the beam is chopped off, as is appropriate for a desorption situation with the gas phase removed. In lowest order perturbation theory for the master equation, which, as discussed in Sect. 5.3, amounts to dropping bound state-continuum transitions from W^{iso}, these weights are equal to one. This follows immediately from (3.29,35,36) because the sum of $W^{iso}(i,i')$ over i vanishes in this approximation. The deviations of S_i from unity due to transitions from states at the top of the surface potential well into the continuum, can be seen in Figs. 5.4,5 for the case of mobile adsorption; simply note that the quantity plotted there is equal to S_i for $n_i(0)=n_i{}^{eq}$.

Invoking detailed balance (3.33) and using (3.34), (7.6) can be rewritten as

$$
\begin{aligned}
S &= \frac{(2\pi mk_BT)^{1/2}N_s}{P} \frac{N_g}{N_sA} \sum_{i,k} S_i W(k,i) n_i{}^{eq} \\
&= \frac{(2\pi mk_BT)^{1/2}N_s}{P} \frac{N_g}{N_sA} \sum_{i,j,k} e_i(0)\, e_j(0) W(k,i) \quad .
\end{aligned}
\tag{7.11}
$$

Brenig (1982) has called (3.68) the prompt sticking coefficient; it includes only single continuum-bound state transitions and implicitly assumes an equilibrium distribution among the bound states maintained by fast bound state-bound state transitions. In contrast the full sticking coefficient (7.11) also includes subsequent bound state – continuum transitions that deplete the higher bound states in the surface potential. Indeed, if we drop $W(k,i)$ from $W^{iso}(k,i)$ in (3.53), i.e., consider only bound state – bound state and continuum – bound state transitions then $\lambda_0 = 0$ and $\varepsilon_i(0) = 1$ so that we find from (3.29) that $S_i = 1$, recovering the prompt sticking coefficient (3.68) from (7.11).

The sticking coefficient is the fraction of particles in an incoming thermal beam trapped in the surface bound states; it is worthwhile to define an energy-dependent sticking coefficient S_k that gives the fraction of trapped particles in a mono-energetic beam. We define

$$
S = \sum_k S_k\, j_k \cdot n \, / \, \sum_k j_k \cdot n \quad ,
\tag{7.12}
$$

where n is the surface normal and j_k is the particle current density. To define it properly we note that

$$
\sum_k j_k \cdot n = \frac{P}{(2\pi mk_BT)^{1/2}}
\tag{7.13}
$$

for a thermal beam. Thus

$$
j_k = \frac{N_g}{V}\, v_k\, n_k{}^{eq}
\tag{7.14}
$$

provided we treat k_z on the same footing as k_x and k_y, i.e., allow k_z to vary from $-\infty$ to $+\infty$. The volume of the gas phase has been denoted by V=AL.

233

Note that, in calculating the flux (7.13), one must restrict the velocity of the particles normal to the surface to negative values $v_k \cdot n < 0$. Thus comparing (7.12) with the first line of (7.11), we can identify

$$S_k = \sum_i S_i \frac{Lm}{\hbar k_z} W(i,k) \tag{7.15}$$

as the energy-dependent sticking coefficient. This equation was first derived by Iche and Nozières (1976). For mobile adsorption, (7.15) reads

$$S_k = \sum_i S_i \frac{Lm}{\hbar k} W(i,k) \tag{7.16}$$

and (7.11) gives

$$
\begin{aligned}
S &= \frac{h}{k_B T} \frac{1}{L} \sum_k S_k v_k e^{-\beta E_k} \\
&= \frac{h}{k_B T} \sum_{k,i} S_i W(i,k) e^{-\beta E_k} \\
&= \int_0^\infty S_{k(E)} e^{-E/k_B T} d(E/k_B T) \quad ,
\end{aligned}
\tag{7.17}
$$

where $S_{k(E)}$ is S_k with $k = (2mE/\hbar^2)^{1/2}$.

It might be useful to clarify the various notations used in the literature on this approach to sticking. Brenig (1982), approaching the problem via scattering theory, modifies the master equations (3.46,47) by introducing an externally controlled flux $j_k(t)$ so that his kinetic equations read

$$\dot{n}_k(t) = \sum_{k'} r_{kk'}^{(B)} j_{k'}(t) + \sum_i r_{ki}^{(B)} n_i(t) \quad ,$$

$$\dot{n}_i(t) = \sum_k r_{ik}^{(B)} j_k(t) + \sum_{i'} r_{ii'}^{(B)} n_{i'}(t) \quad , \tag{7.18}$$

where he identifies the flux, i.e., the number of particles hitting a unit surface area per unit time, as

$$j_k(t) = \frac{1}{L} v_k \cdot n \, n_k(t) \quad , \tag{7.19}$$

where n is the surface normal. In terms of our formalism we can then identify

$$r_{kk'}^{(B)} = \frac{L}{v_{k'} \cdot n} \left[W(k,k') - \delta_{k,k'} \left[\sum_{k''} W(k'',k') + \sum_i W(i,k') \right] \right] \quad , \tag{7.20}$$

$$r_{ki}^{(B)} = W(k,i) \quad , \tag{7.21}$$

234

$$r_{1k}(B) = \frac{L}{v_k \cdot n} \, W(1,k) \quad , \tag{7.22}$$

$$r_{11'}(B) = W^{iso}(1,1') \quad , \tag{7.23}$$

where k and 1 refer to continuum and bound states, respectively. Equation (7.15) is given in Brenig's paper (his equation (3.12)) as

$$S_k(B) = \sum_1 S_1(B) \, r_{1k}(B) \quad . \tag{7.24}$$

Iche and Nozières (1976) write this equation in a one-dimensional model, taking the energy as the independent variable, as

$$P(IN)(\varepsilon) = \int_{-\infty}^{0} W(IN)(\varepsilon,\varepsilon') \, P(IN)(\varepsilon') d\varepsilon' \tag{7.25}$$

for $\varepsilon > 0$, so that

$$P(IN)(\varepsilon) = S_k \qquad \text{for } \varepsilon > 0$$

$$P(IN)(\varepsilon') = S_1 \qquad \text{for } \varepsilon' < 0 \tag{7.26}$$

is their sticking probability. Using classical concepts they call $W(IN)(\varepsilon,\varepsilon')d\varepsilon'$ the probability that the particle hops from ε to the range $(\varepsilon', \varepsilon' + d\varepsilon')$ in one round trip between the two turning points in the surface potential. It is equivalent, up to a density of states factor, to

$$W(IN)(\varepsilon,\varepsilon') = \frac{L}{\sqrt{2m|E_1|}} \, \tilde{W}(1',1) + \delta_{11'} \quad , \tag{7.27}$$

with $\varepsilon = E_1 > 0$ and where $\tilde{W}$ is the matrix of transition probabilities in the master equation (3.46,47). (Note the reverse order: $\varepsilon \rightarrow \varepsilon'$ and $1 \rightarrow 1'$). The normalization

$$\int_{-\infty}^{\infty} W(IN)(\varepsilon,\varepsilon')d\varepsilon' = 1 \tag{7.28}$$

follows automatically from the master equation. For $\varepsilon > 0$ it reads, in our notation,

$$\sum_{1'} [\delta_{11'} + \tilde{W}(1',1)] = 1 + \sum_{1'} \left[W(1',1) - \delta_{11'} \sum_{1''} W(1'',1) \right] = 1 \quad . \tag{7.29}$$

Note that this normalization does not make any statements about the transition probabilities in the master equation. The δ-term in (7.27) does not contribute to the sticking coefficient (7.25).

235

In Sect. 5.3 [see Figs. 5.4,5 and the comments below (7.10)] we found for the weights S_i that for $T_s \rightarrow 0$

$$S_i = P^{(IN)}(\varepsilon') \rightarrow 1 \quad \text{for } \varepsilon' < 0 \quad (7.30)$$

so that the prompt sticking coefficient (7.25) reads in the notation of Iche and Nozières (1976),

$$P^{(IN)}(\varepsilon) = \int_{-\infty}^{0} W^{(IN)}(\varepsilon, \varepsilon') \, d\varepsilon' \quad . \quad (7.31)$$

For this Sedlmeir and Brenig (1980) write

$$S^{(B)}(\varepsilon) = \int_{-\infty}^{0} P^{(B)}(\varepsilon') \, d\varepsilon'$$

$$= \int_{\varepsilon}^{\infty} P^{(B)}(\varepsilon - \Delta\varepsilon) d(\Delta\varepsilon) \quad , \quad (7.32)$$

where we introduced the energy transfer $\Delta\varepsilon$ used by Böheim and Brenig (1981) to produce Fig. 4.6. Recall from (4.71) that

$$P(\Delta\varepsilon) = \frac{L}{v} W(\mathbf{k}', \mathbf{k}) \quad (7.33)$$

was called an energy distribution.

The prompt differential sticking coefficient (7.32) has been evaluated by Sedlmeir and Brenig (1980) with the transition probabilities given in the one-phonon approximation by (4.22-26); see the lowest panel of Fig. 7.12. They also compare the quantum-mechanical calculation with three approximations in which an increasing number of (different) quantum effects are neglected. Next, in the second-lowest panel of Fig. 7.12, the single-particle states in the Morse potential are treated in the WKB approximation. The sticking coefficient then no longer exhibits the correct low-energy behavior but approaches a constant (not unity) at low energy. In the second panel from the top, the adparticle motion is treated completely classically. The dominant effect of this approximation as compared to WKB is the disappearance of the discrete vibrational levels in the final state producing the steps in the dependence of the sticking coefficient as a function of the energy. These steps, however, are a consequence of the use of the Debye approximation for the phonon-spectrum (with a sharp cutoff) and of the neglect of a nonzero width of the vibrational levels of the adparticle, and would be smeared out in more realistic calculations anyway.

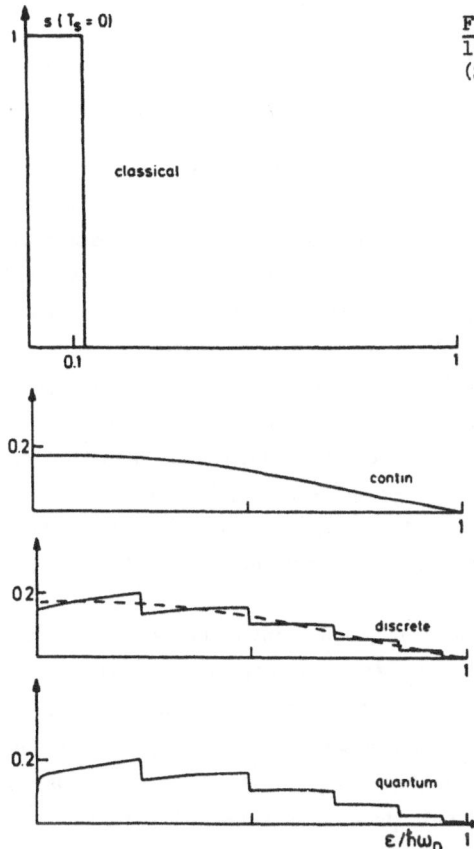

Fig.7.12. Sticking coefficient calculated in various approximations. (Sedlmeir and Brenig 1980)

Finally, in the top panel of Fig. 7.12 the substrate is treated classically as well. This leads to a drastic change in the sticking coefficient. It is now zero down to an energy $\varepsilon = \Delta\varepsilon$ and then rises discontinuously to unity whereas it starts to rise continuously to a value of the order of $\Delta\varepsilon/\omega_D$ in a one-phonon calculation, if the substrate is treated quantum mechanically. The energy integrated sticking coefficients in both approximations are the same, since they are given by the average energy transfer $\Delta\varepsilon$ which is much less sensitive to quantum effects than the differential sticking coefficient itself.

These calculations allow Brenig(1980) and Sedlmeir and Brenig (1980) to shed some light on the apparent contradiction that experimental data seem to confirm the classical result that the sticking coefficient approaches unity at low energies or temperatures, whereas quantum mechanical calculations have S approach zero in this limit. Brenig (1980) argues that this apparent contradiction disappears when considered from a quantitative point

of view. Considering for simplicity a square well for the surface potential, one notes that whereas the exact wave function has an amplitude ratio
between the inside and outside wave function that behaves as (k/k_0) for
small momentum k, its WKB approximation behaves as $(k/k_0)^{1/2}$, where k and k_0
are the wavenumbers outside and inside the attractive well, respectively.
Thus one expects the WKB approximation to deviate from the exact solution,
if the wavelength of the particle becomes comparable to some characteristic
length a of the attractive well, i.e., for qa≈1. Closer inspection shows,
however, that the precise value of qa, where the deviation begins, is very
sensitive to the shape of the potential. For a Morse potential, for instance with a length parameter a, the deviation occurs only below qa ≈
$1/2\pi$. It turns out that the corresponding energies even for light particles
such as helium and hydrogen are of the order of one Kelvin. The addition of
a long-range van der Waals tail shifts the transition region further down to
unobservably low energies of the order of 10^{-5} K (Böheim et al. (1982). Such
an estimate was earlier given by Goodman (1971) and Garcia and Ibañez (1976).
Certainly no data exist at such low energies.

Stutzki and Brenig (1981) have looked in detail into the importance of
Rayleigh waves for the calculation of the sticking coefficient within the
one-phonon approximation in a three-dimensional theory. In the absence of
Umklapp processes they recover the one-dimensional theory of Sedlmeir and
Brenig (1980) if they approximate the surface density of states by a Debye
spectrum. In the fully three-dimensional theory, transfer of parallel momentum including Umklapp processes enhances the sticking probability considerably. Figure 7.13. examines the role of the Rayleigh waves in the
absence of Umklapp processes, i.e., for K_m = 0.

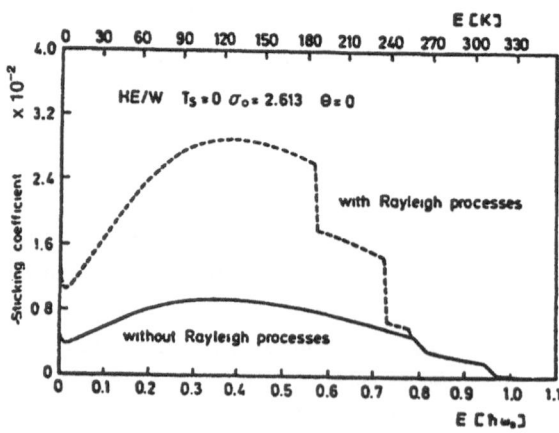

Fig. 7.13. Effect of Rayleigh waves on sticking; ℓ_s =
0.534. (Stutzki and Brenig
1981)

The sharp edges occur when the bound state cannot be reached via Rayleigh phonons. The Debye edge of the one-dimensional model is smeared out, however, over a range $\hbar^2 p_D^2/2m$. The agreement with the one-dimensional model is of course dependent on how well the surface phonon spectrum agrees with a Debye model. Figure 7.14 shows the influence of varying $\ell_s = c_t/c_\ell$, which shifts the Rayleigh dispersion. A soft spectrum has a better overlap with the matrix elements (for He-W) and so raises the sticking coefficient considerably. In particular, the Rayleigh mode is always softer than the longitudinal and transverse modes, so that it contributes more to the sticking ($\approx 50\%$) than one might expect from its contribution to the surface spectrum (25%). This is shown in Fig. 7.13.

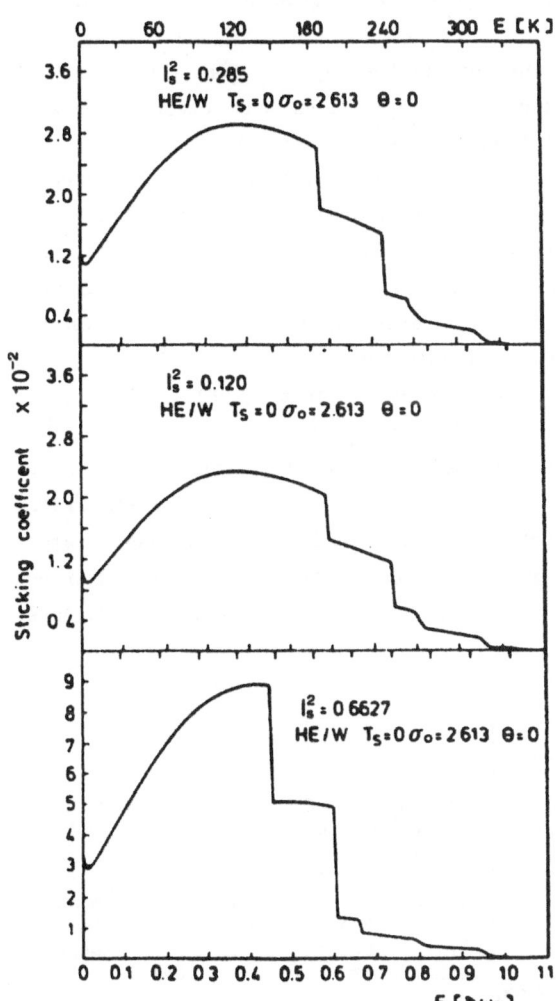

Fig.7.14. The influence of the elastic constants on the sticking coefficient. Parameters as in Fig.7.13. (Stutzki and Brenig 1981)

239

Including Umklapp processes Stutzki and Brenig (1981) find that, as a rough rule, the sticking coefficient gets enhanced by as much as a factor three. This point has recently been examined in great detail by Böheim (1984) for He on metal surfaces. He shows that at low temperatures both the sticking coefficient S and the energy accommodation coefficient α are dominated by "elastic sticking" where the first step is diffraction into bound states. The following transitions may result:

(i) inelastic transitions between bound states leading to sticking;

(ii) inelastic desorption after a few "round trips";

(iii) "elastic desorption", the inverse transition to "elastic sticking". "Elastic desorption" can be more probable than inelastic transitions if the parallel energy of the adparticle is greater than the binding energy.

(iv) One should note that the adparticles may also lose their parallel energy at surface imperfections or collisions with other adsorbates but these effects have not been considered in any theories.

Figure 7.15. gives Böheim's result for He-W(100) with the surface corrugation function and the Morse potential parameters fitted to diffraction scattering experiments. The data points are for the energy accommodation α which according to Leuthäuser (1981) and Brenig (1982) should be equal to S if the transition probabilities in the master equation depend only on the energy difference between initial and final states.

We have evaluated the prompt sticking coefficient (3.68) (Lennard-Jones and Devonshire 1936a,b) and the full sticking coefficient (7.17) (Brenig 1982) for a mobile helium adsorbate using the one-phonon transition probabilities $W(i,k)$ given in (4.22,26). Figure 7.16 gives the results for the He-Ar system for two choices of the surface potential with 6 and 10 bound states, respectively. Note the disturbing fact that the sticking coefficient can get larger than one. There does not seem to be anything in the Iche-Nozières-

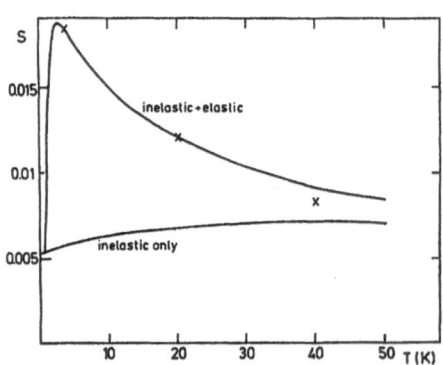

Fig.7.15. Sticking coefficients for He-W(100) with and without inclusion of elastic free-bound transitions. Morse potential parameters: V_0/k_B =161K; $\gamma=1.05$ Å^{-1}. Corrugation amplitude ha$\approx$0.0103 γ^{-1} where a=2.6 Å is the lattice constant. Crosses are experimental results. (Böheim 1984)

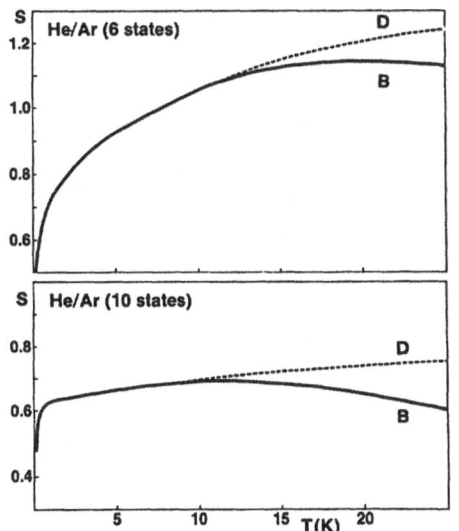

Fig. 7.16. The full (B) and prompt (D) sticking coefficients for two choices of surface potentials (Table 5.1)

Brenig theory to prevent such a situation. The difference between the full (B) and the prompt (D) sticking coefficients is caused by the fact that the coefficients S_i start to deviate from one at higher temperatures.

Following a procedure outlined by Iche and Nozières (1976), Leuthäuser (1981) has developed a Wiener-Hopf approach to solving (7.32). For temperatures such that the energy region around the well top important for sticking is small compared to the well depth, one can assume that the transition probabilities in the master equation, denoted by $P_L(E,E') = W_B(E,E')$, depend only on the energy difference $(E-E')$. For such systems Leuthäuser (1981) gets the interesting result that accommodation and sticking coefficients are equal, i.e. $\alpha = S$. There seems to be some indication of this equality in the low temperature data presented by Leuthäuser.

For a Gaussian transition probability, Leuthäuser (1981) derives an approximate expression for the sticking coefficient

$$S = \mathrm{erf}(\sqrt{\mu_1/4T}) \; \{1 - \{[\frac{\mu_1}{2T} \left[1 - \mathrm{erf}(\sqrt{\mu_1/4T}) \right]$$
$$- \sqrt{\frac{\mu_1}{\pi T}} \; \exp(-\mu_1/4T)]/\mathrm{erf}(\sqrt{\mu_1/4T})\}^2\} \quad , \quad \text{where} \tag{7.34}$$

$$\mu_1 = \int_{-\infty}^{\infty} W(\varepsilon',\varepsilon) \; (\varepsilon - \varepsilon') d\varepsilon \tag{7.35}$$

is the first moment of the transition probability. Using μ_1 as a fitting parameter (7.34) reproduces computer simulations by Tully (1981) based on a classical trajectory approach. In Fig. 7.17 we reproduce his results for

241

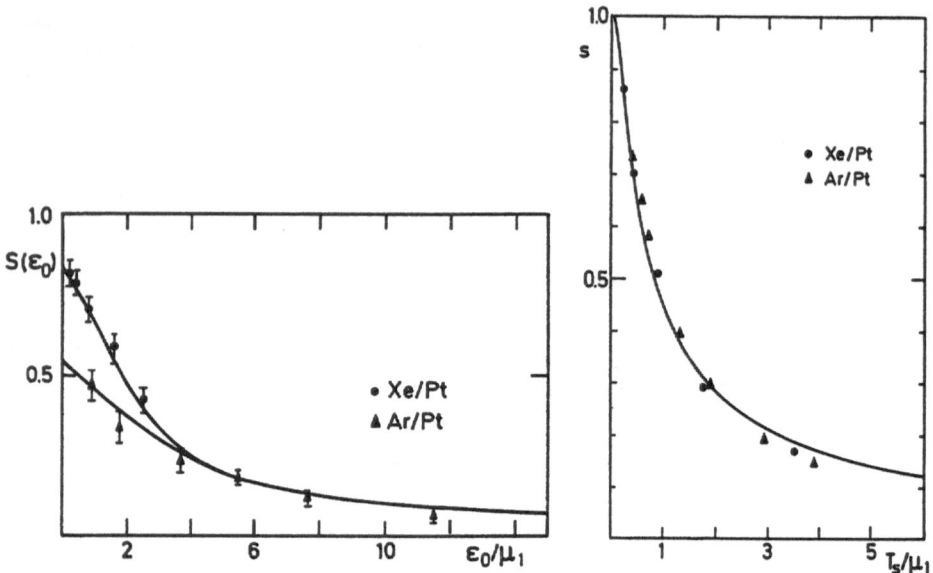

Fig. 7.17. Sticking coefficient as a function of energy, compared to computer simulations by Tully (1981). The first moment (7.35) is fitted to $\mu_1=130$ K for Ar-Pt and to $\mu_1=575$ K for Xe-Pt. The surface temperature is 250 K. (Leuthäuser 1983)

Fig. 7.18. Average sticking coefficient as a function of temperature, compared to computer simulations by Tully (1981). The first moment (7.35) is fitted to $\mu_1=130$ K for Ar-Pt and to $\mu_1=575$ K for Xe-Pt. (Leuthäuser 1983)

the energy dependent sticking coefficient calculated from (7.32). Figure 7.18 then gives the average sticking coefficient as calculated from (7.34). We note in passing that the average desorption energy, plotted in Fig. 6.16, is calculated from the energy-dependent sticking coefficient according to

$$\langle E_d \rangle = \int_0^\infty d\varepsilon_n \int_0^\infty d\varepsilon_t \, \frac{1}{(k_B T_s)^2} \exp[-(\varepsilon_n + \varepsilon_t)/k_B T_s] \, \frac{S(\varepsilon_n)}{S} \, (\varepsilon_n + \varepsilon_t) \quad . \quad (7.36)$$

In closing we note that Kreuzer and Summerside (1981) have discussed the coverage dependence of the sticking coefficient in a simple model based on the master equation (3.107).

7.2.1(b) Classical Theories of Sticking

In classical theories of gas-solid scattering one writes down a set of coupled Newton's equations of motion for a gas particle interacting with a solid. A simple model was studied by Cabrera (1959) and Zwanzig (1960). It has a gas particle interacting via a harmonic force with a semi-infinite, linear harmonic chain as depicted in Fig. 7.19. The equations of motion are

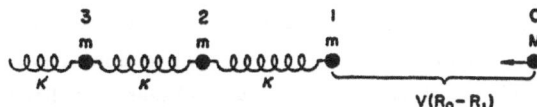

Fig. 7.19. The one-dimensional lattice (M_S) and the gas atom (m). The lattice force constant is k. The interaction of the gas atom with the end of the chain is $V(z_0-z_1)$. (Zwanzig 1960)

$$m \frac{d^2z_0}{dt^2} = - \frac{\partial V(z_0-z_1)}{\partial z_0}$$

$$M_S \frac{d^2z_1}{dt^2} = \frac{\partial V(z_0-z_1)}{\partial z_0} - k(z_1-z_2)$$

$$M_S \frac{d^2z_1}{dt^2} = k(z_{i-1} - 2z_i + z_{i+1}), \qquad i = 2,3,... \tag{7.37}$$

Using generating function techniques (Schrödinger 1914) the set of equations (7.37) can be reduced to a single integro-differential equation in the relative displacement

$$\xi(\tau) = z_0(t) - z_1(t) \quad , \tag{7.38}$$

namely,

$$\frac{d^2\xi(\tau)}{d\tau^2} = \frac{1+\mu}{4\mu} F[\xi(\tau)] - \frac{1}{2} \int_0^\tau ds \frac{J_2(s)}{s} F[\xi(\tau-s)] \quad . \tag{7.39}$$

Here

$$\tau = 2\omega t$$

$$\omega = \sqrt{k/M_S}$$

$$\mu = m/M_S$$

$$F(\xi) = - \frac{1}{k} \frac{dV(z_0-z_1)}{dz_0} \tag{7.40}$$

and $J_2(s)$ is the Bessel function of order two. Equation (7.39) can be solved exactly for V(z) being a truncated harmonic oscillator with depth and with the same force constant k as for the solid, further assuming zero temperature for the solid and either $\mu = 1$ or $\mu = \frac{1}{2}$.

Zwanzig (1960) then shows that trapping always occurs if the initial kinetic energy K of the incoming gas particle is less than some critical value, i.e., for

$$\mu = \frac{1}{2} , \qquad K < K_c = 2.394 \ V_0$$

$$\mu = 1, \qquad K < K_c = 24.54 \ V_0 \quad . \tag{7.41}$$

These results, of course, must be understood in a merely qualitative way indicating that heavier particles get more easily trapped. J.L.Jackson has pointed out, as quoted by Zwanzig (1960), a simple model that demonstrates the main features of energy transfer and trapping. One replaces the harmonic lattice by a linear array of billiard balls, which are initially at rest and do not touch one another. The gas atom interacts with the end of the lattice according to a square well potential whose depth is V_0. In the initial impact, velocity is transferred from the gas atom to the end of the lattice. This velocity is then carried away by successive rigid sphere collisions in the lattice and is forever lost to the gas atom. For $\mu < 1$ the criterion for trapping is $k/V_a < 4\mu/(1-\mu)^2$. If $\mu \geq 1$, the gas atom is always trapped, since it never has its velocity reversed in a collision.

Zwanzig's rather oversimplified model has been reexamined since 1960 over and over again (e.g., McCarroll and Ehrlich 1963; Goodman 1965, 1966; Armand 1968; Beeby and Dobrzynski 1971; Holloway and Beeby 1975; Beeby and Agrawal 1982) extending it to three dimensions, including a more realistic surface potential and accounting for finite temperature effects.

A significant breakthrough in the classical models for gas-solid scattering was achieved by Logan and Keck (1968) with their soft cube model as an improvement of the hard cube model by Logan and Stickney (1966). We introduced it in Sect. 4.3 in connection with the calculations of transition probabilities by Pagni and Keck (1973). The basic assumptions are:

(1) The surface is effectively flat, so that the tangential component of velocity of the gas atom is unchanged;

(2) The gas atom interacts with the surface through a potential which has two parts:

  (a) a stationary attractive part which increases the normal component of velocity of the gas atom before the repulsive collision and decreases it again afterwards;

  (b) an exponential repulsive part which acts between the gas atom and the single surface atom involved in the collision.

(3) The surface atom involved in the collision is connected by a single spring to the remainder of the lattice, the latter being fixed.

(4) The ensemble of oscillators which comprises the surface has an equilibrium distribution of energies at the temperature of the solid.

A sketch of the soft cube model was given in Fig. 4.7. The solution of Newton's equations of motion for the adsorbing gas particle and the surface atom was presented in Sect. 4.3. It can be used to calculate the fraction f_r of atoms scattered into all directions. From this Logan and Keck (1968) define the fraction of particles initially trapped as $S = 1 - f_r$. They find that He and Ne are virtually not trapped on zero temperature Au and Ag surfaces for gas temperatures between 300 K and 2550 K, whereas Xe has $S \approx 0.5$ at 300 K going down to 0.1 to 0.2 at temperatures above 1000 K.

A more complete calculation of the sticking coefficient within the soft cube model has been presented by Modak and Pagni (1973) using the master equation approach with the transition probabilities given by (4.93). They define the sticking coefficient as the fraction of particles whose energy, after collision, is less than the surface well depth $v = V_0/k_BT$. Note that they define their energy zero at the bottom of the surface potential and that they include a Boltzmann factor in their transition probabilities. We thus have, with, $\varepsilon = E/V_0$

$$S = \int_v^\infty d\varepsilon' \int_0^v d\varepsilon R(\varepsilon,\varepsilon') / \int_v^\infty d\varepsilon' \int_0^\infty d\varepsilon R(\varepsilon,\varepsilon') \quad . \tag{7.42}$$

This is the prompt sticking coefficient. It is plotted in Figs. 7.20,21 for the heavier rare gases scattering off tungsten. Taking surface potential

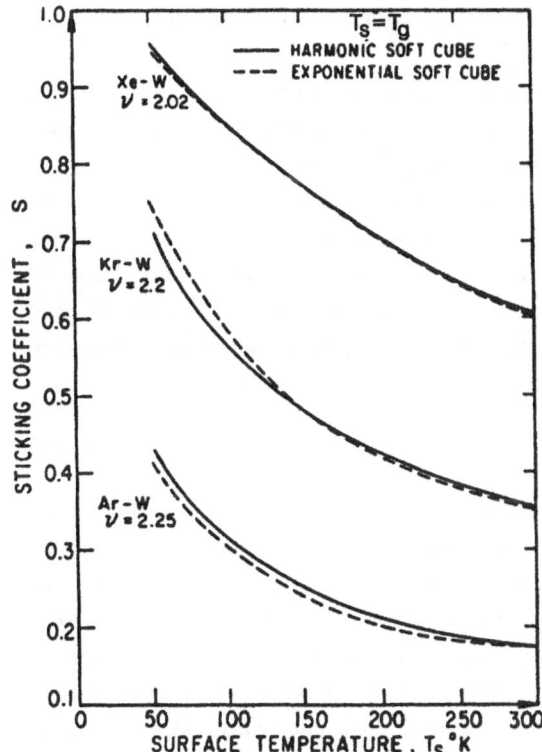

Fig. 7.20. Sticking coefficient S vs. the temperature of the solid T_S, for a thermal incident beam with $T_g = T_S$ of xenon, krypton and argon on tungsten. The dashed lines are soft cube calculations with a repulsive exponential potential. The heavy lines are the results for the harmonic soft cube model. The frequency parameter ν is defined below (4.79). (Modak and Pagni 1973)

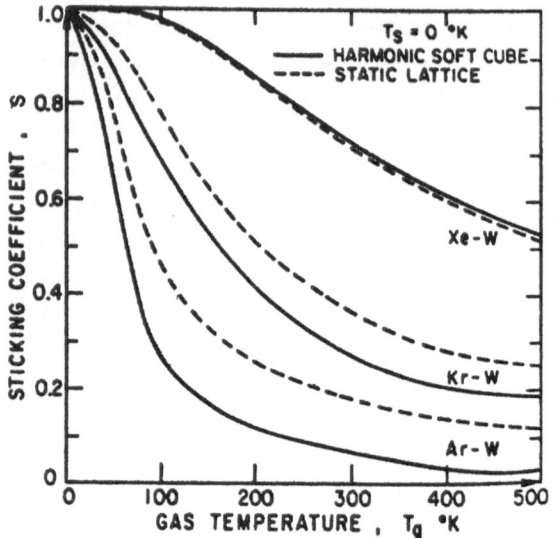

Fig.7.21. Sticking coefficient vs. gas temperature T_g for a thermal beam of rare gas atoms on tungsten at $T_S=0$ K. The dashed lines represent static lattice results. The heavy lines are the predictions of the harmonic soft cube model. (Modak and Pagni 1973)

parameters that were independently determined by Logan (1969) to fit accommodation coefficient experiments, Modak and Pagni (1973) find good agreement with the sticking coefficients measured by Stoll, Smith, and Merrill (1971) for Ar, Kr, and Xe on Ag, Pt, and W.

Weinberg and Merrill (1971) have developed a heuristic classical model based on an assumed attractive square well and an impulsive repulsive potential that enables one to calculate the prompt sticking coefficient from thermal accommodation coefficient data. They consider a gas particle with kinetic energy E impinging normally onto a surface. Once within the range of the attractive surface potential (square) well of depth V_0 its energy is $E_i = V_0 + E$. It will be trapped if after colliding with the repulsive hard wall its energy $E_f \leq V_0$. Treating E_i and E_f as average energies of incoming and outgoing beams, one defines the accommodation coefficient as

$$\alpha(T) = \frac{E_i - E_f}{E_i - E_S} , \qquad (7.43)$$

where E_S is the thermal energy of the solid. Since trapping is already included in the experimentally measured accommodation coefficient (the use of observed accommodation coefficients emphasizes the empirical nature of the model), it might seem superficially that it is improper to use an experimentally measured accommodation coefficient in conjunction with the present model. However, the assumption is made that a gas atom at thermal energy interacting with a solid through a relatively deep potential well is equivalent to a gas atom at a very high effective temperature interacting with the

246

solid only through a repulsive potential. The energy of the gas atom at the surface is $E_i + V_0$ in both cases, and it is presumed that only a negligible amount of trapping occurs in connection with this high-temperature accommodation coefficient, justifying its use in the present model. If the model is used for cases where this assumption is not valid then the estimated trapping probability will be concomitantly large.

Equation (7.43) leads to the following result for the trapping condition:

$$\frac{V_0}{E_i} \geq \frac{1-\alpha(T)}{1-(E_s/V_0)\alpha(T)} \quad , \qquad \text{where} \tag{7.44}$$

$$T = (V_0+E)/k_B \tag{7.45}$$

is the effective temperature of the gas atom as it interacts with the repulsive potential of the solid. Taking the equality in (7.44) gives the critical trapping condition ($E_i=E_{ic}$, $E=E_c$), which is represented in Fig. 7.22, where V_0/E_{ic} is plotted versus $\alpha(T_c)$ with E_s/V_0 as a parameter. The region of trapping based on the empirical model lies between the appropriate curve (a function of surface temperature) and the horizontal line $V_0/E_{ic} = 1$. It is clear that the region of trapping is decreased as the surface temperature is incresed, and trapping is forbidden according to the model calculation if $T_s \geq V_0/k_B$. It is recognized that E_i and E_f in Equation (7.43) are thermally averaged quantities, and thus E_s should be treated in a similar manner. However, the association of E_s with $k_B T_s$ is a much better approximation than merely putting $T_s = 0$ K.

The region of trapping can be reduced further if it is noted that the critical temperature for trapping is high enough so that $\alpha(T_c) \leq \alpha(\infty)$, where

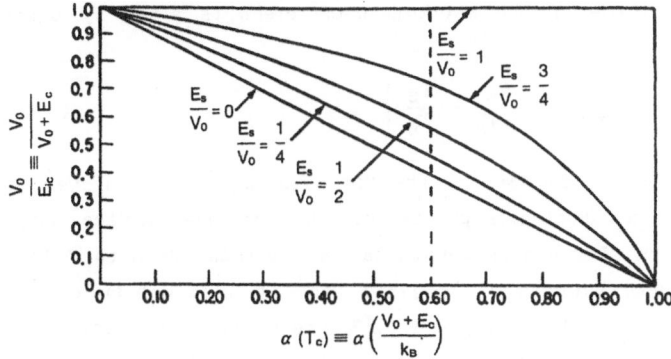

Fig.7.22. Critical dimensionless well depth vs. critical accommodation coefficient with the dimensionless surface temperature as a parameter. The trapping region lies between the appropriate curve and the horizontal line $V_0/E_{ic}=1$. (Weinberg and Merrill 1971)

$\alpha(\infty)$, the high-temperature limit of $\alpha(T)$, is given by (Goodman and Wachman 1976)

$$\alpha(\infty) = 2.4\mu^* /(1+\mu^*)^2 \qquad (7.46)$$

where μ^* is the ratio of the mass of the gas species m to the mass of the surface atom M_s. It now lies between the appropriate curve (a function of surface temperature), the horizontal line $V_0/E_{ic} = 1$, and the vertical line (shown dotted in Fig. 7.22) $\alpha(T_c) = 0.60$ which is the maximum value of $\alpha(\infty)$ (where $\mu^* = 1$) and thus the maximum value of $\alpha(T_c)$.

Using (7.44,46), it is possible to calculate the minimum dimensionless well depth for trapping (a maximum critical initial kinetic energy). This calculation is useful as a qualitative guide in determining the importance of trapping in the absence of accommodation coefficient data and becomes exact for large T_g.

The sticking coefficient S may be obtained by integrating a one-dimensional Maxwell-Boltzmann velocity distribution function from zero velocity to the critical velocity for trapping, v_c, and normalizing by dividing by the mean incident velocity. The critical trapping velocity may be calculated from (7.44) if accommodation coefficient data are available, and

$$v_c = [(2E_{ic} - 2V_0)/m]^{1/2} \quad . \qquad (7.47)$$

The sticking coefficient is thus found to be

$$S = 1 - \exp[-(E_c/k_BT_g)] \quad \text{or} \qquad (7.48)$$

$$S = 1 - \exp\left[\mu \frac{\alpha(T_c)}{1-\alpha(T_c)} \left(\frac{V_0}{k_BT_g} - \frac{T_s}{T_g}\right)\right] \quad . \qquad (7.49)$$

The maximum possible value of the sticking coefficient is given by using (7.46) for $\alpha(T_c)$

$$S_{max} = 1 - \exp\left[-\frac{2.4\mu^*}{1-0.4\mu^*+\mu^{*2}} \left(\frac{V_0}{k_BT_g} - \frac{T_s}{T_g}\right)\right] \quad . \qquad (7.50)$$

The sticking coefficient for xenon, krypton, and argon on tungsten is plotted in Fig. 7.23 as a function of gas temperature with the surface temperature as a parameter. For gas temperatures ranging from 300 K to 2000 K and surface temperatures between 0 and 775 K, the sticking coefficient of xenon varies between essentially unity and 0.67, the sticking coefficient of krypton varies between 0.90 and 0.21, and the sticking coefficient of argon varies between 0.43 and 0.025. At a surface temperature of 0 K and gas temperatures between 300 and 2000 K, the sticking coefficient of neon varies

248

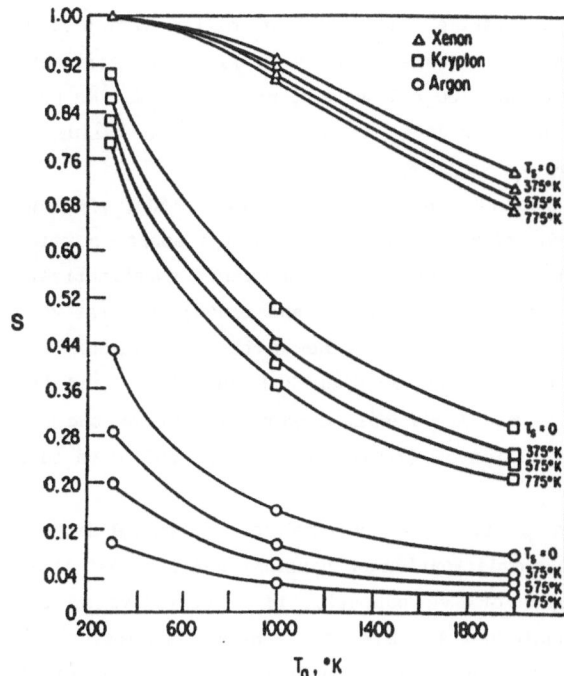

Fig. 7.23. Sticking coefficient of rare gases on tungsten as a function of gas temperature with surface temperature as a parameter. (Weinberg and Merrill 1971)

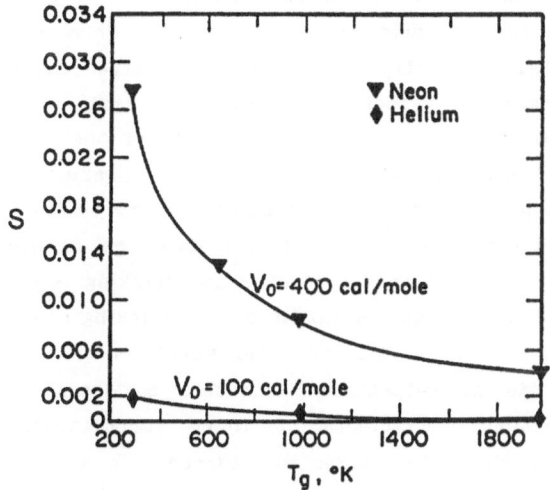

Fig. 7.24. Sticking coefficient of helium and neon on tungsten at $T_S = 0$ K as a function of gas temperature. (Weinberg and Merrill 1971).

between 0.027 and 0.004, and the sticking coefficient of helium varies from 0.0019 to 0.0003. These calculated data are shown in Fig. 7.24 for well depths of neon and helium of 400 and 100 cal mole^{-1}, respectively. No neon is trapped at a surface temperature greater than 200 K, and no helium is trapped at a surface temperature above 50 K.

Weinberg and Merrill (1971) conclude that the square well model predicts trends in the sticking coefficient of the rare gases on metal surfaces which are qualitatively correct in all cases and which are nearly quantitatively correct except for the heavier rare gases (krypton and xenon) at relatively low gas temperatures. The fact that the model seems to overestimate the xenon trapping probability at gas temperatures of 300 and 1000 K and the krypton trapping probability at 300 K may result because a part of the gas is trapped according to the definition of (7.44) and is still scattered in a directed manner.

To model the gas-surface interaction by a square well potential may at first seem like an unacceptable simplification of physical reality, but it is argued that the attractive region of the potential is conservative so the shape and width of the well should not be important, and an impulsive potential may be a reasonable, though crude, approach to reality for the steep repulsive portion of the potential. Using the three-dimensional accommodation coefficient in connection with a one-dimensional theory (via experimental data or the semiempirical formula) is based on the premise that the preponderance of energy exchange in gas-solid collisions results from the normal component of the incident gas velocity and the fact that after entry into the potential well the transverse velocity components are small compared to the velocity perpendicular to the surface. The success of the cube theories in predicting the scattering distributions support the assertion that energy exchange occurs primarily in the normal component of velocity.

It is thus concluded by Weinberg and Merrill (1971) that the simple square well model can predict qualitative trends in rare gas sticking coefficients and it can perhaps predict quantitative values of the sticking coefficient for relatively light gas species (small potential well) and/or high gas temperature. The model has no adjustable parameters but requires experimental thermal accommodation coefficients at lower temperatures. Where these data are available, the calculations are exceedingly simple. Even if accommodation coefficient data are not available, maximum trapping probabilities can still be calculated using values of $\alpha(\infty)$ and (7.50).

7.2.2 Energy Accommodation

7.2.2(a) Quantum Theories of Accommodation

The energy accommodation coefficient α has been introduced by Knudsen to characterize the efficiency of energy transfer between a gas and a solid. Assume that the gas is at a temperature T_g and the solid at T_s. Let E_g be the mean energy of a gas particle at T_g, E_s its mean energy at T_s and E_g' its energy after scattering off the surface from an initial energy E_g. We then define

$$\alpha = \lim_{T_s \to T_g} \alpha(T_g, T_s) = \lim_{T_s \to T_g} \frac{E_g' - E_g(T_g)}{E_s(T_s) - E_g(T_g)} = \frac{\Delta E}{\Delta E^{id}} \qquad (7.51)$$

Thus if $E_g' = E_s$ we get perfect accommodation with $\alpha = 1$. For the calculation of α we first outline the master equation approach.

We can relate the energy change ΔE to the rate of energy change $\Delta \dot{E}$ per unit time if we divide by the flux F of particles hitting the surface area A per unit time, i.e.,

$$\Delta E = \frac{\Delta \dot{E}}{AF} \quad . \qquad (7.52)$$

This allows us to make the connection with the transition probabilities $W(\imath, \imath')$ by multiplying the master equation (3.47) with E_k and summing over k, so that we get

$$\Delta \dot{E} = N_g \sum_{k,\imath} W(k, \imath) E_k n_\imath - N_g \sum_{\imath,k} W(\imath, k) E_k n_k \quad , \qquad (7.53)$$

where $\imath$ denotes, as usual, both bound states and continuum states of gas particles. The total number of gas particles N_g appears explicitly due to the normalization of

$$n_k = n_k^{eq} = \frac{1}{V} \left[\frac{2\pi\hbar^2}{mk_B T_g} \right]^{3/2} e^{-\beta_g E_k} \qquad (7.54)$$

to one, see (3.49). Note that the bound states contribute negligibly to the normalization. The flux is, of course, given by $F = P/(2\pi m k_B T_g)^{1/2}$. Note that the probability densities n_k^{eq} depend on the gas temperature $T_g = 1/k_B \beta_g$ whereas the transition probabilities $W(k, \imath)$ are controlled by the solid temperature T_s.

Strachan (1937) kept only the continuum-continuum contributions in (7.53) whereas Goodman (1971) also kept the transitions to and from bound states, although in his review (Goodman 1975) he rejects his one-dimensional model and does not include the bound state transitions in a three-dimensional theory either. For now we follow Strachan's (1937) approach and discuss the more general view presented by Brenig (1982) at the end of the section. Let us then split the first term in (7.53) into contributions with $E_k < E_{k'}$ and $E_k > E_{k'}$ and eliminate the former by (i) renaming summation indices and (ii) applying detailed balance, (3.33) at $T_s = 1/(k_B \beta_s)$. We get

$$\Delta \dot{E} = \frac{N_g}{V} \left[\frac{2\pi \hbar^2}{m k_B T_g} \right]^{3/2} \sum_{k,k'} \theta(E_k - E_{k'})$$

$$* W(k',k) \, (E_k - E_{k'}) e^{-\beta_g E_k} [e^{(\beta_s - \beta_g)(E_{k'} - E_k)} - 1] \quad . \tag{7.55}$$

Because we eventually want to take the limit $T_s \to T_g$, we expand the square brackets and get for $T_s \approx T_g$

$$\Delta \dot{E} = \frac{N_g}{V} \left[\frac{2\pi \hbar^2}{m k_B T_g} \right]^{3/2} \frac{T_s - T_g}{k_B T_g^2} \sum_{k,k'} \theta(E_k - E_{k'})$$

$$* W(k',k) \, (E_k - E_{k'})^2 e^{-\beta_g E_k} \quad . \tag{7.56}$$

Physisorption kinetics of mobile adsorbates can be described adequately in one-dimensional models. The transition probabilities are then given by (4.4) which allows us to perform the summation over the lateral components K of the wavevector k. This effectively implies that the preexponential factor in (7.54) arising from the chemical potential, be raised only to the power 1/2 rather than 3/2. The resulting mean energy change for mobile adsorption then follows from (7.56) as

$$\Delta \dot{E} = \frac{N_g}{L} \left[\frac{2\pi \hbar^2}{m k_B T_g} \right]^{3/2} \frac{T_s - T_g}{k_B T_g^2} \sum_{k > k'} W(k',k) \, (E_k - E_{k'})^2 e^{-\beta_g E_k} \quad , \tag{7.57}$$

where momenta k and energies E_k refer to the motion in the z-direction only.

To calculate α, we must divide by the ideal energy transfer rate $\Delta \dot{E}_{id}$ for perfect accommodation. Let us first look at a flux of particles with particle current density

$$\frac{p}{m} f(r,p) \quad . \tag{7.58}$$

The energy current density associated with (7.58) is

$$\frac{p^2}{2m} \frac{p}{m} f(r,p) \quad . \tag{7.59}$$

Let us assume that the gas is in thermal equilibrium at temperature T_g so that

$$f(\mathbf{r},\mathbf{p}) = \frac{N_g}{V^2}\left(\frac{2\pi\hbar^2}{mk_BT_g}\right)^{3/2} e^{-\beta_g p^2/2m} = \frac{N_g}{V}\, n_k^{eq} \quad , \tag{7.60}$$

where $\mathbf{p} = \hbar\mathbf{k}$ assumes discrete values for a particle in a box. The energy deposited per unit time on a surface area A is then

$$\dot{E}_g = \int_A dA \sum_k \frac{\hbar\mathbf{k}\cdot\mathbf{n}}{m}\, f(\mathbf{r},\hbar\mathbf{k}) = 2k_BT_g\frac{AP}{(2\pi mk_BT_g)^{1/2}} \quad , \tag{7.61}$$

where $\mathbf{n}$ is the surface normal and $PV=N_gk_BT_g$. A beam of thermal particles leaving a surface at temperature T_s will carry away an energy flux like (7.61) but with T_g replaced by T_s. Thus the ideal energy exchange rate is given by

$$\Delta\dot{E}_{id} = 2k_BT_sA\frac{P}{(2\pi mk_BT_s)^{1/2}} - 2k_BT_gA\frac{P}{(2\pi mk_BT_g)^{1/2}} \quad . \tag{7.62}$$

The standard argument assumes that incoming and outgoing particle fluxes are the same so that T_s in the denominator of the second term is replaced by T_g and

$$\Delta\dot{E}_{id} = 2k_B(T_s-T_g)\frac{PA}{(2\pi mk_BT_g)^{1/2}} \quad . \tag{7.63}$$

Combining (7.55) with (7.60) we get the energy accommodation coefficient to be

$$\alpha = \frac{\Delta\dot{E}}{\Delta\dot{E}_{id}} = \frac{1}{2}\frac{\hbar^3}{m}\frac{(2\pi)^2}{(k_BT)^4}$$

$$* \frac{1}{A}\sum_{k,k'}\theta(E_k - E_{k'})\, W(k',k)\,(E_k - E_{k'})^2\, e^{-E_{k'}/k_BT} \quad , \tag{7.64}$$

where $T=T_g=T_s$. For a mobile adsorbate one gets, using (7.57),

$$\alpha\big|_{(mobile)} = \frac{\pi\hbar}{(k_BT)^3}\sum_{k>k'} W(k',k)\,(E_k-E_{k'})^2\, e^{-E_k/k_BT} \quad , \tag{7.65}$$

Brenig (1982) has reexamined this expression of the accommodation coefficient and, based on an analysis of the master equation for times large compared to t_d, concludes that bound state – bound state transitions modify the "prompt" accommodation coefficients (7.63,65) giving for mobile adsorption

$$\alpha\big|_{(mobile)} = \frac{\pi\hbar}{(k_BT)^3}\sum_{k>k'}(E_k-E_{k'})^2\, e^{-E_k/k_BT}$$

$$* \left[W(k',k) - \sum_{i,i'} W(k',i')(W^{iso})^{-1}(i',i)W(i,k) \right] \quad . \tag{7.66}$$

253

Expressions (7.64,65) were first obtained by Devonshire (1937a) for the spe-
cial case where the transition probabilities are calculated for transitions
in a Morse potential within the one-phonon approximation. Using (4.26) one
gets for mobile adsorbates

$$\alpha|(\text{mobile}) = \frac{3\pi^2}{2r^6} \frac{m}{M_S} \delta^{-3} \int_0^\infty d\eta \int_\eta^\infty d\eta' \; \theta(r-\eta'^2+\eta^2)$$

$$* \; \frac{\eta\eta'\sinh(2\pi\eta)\sinh(2\pi\eta')}{\sinh^2(\pi(\eta+\eta'))\sinh^2(\pi(\eta-\eta'))} \left[\left| \frac{\Gamma(\frac{1}{2} - \sigma_0 + i\eta)}{\Gamma(\frac{1}{2} - \sigma_0 + i\eta')} \right| + \left| \frac{\Gamma(\frac{1}{2} - \sigma_0 + i\eta')}{\Gamma(\frac{1}{2} - \sigma_0 + i\eta)} \right| \right]^2$$

$$* \; (\eta'^2-\eta^2)^5 \left[\left[\exp\left[\frac{\delta}{r} (\eta'^2-\eta^2) \right] - 1 \right]^{-1} + 1 \right] \exp(-\frac{\delta}{r} \eta'^2) \quad . \tag{7.67}$$

Here we also used a bulk Debye model with a single Debye temperature. We
note that the high temperature limit of (7.67) is (Gilbey 1961),

$$\alpha \rightarrow \frac{4m}{M_S} , \tag{7.68}$$

whereas inclusion of recoil effects suggest (Baule 1914) that

$$\alpha \rightarrow \frac{4mM_S}{(m+M_S)^2} , \tag{7.69}$$

which is larger than Goodman's result (7.46) by a factor of 5/3. Gilbey
(1961) argued that Devonshire's simplification of using bulk rather than surface phonons overestimates the ease of excitation of high-frequency lattice
modes and underestimates energy transfer to low-frequency modes. Manson
(1972) has calculated the accommodation coefficient starting from an expression like (7.65) but taking proper surface modes into account. His calculations for the scattering off a tungsten surface fit the data by Thomas
(1967) and Kouptsidis and Menzel (1969, 1970) for T > 100 K.

However, using only a repulsive surface potential, his results are inade-
quate at low temperatures where, not surprisingly, he finds disagreement
between theory and experiment. Goodman (1971) reviewed the theory of the
accommodation coefficient up to 1971. With Gillerlain (Goodman and Gillerlain, 1971) he attempted to include bound state - bound state and bound
state - continuum transitions. This theory is also supposed to improve the
high-temperature fit to experiment. Goodman (1972) also produced a three-
dimensional one-phonon theory, including neither surface modes nor an
attractive potential nor bound state transitions, again finding agreement
with the high-temperature results. The confusing state of the theory was

described in a review by Goodman (1975). In a note Garcia et al. (1980) [with no reference to the earlier paper by Manson (1972)] argue that agreement at high temperatures can be achieved by inclusion of two-phonon processes. The presence of a van der Waals long-range attraction ($\approx$ C_{3z}^{-3}) and inclusion of bound state – continuum and continuum – bound state transitions leads to a low-temperature behavior

$$\alpha(T \to 0) = c_0 + c_1 T + c_2 T^2 + \ldots \tag{7.70}$$

The constant c_0 is, however, so small that the experimental data cannot be fitted. The same conclusion was reached shortly thereafter by Goodman (1980); another review was presented by him in 1981. The contributions of Brenig's group to the theory of the sticking coefficient start with a simple model by Müller and Brenig (1979) who assume a Gaussian transition probability (4.72) where the energy loss Δ is related to a friction coefficient. We will outline this approach via nonequilibrium thermodynamics in Chap. 8.

We reported in (7.34) a result by Leuthäuser (1981) which shows that the following relation between partial derivatives should hold:

$$\frac{\partial \alpha}{\partial T_s} = \lim_{T_g \to T_s} \left[\left[\frac{\partial \alpha(T_s, T_g)}{\partial T_s} \right]_{T_g} + \left[\frac{\partial \alpha(T_s, T_g)}{\partial T_g} \right]_{T_s} \right] \ , \tag{7.71}$$

where $\alpha(T_s, T_g)$ is the nonequilibrium accommodation coefficient given in

Table 7.1. Experimental values of the different partial derivatives of the accommodation coefficient relation (7.71). (Leuthäuser 1981)

T_s	$\dfrac{\partial \alpha(T_s, T_g)}{\partial T_s}$	$\dfrac{\partial \alpha(T_s, T_g)}{\partial T_g}$	$\dfrac{\partial \alpha}{\partial T_s} + \dfrac{\partial \alpha}{\partial T_g}$	$\dfrac{\partial \alpha}{\partial T}$
[K]	$[10^3 K^{-1}]$	$[10^3 K^{-1}]$	$[10^3 K^{-1}]$	$[10^3 K^{-1}]$
Ar/W				
90	(1.8)	2.6	4.4	4.0
120	0.6	1.7	2.3	2.3
173	0.4	0.6	1.0	1.1
211	0.3	0.7	1.0	0.9
Kr/W				
90	(5.4)	2.7	8.1	6.0
120	0.6	0.9	1.5	1.6
173	0.6	0.9	1.5	1.4
211	0.6	0.3	0.9	1.1

(7.51). Table 7.1 shows that this relation is rather well satisfied for Ar and Kr on W. We point out once more that for Gaussian transition probabilities Leuthäuser (1983) has shown that sticking and accommodation coefficients are equal. A detailed fit to Tully's (1981) computer simulations for Ar-Pt and Xe-Pt has been given by Leuthäuser (1984), see also Brako (1983). As mentioned already in Sect. 7.2.1(a) Böheim (1984) has recently included diffraction into bound states with subsequent inelastic transitions in the calculation of sticking and accommodation coefficients, assumed equal according to Leuthäuser's argument. These effects lead to a dramatic increase in both S and α at low temperatures. We have already reproduced in Fig. 7.15 his low temperature results for the sticking coefficient S for He-W(100). In Fig.7.25 we now give for completeness his calculations for S = α for the same system but over a larger temperature regime. Hopefully there will be more calculations along these lines in the near future.

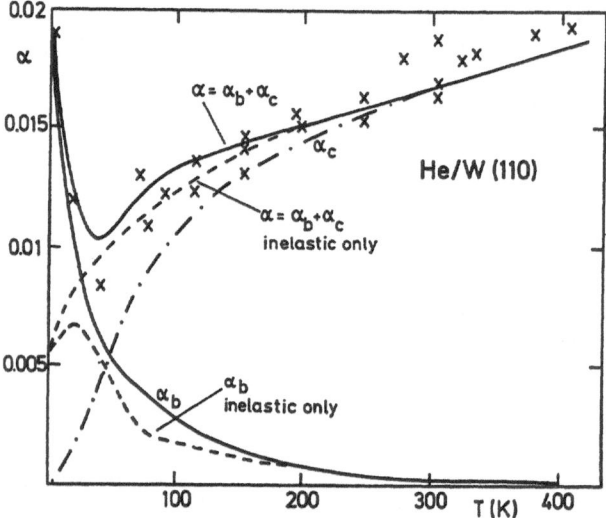

Fig.7.25. Energy accommodation coefficient α and its bound state contribution α_b for He-W(110). Solid line: elastic and inelastic contributions; dashed line: inelastic transitions only; dash-dotted line: continuum contribution α_c. (Böheim 1984)

7.2.2(b) Classical Theories of Accommodation

As we have seen in Chap. 5 on desorption and in Sect. 7.2.1(b) on sticking, classical theories start by writing down Newton's equations of motion for the adsorbing gas particle interacting with a limited number of surface atoms, the latter in addition being coupled to the vibrational degrees of freedom of the solid. With realistic interatomic forces such a programme can only be carried out numerically. This was done successfully by Tully and his group (1980, 1981) for the calculation of desorption times, time of flight spectra, and sticking coefficients, as mentioned already in the respective chapters. These results can be compared with experimental data to test the underlying physical assumptions about interaction potentials and the mechanism of energy dissipation. They also serve as testing grounds for simpler models for which explicit analytical results can be obtained to hopefully lead to qualitative understanding of physisorption kinetics. The stochastic classical trajectory theory was developed by Adelman (1980) and his coworkers. For related work see also Masel, Merrill, and Miller (1974), Lin and Wolken (1976a,b), Lin, Adelman, and Wolken (1977), Doll (1980), Nitzan and Carmeli (1982).

The first classical model to explain energy accommodation was developed by Baule (1914) who considered a line of hard spheres on which a lighter particle impinges. He found for the accommodation coefficient, averaged over all impact parameters,

$$\alpha = \frac{2\mu}{(1+\mu)^2} \qquad \text{and} \tag{7.72}$$

$$\alpha = \frac{4\mu}{(1+\mu)^2} \tag{7.73}$$

for head-on collisions. Here μ is the reduced mass. Landau (1935) developed the first "continuum" theory treating the solid as an elastic continuum with which a gas atom interacts via a repulsive potential. He argues that except for He and H_2, the theory of energy accommodation should be based on classical mechanics. At low temperatures he finds that α is proportional to $T^{3/2}$ provided $\alpha \ll 1$.

Foremost among the classical models developed to understand the phenomenon of energy accommodation is again a gas atom interacting via a harmonic force with a one-dimensional harmonic chain as introduced by Zwanzig (1960) and Cabrera (1959). Goodman (1962, 1970, 1975, 1976, 1980, 1981), McCarroll and Ehrlich (1963), Chambers and Kinzer (1963, 1966) have done a considerable amount of analytical work on this model, in particular extending it to three dimensions. Beeby and coworkers (1971, 1975) and Jewsbury et al. (1977) are continuing along these lines. Steinbrüchel (1977) has looked at the importance of multiple collisions on energy accommodation (see also Füstöss 1982), and Richard and Depristo (1983) have developed a classical scaling theory.

The hard and soft cube models have also been analysed to provide information on energy accommodation (Logan 1969). Within the hard cube model Grimmelmann et al. (1980) found an interesting relation

$$\langle E_{g'} \rangle = \alpha_g \langle E_g \rangle + \alpha_s \langle E_s \rangle \tag{7.74}$$

between the initial and final energies of the scattered gas particle $\langle E_g \rangle$ and $\langle E_{g'} \rangle$ respectively and $\langle E_s \rangle = 2k_B T_s$. For a purely repulsive potential they find

$$\alpha_g = \left[\frac{1-\mu}{1+\mu} \right]^2$$

$$\alpha_s = \frac{\mu(2-\mu)}{(1+\mu)^2} \ . \tag{7.75}$$

A fit to data on Ar-W by Janda et al.(1980) produces an "effective" mass of the cube representing the impact site on the solid surface that has a mass of about 4 W atoms, indicating that this is the size of a relevant surface cluster participating in the collision at incident energies between 300 K and 1000 K. At higher energies this effective mass is expected to decrease to that of a single W atom.

8. Kramers Equation

The master equation approach, on which all our discussion of physisorption kinetics has been based so far, requires a detailed knowledge of the underlying dynamics for the calculation of the transition probabilities unless some phenomenology such as a Gaussian ansatz is employed. Once the microscopic approach is worked out and understood it is imperative to extract a simplified description that entails the essential features of the phenomenon. This is particularly useful to (i) derive earlier phenomenological models and (ii) to get a simple starting point to develop a theory for the more complicated kinetics of chemisorption.

A scheme to extract simpler kinetic equations from the master equation has been worked out in nonequilibrium statistical mechanics. Using a Kramers-Moyal expansion, one rewrites the integro-differential master equation as an infinite order differential equation. Under well-defined conditions, the latter can be truncated to yield a second-order differential equation of the Fokker-Planck type. For physisorption and chemisorption kinetics, the particular variant obtained is called the Kramers equation. It was originally conceived by Kramers (1940) in his attempt to develop a diffusion model of chemical reactions. Its use in chemisorption kinetics has been very promising (see for instance d'Agliano et al. 1975).

To understand the physics behind the Kramers equation in the context of physisorption kinetics, let us again describe the gas particle interacting with a solid using classical statistical mechanics. As long as the interaction between particles in the dilute gas phase or in the low-coverage adsorbate is negligible, it suffices to study the single particle distribution function $f(\mathbf{r}, \mathbf{v}, t)$ introduced in (6.9) subject to a kinetic equation (6.11) with a collision term dropped. In the theory of Brownian motion, one assumes that the right-hand side of (6.11) summarizes the influence of all degrees of freedom of a gas in which a Brownian particle moves. With Kramers (1940) one gets in one dimension

$$\frac{\partial}{\partial t} f(z, v, t) + v \frac{\partial f}{\partial z} + \frac{F(z)}{m} \frac{\partial f}{\partial v} = \eta \left[\frac{\partial}{\partial v} (vf) + \frac{k_B T}{m} \frac{\partial^2 f}{\partial v^2} \right] . \qquad (8.1)$$

Here

$$F(z) = -\frac{dV_S(z)}{dz} \qquad (8.2)$$

is the force acting on the gas particle due to the surface potential V_S. The terms on the right-hand side represent a drift and a diffusion term, respectively: the interpretation of the latter is obvious if we recall Einstein's relation between the friction coefficient η and the diffusion coefficient D, namely

$$D = \frac{k_B T}{m \eta} \qquad . \qquad (8.3)$$

In the Kramers equation, all the details about energy loss mechanisms are summarized in a single transport coefficient, namely the friction coefficient η.

A general solution of (8.1) is not known, but Kramers (1940) has shown that for small friction, i.e., as long as

$$\eta < \omega_0 \frac{k_B T}{V_0} \qquad , \qquad (8.4)$$

the desorption rate constant is approximately given by

$$r_d = t_d^{-1} = \eta \frac{V_0}{k_B T} e^{-V_0/k_B T} \qquad . \qquad (8.5)$$

Here V_0 is the depth of the surface potential well and ω_0 is the oscillation frequency near the bottom of the surface potential which for a Morse potential is given by

$$\omega_0 = \sqrt{2\gamma^2 V_0/m} \qquad . \qquad (8.6)$$

In the limit of large friction, the Kramers equation (8.1) can be reduced to a quasi-linear Fokker-Planck equation [see, e.g., van Kampen (1981) Sect. VIII.7]

$$\frac{\partial}{\partial t} \tilde{f}(z,t) = -\frac{\partial}{\partial z} \left[\frac{F(z)}{m \eta} \tilde{f}(z,t) \right] + D \frac{\partial^2 \tilde{f}(z,t)}{\partial z^2} \qquad , \qquad (8.7)$$

where $\tilde{f}(z,t)$ is the integral of $f(z,v,t)$ over all velocities. In the next section, we will sketch a derivation of the Kramers equation (8.1) from the master equation utilizing a Kramers-Moyal expansion (Gortel et al. 1981; Kreuzer and Teshima 1981; Jack and Kreuzer, 1982,1985). The next section will then develop a yet more coarse-grained theory by deriving the balance equations of nonequilibrium thermodynamics for macroscopic quantities such
260

as particle number density and energy density; a set of transport coefficients such as the friction coefficient η and the sticking coefficient S will be all that remains from the dynamical gas-solid interaction.

8.1 Derivation from the Master Equation

To derive the Kramers equation (8.1) from the master equation (3.46) for physisorption, we follow the approach by Gortel et al. (1981). It is advantageous to rewrite the sum over bound states as an integral over energy E such that $E=E_i$ at the bound states. Gortel et al.(1981) choose a dimensionless variable $\varepsilon = E/\hbar\omega_D$; Jack and Kreuzer (1985) have recently argued that the derivation is yet more transparent if one uses a variable $e=E/V_0$ instead, where V_0 is the depth of the surface potential. The master equation then reads

$$\frac{\partial n(e,t)}{\partial t} = \int_{-1}^{\infty} de' \, \rho(e') [\, W(e,e')n(e',t) - W(e',e)n(e,t) \,] \quad . \tag{8.8}$$

The density of states is given for negative energies by

$$\rho(e) = \frac{di}{de} \tag{8.9}$$

and for positive energies by

$$\rho(e) = \frac{dk}{de} \tag{8.10}$$

where i and k label bound states and continuum states, respectively. For a Morse potential we set

$$\rho(e) = 1/\sqrt{|e|} \tag{8.11}$$

absorbing a factor $\sigma_0/2$ into the definition of $W(e,e')$. To rewrite the various transition probabilities such as (4.22-26) in the energy variable e is a straightforward matter.

Van Kampen (1961, 1981) has developed a systematic expansion of the master equation in terms of the inverse of a large parameter such that the lowest two terms yield a Fokker-Planck equation. This procedure seems to cause some inconsistencies if the independent variable is bounded as it is in our case: $-1 < e < \infty$. It is then more convenient to start from the equivalent Chapman - Kolmogorov equation (3.11) (Chandrasekhar 1943; Kac and Logan 1979). We multiply the latter with the density of states $\rho(e)$ and a test function $\tau(e)$ which is infinitely differentiable in the interval $-1 < e < \infty$

and vanishes at its endpoints. After integration we get

$$\int_{-1}^{\infty} \tau(e)\rho(e)P_1|_1(e,t;e_1,t_1)de$$

$$= \int_{-1}^{\infty}\int_{-1}^{\infty} \tau(e)\rho(e)P_1|_1(e,t;e_2,t_2) \; P_1|_1(e_2,t_2;e_1,t_1)dede_2 \qquad (8.12)$$

On the right-hand side we now expand

$$\tau(e) = \sum_{n=0}^{\infty} \frac{1}{n!} (e-e_2)^n \tau^{(n)}(e_2) \quad , \qquad (8.13)$$

where $\tau^{(n)}$ is the n-th derivative of $\tau(e)$. We can then use (3.13,15), suitably modified with density of states factors, to rewrite the right-hand side of (8.12) for small times $\Delta t=t-t_2$.

$$\sum_{n=0}^{\infty} \frac{1}{n!} \int_{-1}^{\infty}\int_{-1}^{\infty} de \, de_2 \; \rho(e)(e-e_2)^n \, P_1|_1(e,t;e_2,t_2)\tau^{(n)}(e_2) \, P_1|_1(e_2,t_2;e_1,t_1)$$

$$= \int_{-1}^{\infty} de \, \rho(e)\tau(e)P_1|_1(e,t_2;e_1,t_1)$$

$$+ \Delta t \sum_{n=1}^{\infty} \frac{1}{n!} \int_{-1}^{\infty} de_2\bar{\alpha}_n(e_2) \, P_1|_1(e_2,t_2;e_1,t_1)\tau^{(n)}(e_2) \quad . \qquad (8.14)$$

Here we defined the moments for $n\geq 1$,

$$\bar{\alpha}_n(e_2) = \rho(e_2)\int_{-1}^{\infty} de \, \rho(e)(e-e_2)^n \, W(e,e_2) \qquad (8.15)$$

The n-th term in (8.14) after n partial integrations gives

$$\Delta t \sum_{n=1}^{\infty} \frac{(-1)^n}{n!} \int_{-1}^{\infty} de \, \tau(e) \frac{\partial^n}{\partial e^n} [\bar{\alpha}_n(e)P_1|_1(e,t+\Delta t;e_1,t_1)] \quad . \qquad (8.16)$$

The first term in (8.14) is next taken to the left-hand side of (8.12). After division by Δt we take the limit $\Delta t\to 0$. Next we multiply by $P_1(e_1,t_1)=n(e_1,t_1)$ and integrate over e_1 to obtain, after using (3.12),

$$\int_{-1}^{\infty} de \, \tau(e) \left[\frac{\partial}{\partial t} [\rho(e)n(e,t)] - \sum_{n=1}^{\infty} \frac{(-1)^n}{n!} \frac{\partial^n}{\partial e^n} [\bar{\alpha}_n(e)n(e,t)] \right] = 0 \quad . \qquad (8.17)$$

As this relation must hold for any nonzero test function $\tau(e)$, we conclude that the expression in braces must vanish. Thus, we have arrived at an equivalent form of the master equation, usually referred to as the Kramers-Moyal expansion,

$$\frac{\partial}{\partial t}\,[\rho(e)n(e,t)] = \sum_{n=1}^{\infty} \frac{(-1)^n}{n!} \frac{\partial^n}{\partial e^n}\,[\tilde{\alpha}_n(e)n(e,t)]$$

$$= \sum_{n=1}^{\infty} \frac{(-1)^n}{n!} \frac{\partial^n}{\partial e^n}\,[\alpha_n(e)\rho(e)n(e,t)] \quad , \quad \text{where} \tag{8.18}$$

$$\tilde{\alpha}_n(e) = \rho(e)\,\alpha_n(e) \quad . \tag{8.19}$$

The hope is to approximate (8.18) by the first two terms

$$\frac{\partial \rho(e)n(e,t)}{\partial t} = -\frac{\partial}{\partial e}\,[\alpha_1(e)\rho(e)n(e,t)] + \frac{1}{2}\frac{\partial^2}{\partial e^2}\,[\alpha_2(e)\rho(e)n(e,t)] \quad . \tag{8.20}$$

The validity of (8.20) obviously hinges on (i) how fast the α_n decrease as a function of n, and (ii) how smooth they are. In Figs. 8.1,2, we reproduce α_1 and α_2 for negative energies for the Xe-W system with the transition probabilities given in the one-phonon approximation (4.22-26). At lower temperatures α_2 is substantially smaller than α_1.

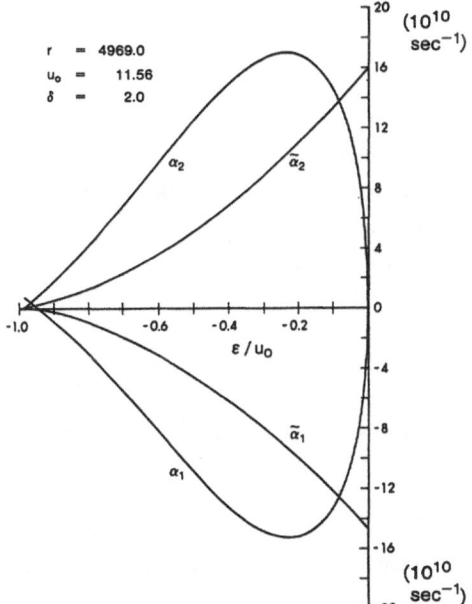

Fig.8.1. Moments $\tilde{\alpha}_n(e)$ and $\alpha_n(e)$ from (8.15,19) for the Xe-W system. Potential parameters $u_0=V_0/\hbar\omega_D$, $r=2m\omega_D/\hbar\Upsilon^2$; $\delta=\hbar\omega_D/k_BT$. Debye temperature $T_D=405$ K. (Gortel et al. 1981)

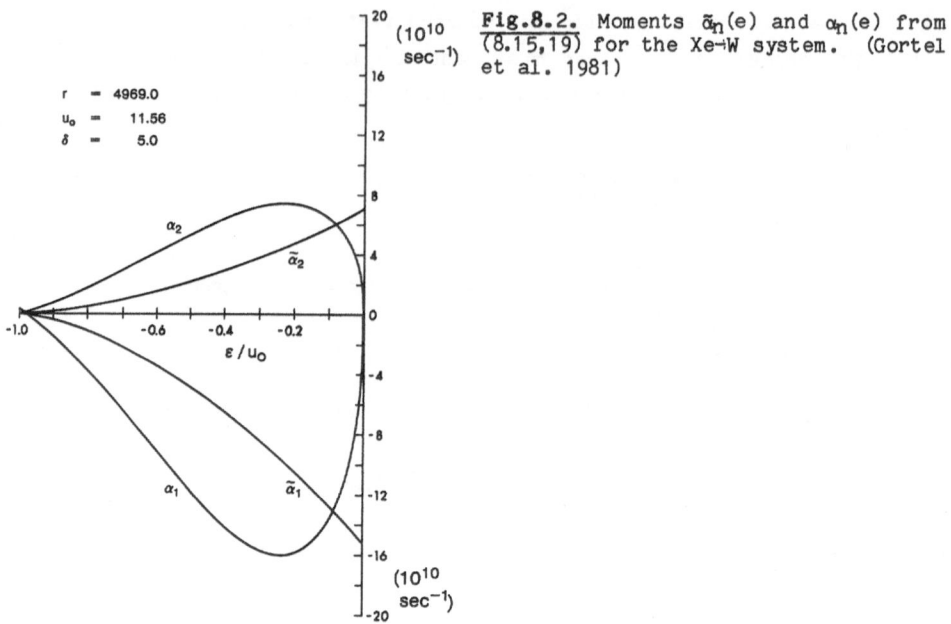

Fig.8.2. Moments $\tilde{\alpha}_n(e)$ and $\alpha_n(e)$ from (8.15,19) for the Xe→W system. (Gortel et al. 1981)

The connection between the moments defined here as a function of the variable e, and those defined by Gortel et al. (1981) in terms of $\varepsilon = eu_0$, where $u_0 = V_0/\hbar\omega_D$, is given by

$$\alpha_n(\varepsilon) = \alpha_n(e)u_0^n$$

$$\tilde{\alpha}_n(\varepsilon) = \tilde{\alpha}_n(e)u_0^{n-1/2} \quad . \tag{8.21}$$

The powers of u_0 guarantee a faster convergence for the Kramers-Moyal expansion. We should also comment on the preference of working with the nonsymmetric kernel $W(e',e)$ instead of starting from the symmetrized version (5.24) (Pagni and Keck 1973). For the latter we would find

$$\frac{\partial}{\partial t}[\rho(e)\chi(e,t)] = \sum_{n=1}^{\infty} \frac{(-1)^n}{n!} \frac{\partial^n}{\partial e^n}[\Delta_n(e)\rho(e)\chi(e,t)] \quad , \quad \text{where} \tag{8.22}$$

$$\Delta_n(e) = \exp[-E/2k_BT]\int_{-1}^{\infty} de' \, \rho(e')(e'-e)^n \, W(e',e)\exp[E'/2k_BT] \quad . \tag{8.23}$$

This symmetrization changes nothing in the physics. However, it turns out that for most systems the exponential under the integral in (8.23) can be neglected, so that

$$\Delta_n(e) = \alpha_n(e)\exp[-E/2k_BT] \quad . \tag{8.24}$$

Because α_n is only weakly dependent on e as compared to the exponential factor, the behavior of Δ_n will be dominated (rather artificially) by the latter. This might have led to some erroneous conclusions about the relative worths of the diffusion and the iteration approximations.

The smooth shape of the moments α_n suggests that there should be simple approximations, obtainable within the one-phonon approximation, to the transition probabilities as given by (4.22-26). Starting from an asymptotic expansion for $\ln\Gamma(z)$ (Bateman 1953), one can derive an expansion

$$\frac{\Gamma(x+1+a)}{\Gamma(x+1)} = x^a \exp\left[\sum_{n=2}^{\infty} \frac{B_n(-a) - B_n(0)}{n(n-1)x^{n-1}}\right] \quad , \tag{8.25}$$

where $B_n(y)$ are Bernoulli polynomials (Gradshteyn and Ryzhik 1965). Likewise, one expands the factor in (4.25)

$$\frac{\sinh\left[2\pi\sqrt{ru_0e}\right]}{\sinh^2\left[\pi\sqrt{ru_0e}\right] + \cos^2\left[\pi\sqrt{ru_0}\right]}$$

$$= 2 + 4\sum_{n=1}^{\infty} (-1)^n \cos\left[2\pi n\sqrt{ru_0}\right]\exp\left[-2\pi n\sqrt{ru_0e}\right] \quad . \tag{8.26}$$

Up to terms of order σ_0^{-4}, one then finds for $-1 < e < 0$ (Gortel et al. 1981; Jack and Kreuzer 1985)

$$\alpha_1(e) = 5\eta_0 \frac{k_BT}{V_0} \frac{1}{\sqrt{-e}} \left[\frac{1}{1+e} - \frac{\tanh^{-1}(\sqrt{-e})}{\sqrt{-e}}\right] F_6\left[\frac{\tanh^{-1}(\sqrt{-e})}{\sqrt{-e}}\right]$$

$$- 2\eta_0\sqrt{-e} \, F_5\left[\frac{\tanh^{-1}(\sqrt{-e})}{\sqrt{-e}}\right] + O(\sigma_0^{-6}) \quad \text{and} \tag{8.27}$$

$$\alpha_2(e) = 4\eta_0 \frac{k_BT}{V_0} \sqrt{-e} \, F_5\left[\frac{\tanh^{-1}(\sqrt{-e})}{\sqrt{-e}}\right] + O(\sigma_0^{-6}) \quad , \quad \text{where} \tag{8.28}$$

$$F_n(s) = s^{-n} \left[1 - \exp\left[-\sigma_0 \frac{\hbar\omega_D}{V_0} s\right] \sum_{m=0}^{n-1} \frac{s^m}{m!} \left[\sigma_0 \frac{\hbar\omega_D}{V_0}\right]^m\right] \quad . \tag{8.29}$$

Here we have introduced a friction coefficient

$$\eta_0 = \frac{9\pi V_0{}^2\gamma^4}{mM_S\omega_D{}^3}$$

$$= 36\pi\omega_D \ \frac{m}{M_S} \ \frac{u_0{}^4}{\sigma_0{}^4}$$

$$= \frac{9\pi}{4} \ \frac{m}{M_S} \left[\frac{\omega_0}{\omega_D}\right]^3 \omega_0 \tag{8.30}$$

in terms of the relevant parameters for the gas-solid system. Here ω_0 is the classical oscillation frequency (8.6) of an adparticle at the bottom of the surface potential. Equations (8.27,28) are valid for $\sigma_0 > V_0/k_BT \gg 1$. It thus seems appropriate to identify σ_0 as the large parameter to be used in van Kampen's expansion of the master equation (van Kampen 1961,1981).

To complete the derivation of the Kramers equation (8.1), it is expedient to rewrite it in terms of the action variable

$$J = \oint \ pdr \tag{8.31}$$

In situations of low friction as defined by (8.4), the particle orbits in phase space are nearly closed for negative energies, so that (8.1) can be averaged over one period to yield (Kramers,1940)

$$\frac{\partial \bar{n}}{\partial t} = \eta \ \frac{\partial}{\partial J} \left[J\bar{n} + \frac{k_BTJ}{\nu} \ \frac{\partial \bar{n}}{\partial J}\right] \tag{8.32}$$

where $\bar{n}$ is the occupation function $n(e,t)$ averaged over one period and

$$\nu = \frac{\partial E}{\partial J} \tag{8.33}$$

is the frequency of oscillation. For a Morse potential, the action can be given explicitly as

$$J = \frac{2\pi}{\gamma} \left[\sqrt{2mV_0} - \sqrt{2mE}\right]$$

$$= h\sigma_0 \left[1 - \sqrt{-e}\right] \quad , \tag{8.34}$$

with $J=0$ at the bottom of the well and $J_0=h\sigma_0$ at the top. The frequency of oscillation is then

$$\nu = \frac{\gamma}{\pi} \sqrt{\frac{V_0}{2m}} \ |e| \quad , \tag{8.35}$$

being equal to $\omega_0/2\pi$ at the bottom of the potential well. Introducing a dimensionless action variable

266

$$x = \frac{J}{\hbar \sigma_0} = 1 - \sqrt{-e} \quad , \tag{8.36}$$

in terms of which the frequency becomes

$$\nu = \frac{\omega_0}{2\pi} (1 - x) \quad , \tag{8.37}$$

we can rewrite the Kramers equation (8.32) as

$$\frac{\partial n(x,t)}{\partial t} = \eta \, \frac{\partial}{\partial x} \left[\frac{k_B T}{2V_0} \, \frac{x}{1-x} \, \frac{\partial n(x,t)}{\partial x} + xn(x,t) \right] \quad . \tag{8.38}$$

Returning to (8.27,28), we keep only terms to order σ_0^{-4}, neglect the exponential term in $F_n(s)$, (8.29), and insert them into (8.20). We obtain a generalized Kramers equation

$$\frac{\partial n(x,t)}{\partial t} = \frac{\partial}{\partial x} \left[\eta(x) \left[\frac{k_B T}{2V_0} \, \frac{x}{1-x} \, \frac{\partial n(x,t)}{\partial x} + xn(x,t) \right] \right] \quad , \tag{8.39}$$

where we introduced a friction function

$$\eta(x) = \eta_0 \, \frac{1}{x} \, F_s \left[\frac{\tanh^{-1}(1-x)}{1-x} \right] \quad , \tag{8.40}$$

with η_0 given in (8.30).

It is important to note that (8.39) has the correct equilibrium solution, namely

$$n^{eq}(x) = c \, \exp \left[\frac{V_0}{k_B T} \, (1-x)^2 \right] = c \, \exp \left[-\frac{eV_0}{k_B T} \right] \quad . \tag{8.41}$$

This is not the case for (8.20), i.e., before the above expansion in terms of σ_0^{-1} is done, because then the equilibrium condition

$$\frac{2\alpha_1(e)}{\alpha_2(e)} = -\frac{V_0}{k_B T} + \frac{d\ln(\rho \alpha_2)}{de} \tag{8.42}$$

is not satisfied, particularly not at the bottom of the surface potential well. This is, in part, a reflection of the fact that the Kramers-Moyal expansion breaks down within a fluctuation of the end points of the range of the independent variable E. Also note that the moments α_n may change sign in this region. An attempt to tackle this problem has been made by Pagni (1973). Indeed, to keep only the lowest terms in (8.25) requires that $x \ll a$; this in turn imposes the condition

$$1 - \sqrt{-e} \gg \frac{1}{2\sigma_0} \quad , \tag{8.43}$$

implying that the Kramers equation can only be acceptable to within a fluctuation of the bottom of the surface potential.

To reduce the friction function (8.40) to a constant, we note that for

$$\frac{\sigma_0}{u_0} = \frac{2\omega_D}{\omega_0} \gg 1 \tag{8.44}$$

the exponential term in (8.29) can be neglected. Also the power law can be approximated as a linear function of x, so that the friction function becomes

$$\eta(x) \approx \eta \approx a \, \eta_0 \tag{8.45}$$

where the constant a can be determined, e.g., by a Legendre fit, to be $a \approx 1.15$ (Jack and Kreuzer 1985). This concludes the derivation of the Kramers equation for physisorption.

It is instructive to interpret the various conditions that had to be imposed in the course of deriving the Kramers equation from the master equation. The inequality (8.44) suggests that the time scale of the phonon bath, as measured by the Debye frequency ω_D, must be much faster than that of the particle motion as measured by the classical oscillation frequency ω_0. Moreover, because the energy difference between the lowest two energy levels in a Morse potential is proportional to ω_0, we see that the energy difference between adjacent energy levels must be much less than the maximum energy of the phonons that actually facilitate transitions between them; this is, of course, also necessary for the one-phonon approximation to be employed. Turning next to the condition (8.43) we rewrite it as

$$V_0 + E \gg \tfrac{1}{2} \hbar\omega_0 \tag{8.46}$$

implying that the kinetic energy of an adsorbed particle must be large compared to its zero-point energy in the surface potential. In terms of the action variable (8.34), this reads

$$J \gg \tfrac{1}{2} h \qquad \text{or} \tag{8.47}$$

$$x = \frac{J}{J_0} \gg \frac{1}{2\sigma_0} \quad . \tag{8.48}$$

The first of these relations clearly states that all the manipulations aimed at deriving the Kramers equation were appropriate to extract the classical limit as well. It is therefore not surprising that Jack and Kreuzer (1982) obtained the same expression for the friction coefficient, namely (8.30), as resulted from a purely classical approach advanced by Caroli et al. (1978). One may therefore also conclude that for weakly coupled gas-solid systems for which σ_0 is large, an incoherent sequence of one-phonon processes (rather than coherent multi-phonon processes) suffices to extract the classical limit from a kinetic theory based on quantum mechanics. The reason, presumably, is simply that for weakly coupled systems, coherent multi-phonon processes are negligible.

The condition (8.48) again implies that σ_0 be large. This, in turn, produces a large number of bound states in the surface potential, a necessary feature to take the continuous limit (8.8) in the master equation.

To describe desorption with the Kramers equation (8.1), we have to impose radiative boundary conditions at the top of the surface potential well to account for the loss of particles into the gas-phase continuum. This can be done simply in the weak coupling limit (8.38) with an ansatz

$$n(x,t) = [n^{(eq)}(x) - 1] \, e^{-t/t_d} \quad, \tag{8.49}$$

where (8.41) is used. Inserted into (8.38) and integrated over x, this yields immediately (8.5). Jack and Kreuzer (1985) go beyond this approximation and show that starting from (8.39) and using the systematic expansions (8.27,28) for the moments, one gets correction factors to (8.5), namely

$$t_d = \tfrac{1}{2} \, n_0^{-1} \, \frac{k_B T}{V_0} \, e^{V_0/k_B T} \, [1 - e^{-\sigma_0 \hbar \omega_D/V_0}$$

$$*[1 + \sigma_0 \frac{\hbar \omega_D}{V_0} + \frac{1}{2} \left[\sigma_0 \frac{\hbar \omega_D}{V_0} \right]^2$$

$$+ \frac{1}{6} \left[\sigma_0 \frac{\hbar \omega_D}{V_0} \right]^3 + \frac{1}{24} \left[\sigma_0 \frac{\hbar \omega_D}{V_0} \right]^4]] \quad. \tag{8.50}$$

This expression leads to good agreement with the exact desorption times calculated directly from the master equation, even for shallow surface potentials with only a few bound states.

8.2 Macroscopic Laws

To establish a macroscopic law of friction for an adsorbed particle, we first derive an equation for its total energy (Jack and Kreuzer 1982)

$$\langle e \rangle = \int_{-1}^{0} e \rho(e) \, n(e,t) \, de \tag{8.51}$$

by multiplying (8.8) by e and integrating. Neglecting adsorption and desorption we get

$$\frac{d\langle e \rangle}{dt} = \langle \alpha_1(e) \rangle \quad. \tag{8.52}$$

Putting a particle into a state e<0, we can use this equation to follow its journey down to the bottom of the potential well. Let us assume that the mean square deviations

$$\sigma^2(\langle e \rangle) = \langle e^2 \rangle - \langle e \rangle^2 \qquad (8.53)$$

remain small. They satisfy an equation

$$\frac{d\sigma^2}{dt} = \langle \alpha_2(e) \rangle + 2[\langle e\alpha_1(e) \rangle - \langle e \rangle \langle \alpha_1(e) \rangle] \ . \qquad (8.54)$$

Expanding the moments around $\langle e \rangle$, we get (van Kampen 1961, 1981)

$$\frac{d\langle e \rangle}{dt} = \alpha_1(\langle e \rangle) + \tfrac{1}{2} \alpha_1''(\langle e \rangle) \ \sigma^2$$

$$\frac{d\sigma^2}{dt} = \alpha_2(\langle e \rangle) + 2\alpha_1'(\langle e \rangle) \ \sigma^2 + \tfrac{1}{2} \alpha_2''(\langle e \rangle)\sigma^2 \ , \qquad (8.55)$$

where the primes indicate differentiation with respect to $\langle e \rangle$. With the moments approximated by (8.27,28) (and after some further approximations), we get in the variables $x = 1 - \sqrt{-\langle e \rangle}$ and $y = \sigma^2$,

$$\frac{dx}{dt} = -\eta_0 \left[x - \frac{k_B T}{2V_0} \right] + \frac{\eta_0}{8} \left[1 - \frac{k_B T}{2V_0} \right] \frac{y}{(1-x)^4}$$

$$\frac{dy}{dt} = 4\eta_0 \frac{k_B T}{V_0} x(1-x) \ -2\eta_0 \left[1-2x + \frac{k_B T}{2V_0} \right] \frac{y}{1-x} \ . \qquad (8.56)$$

These are the macroscopic equations, i.e., the friction law governing the descent of an adsorbed particle to the bottom of the surface potential well (Jack and Kreuzer 1982). Note that the time scale is set by the friction coefficient η_0. Equations (8.56) have an equilibrium solution for $dx/dt = dy/dt = 0$ to lowest order in the fluctuations $k_B T/V_0$,

$$x(eq) \approx \frac{k_B T}{2V_0}$$

$$e(eq) \approx -1 + \frac{k_B T}{V_0}$$

$$\sigma^2 \approx \left[\frac{k_B T}{V_0} \right]^2 \ , \qquad (8.57)$$

in agreement with a Maxwell-Boltzmann distribution. Jack and Kreuzer (1982) have analyzed the stability of this equilibrium solution by establishing the domain of initial conditions $x(t=0)$ and $y(t=0)$ for which the solutions of (8.56) approach (8.57) as $t \to \infty$. A sufficient condition (van Kampen, 1961, 1981) for this to occur is that $\alpha_1' < 0$ or $x < 0.5$. However, note from the second equation of (8.56) that the fluctuations $y=\sigma^2$ always remain positive for all $0 < x < 1$ as long as $y(t=0) > 0$. We also note that (8.56) has a second stationary solution in addition to (8.57) which, however, is not stable but a saddle point. The domain of attraction for the stable solution (8.57) is bounded by a solution of

$$\frac{dy}{dx} = \frac{dy/dt}{dx/dt} \tag{8.58}$$

that starts at the saddle point and has an initial slope there that is equal to the right-hand side, evaluated from (8.56) using l'Hospital's rule. In Fig. 8.3 are plotted the curves $dx/dt=0$ and $dy/dt=0$ in the $(z,\log_{10}y)$-plane, together with the solution of (8.58). The latter intersects $y=0$ for $k_BT/V_0\approx0.04$ at $x_m = 0.86737$ and for 0.09 at $x_m = 0.81125$, both considerably higher than $x_m = 0.5$. In Fig. 8.3, we also reproduce four solutions of (8.56) starting from initial data with zero fluctuations $y(t=0)=0$. It should be obvious that the solution starting at precisely $x(t=0) = x_m$ is unstable. Note also that the boundary of the domain of attraction obtained by integrating (8.58), starting from the saddle point, is the locus of those points (x,y) that lie roughly within a thermal fluctuation of the line $dx/dt=0$.

An analytic solution of (8.56) can be found for $y(t=0) = 0$ and $x(t=0)\ll1$ after linearization which gives

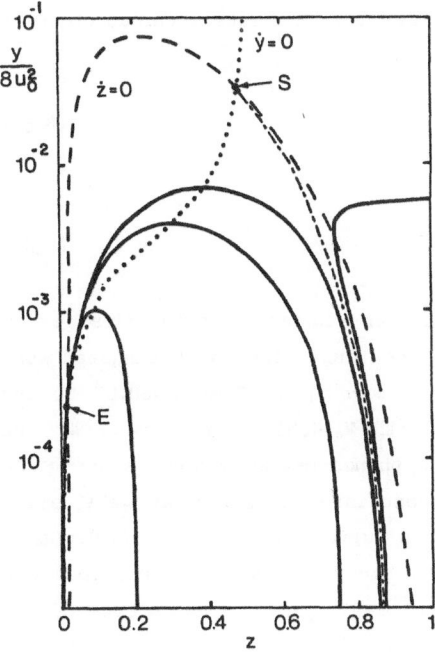

Fig. 8.3. Analysis of (8.56): $\dot{x}<0$ below the dashed curve $\dot{x}=0$; $\dot{y}>0$ to the right of the dotted curve $\dot{y}=0$. E and S indicate the equilibrium and saddle points with $\dot{x}=\dot{y}=0$. The dash-dotted curve through S is a solution of (8.58) limiting the domain of attraction of E. The remaining four curves are trajectories calculated from (8.56). (Jack and Kreuzer 1982)

$$\frac{dx}{dt} = -n_0\left[x - \frac{k_BT}{2V_0}\right]$$

$$\frac{dy}{dt} = 4n_0 \frac{k_BT}{V_0} x - 2n_0y \quad , \tag{8.59}$$

so that

$$x(t) = \left[x(0) - \frac{k_BT}{2V_0}\right]e^{-n_0t} + \frac{k_BT}{2V_0}$$

$$y(t) = \left[\frac{k_BT}{V_0}\right]^2 [1 - e^{-2n_0t}] + 4\frac{k_BT}{V_0}\left[x(0) - \frac{k_BT}{2V_0}\right][e^{-n_0t} - e^{-2n_0t}] \quad . \tag{8.60}$$

Note that $y(t)$ has a maximum when

$$n_0t = \ln 2 + \ln\left[1 + \frac{k_BT}{4V_0} \frac{1}{x(0) - k_BT/2V_0}\right] \quad . \tag{8.61}$$

Moreover,

$$y(t \to \infty) = \left[\frac{k_BT}{V_0}\right]^2$$

$$x(t \to \infty) = \frac{k_BT}{2V_0} \tag{8.62}$$

in accordance with the equilibrium solution (8.57).

Before we proceed to test (8.56), we discuss some data obtained directly by calculating the average energy per particle

$$\langle e \rangle = \sum_i e_i n_i(t) / \sum_i n_i(t) \tag{8.63}$$

and the fluctuations

$$\sigma^2 = \sum_i [e_i - \langle e \rangle]^2 n_i(t) / \sum_i n_i(t) \quad , \tag{8.64}$$

with $n_i(t)$ obtained by diagonalizing the master equation. In the first example, we put a xenon atom initially into the bound state on a tungsten surface at an energy $\langle e(t=0) \rangle = -0.65$ which corresponds for $\sigma_0=80.07$ to the
16th bound state in a Morse potential with $V_0=4682$ K and $\gamma^{-1}=0.5$ Å. We
choose a temperature $T = 202$ K. Owing to phonon-mediated bound state-bound
state transitions, the particle will descend into the potential well, emitting non-equilibrium phonons into the solid (upper panel in Fig. 8.4). Large
nonequilibrium fluctuations build up that decay to thermal ones, once the
particle has reached its equilibrium energy $e \approx -1 + k_BT/V_0$ at the bottom of the

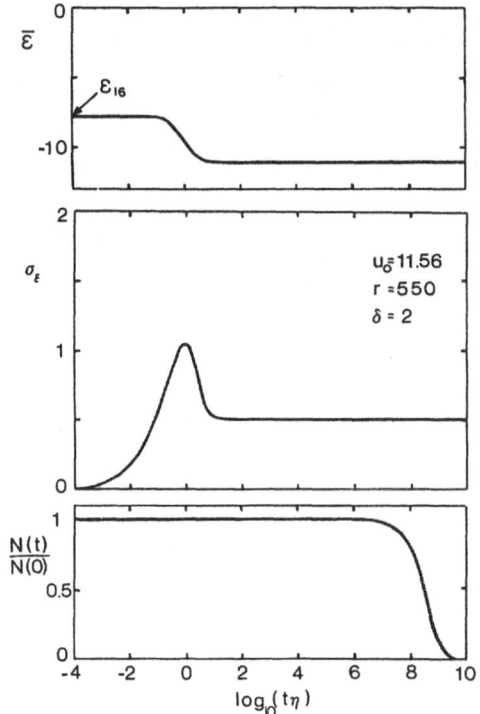

Fig. 8.4. Time evolution of a particle put into the surface potential well. Upper panel: average energy (8.62); center panel: fluctuations (8.63) as a function of $\log_{10}(\eta_0 t)$. The curves are exact results from the master equation, and also from the macroscopic laws (8.56). The lower panel represents the normalized occupation probability (5.5) from the master equation with the desorption process taken into account. Other parameters (typical for Xe-W): $m/M_S=0.714$, $\omega_D=5.3 \times 10^{13}$ s^{-1} so that $\eta_0=1.89 \times 10^{12}$ s^{-1}. (Jack and Kreuzer 1982)

potential well after a relaxation time $\eta_0^{-1} = 5.3 \times 10^{-13}$ s. Its further fate depends on the boundary conditions. With a gas phase present in front of the solid or with bound state-continuum and continuum-bound state transitions switched off, the particle has reached its final state. However, under conditions of thermal desorption, i.e. with the gas phase removed and phonon-mediated bound state-continuum transitions operative, the particle will eventually desorb, which happens after a desorption time t_d, which at T=202 K is about eight orders of magnitude larger than η_0^{-1}, the internal relaxation time. This remarkable separation of time scales can be seen in the lower panel of Fig. 8.4. We note parenthetically that, although the distribution function $n_i(t)$ is for $t \gg \eta_0^{-1}$ very nearly a Maxwell-Boltzmann distribution near the bottom of the well, this is not the case within a few Debye energies from the top of the well from where desorption takes place. Returning now to the macroscopic equations (8.56), we simply note that with initial conditions e(t=0)=-0.65, i.e., x(t=0)=0.2 and y(t=0)=σ^2(t=0)=0, the solutions of (8.56) agree within the accuracy of the graph with the "exact data".

In Fig. 8.5 we present data on a situation where a particle is initially put into a bound state at an energy e = -0.065, corresponding to the 60th

273

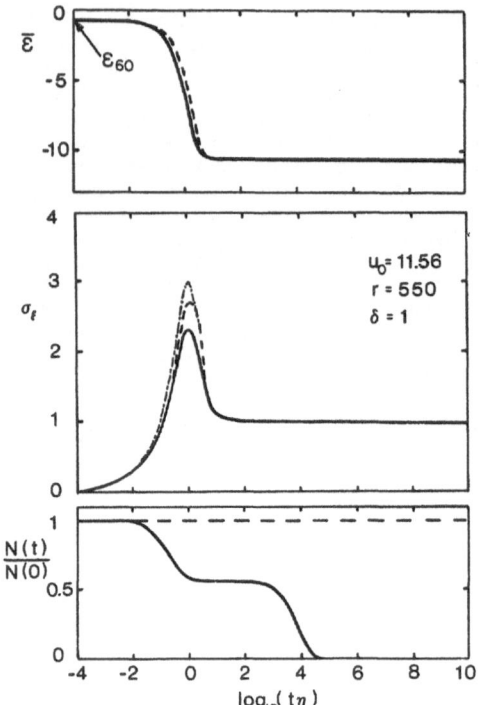

Fig. 8.5. Same as Fig.8.4, but for different initial conditions. In addition, dashed lines result from the master equation with bound state-continuum transitions suppressed. Dashed-dotted lines from macroscopic laws (8.56). (Jack and Kreuzer 1982)

energy level in the surface potential well, with the temperature such that $T = T_D = 405$ K. We start with the "exact" results from the master equation and look first at a situation where desorption is possible. Again the particle will, within a time of order η_0^{-1}, drift to the bottom of the potential well with substantial fluctuations building up. However, the probability for the particle to remain adsorbed drops from one to about 0.56, with final desorption taking place at a time about $10^4 \eta_0^{-1}$ later. If the desorption process is suppressed by dropping the bound state-continuum transitions, larger fluctuations can build up (dashed lines in Fig. 8.5) that would, in the previous situation, lead to desorption. The macroscopic equations (8.56) yield somewhat larger fluctuations (dash-dotted line in Fig. 8.5), but still reproduce the exact data rather well. To see the general trend, we plot in Fig. 8.6 the maximum of σ versus the initial bound-state energy $e(t=0)$ of the particle. The macroscopic equations (8.56) reproduce the exact results from the master equation (given for $\gamma^{-1} = 0.5$ Å and 1.0 Å) very well up to about $e(t=0) \approx -0.1$. For larger $e(t=0)$, marked deviations start as one approaches the stability limit $x_m = 0.811$ or $e(t=0) \approx -0.036$ for (8.56).

Finally, in Fig. 8.7, we give two perspective views of the distribution function $n_i(t)$, obtained from the master equation, plotted first over the

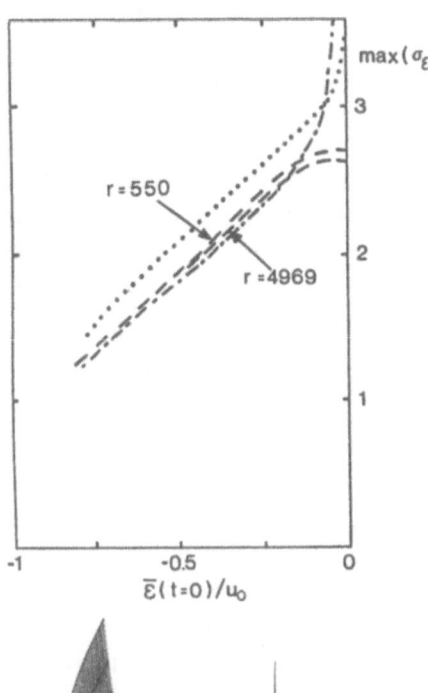

Fig. 8.6. Maximum in the nonequilibrium energy fluctuations σ for a particle that is initially put into a bound state at energy e(t=0). Dashed lines from the master equation with bound state-continuum transitions suppressed. Dashed-dotted line from the macroscopic laws (8.56). Dotted line from van Kampen's equation as discussed in the reference. (Jack and Kreuzer 1982)

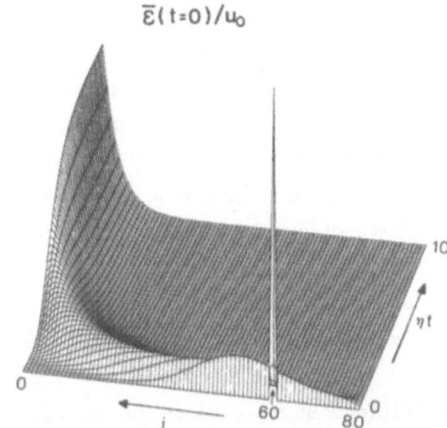 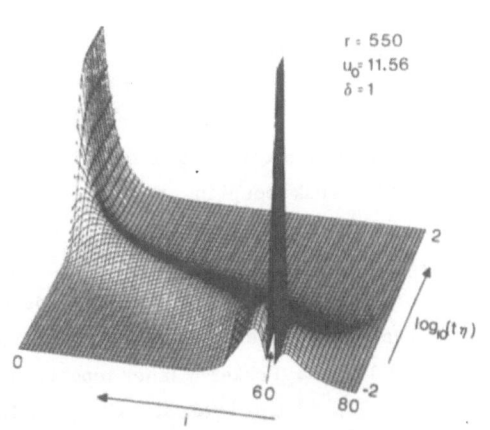

Fig. 8.7. Two views of the time evolution of the distribution function $n_i(t)$ after a particle is put into the i=60 bound state. Along the i axis we plot the index of the bound state, not its energy. The log scale enhances the initial evolution. (Jack and Kreuzer 1982)

$(i, n_0 t)$ plane and then over the $(i, \log_{10} n_0 t)$ plane, the latter to enhance the initial evolution for which a highly non-Gaussian distribution develops. A particle put into the i=60 state will initially undergo transitions into higher bound states and, to a greater extent, into lower bound states, the latter eventually overwhelming, leading to the overall drift to the bottom of the potential well accompanied by large transient nonequilibrium fluctuations. The latter are bigger for higher temperatures and deeper down in the potential well.

8.3 Friction Coefficient

In deriving the weak coupling limit (8.32 or 38) of the Kramers equation from the master equation, we obtained a microscopic expression (8.30) for the friction coefficient η_0. The same coefficient also appears in the Langevin approach to Brownian motion. The equivalent theory for physisorption kinetics has been worked out by Caroli et al. (1978). Starting from the classical Liouville equation for the coupled gas-solid system, they derive in the heavy adatom limit, i.e., for $m/M_S \gg 1$, a Langevin equation governing the time evolution of the momentum of the adsorbing particle, namely

$$\frac{d}{dt}\ p = -\frac{\partial}{\partial r}\ V_S(r) + F(r,t) - p(t)\cdot\hat{\eta}_0(r) \quad , \quad \text{where} \tag{8.65}$$

$$\hat{\eta}_0(r) = \frac{1}{k_B T m}\int_0^\infty d\tau\ \langle F(r,0)F(r,\tau)\rangle \tag{8.66}$$

is a space dependent friction tensor to be calculated from the correlations of the random forces acting on the adsorbing particles due to the thermal motion of the solid. Here, V_S is, as usual, the average surface potential. Employing a series of approximations, such as a linear chain of atoms interacting via equivalent local harmonic potentials, Caroli et al (1978) derive, in the weak coupling regime at low temperatures, an expression for η_0 that is identical to (8.30) up to a factor 9π. Some discussion is presented anticipating possible difficulties in going beyond the weak coupling limit.

D'Agliano et al. (1975) have developed a Brownian motion model of the interaction between adsorbing particles and metal electrons. They base their model on a Fokker-Planck equation [derived by Schaich (1974) in the large mass limit] and evaluate the friction coefficient using a bootstrap approach that relies heavily on linear response theory. Although this is a very promising approach to chemisorption kinetics, it seems of little importance for the physisorption case where electronic degrees of freedom are far less effective than phonons (see, e.g., Schönhammer and Gunnarson 1982, 1983a,b). D'Agliano et al (1975) have applied their method to the latter situation as well. They start from the work of Pagni and Keck (1973) who also approximated their master equation by a diffusion equation which is nothing but the stationary part of the Fokker-Planck equation. D'Agliano et al. argue that the moment Δ_2 given in (8.23) in terms of the symmetrized kernel of the master equation, determines the friction coefficient as

$$\eta_0 = \frac{\Delta_2(E)}{2k_B T}\ E\ n^{eq}(E) \quad . \tag{8.67}$$

They further argue that for a light adparticle such as hydrogen, this can be explicitly evaluated and find as an estimate for the friction coefficient

$$\eta_0 \approx \frac{4}{\pi} \omega_0 \frac{m}{M_S} [1 + O(m/M_S) + ...] \quad , \qquad (8.68)$$

which differs from (8.30) by the factor $9\pi^2/16(\omega_0/\omega_D)^3$, which, by virtue of (8.44), must be small. The fact that no characteristic frequency of the solid appears in (8.68), is a consequence of the light adatom limit being employed, for which the lattice is essentially rigid.

8.4 Hydrodynamics of Adsorption

The description of a macroscopic system such as a gas interacting with a solid, can be attempted at several levels of complexity, each designed to provide a certain amount of detail of its time evolution. Most encompassing is, of course, a description in terms of the microscopic dynamics of the interacting many-body system; a starting point would be the classical Liouville equation for the many-body distribution function, or the quantum-mechanical Liouville-von Neumann equation for the density matrix. Such details are rarely needed or comprehensible, so that one usually proceeds to coarse-grain the time evolution and the spatial structure over time and length scales that reflect the underlying collision processes, ultimately leading to thermal equilibrium. The results are approximate kinetic equations like the master equation, the Fokker-Planck equation, or the Langevin equation for single particle distribution functions, and perhaps similar ones for low-order correlation functions. If the processes studied are slow on the kinetic time scale (e.g., given by the inverse of the friction coefficient η_0), it is usually advantageous to attempt a hydrodynamic description. Its ingredients are macroscopic densities, such as mass or particle number density, momentum density, energy density, each one subject to a differential conservation law. Such a set of balance equations for the above mechanical quantities can be rigorously derived from the Liouville equation (e.g., Kreuzer 1981). Continuing our discussion of kinetic equations, we prefer here to start such a derivation from the Kramers equation (8.1).

We again concentrate on a highly mobile adsorbate for which (i) the surface force F and (ii) the friction terms on the right hand side of (8.1) effect only the motion of the adparticle normal to the surface. The Kramers equation then takes on the above one-dimensional form. The three-dimensional single-particle distribution function then factorizes into

$$f^{(3)}(\mathbf{r},\mathbf{p},t) = f(z,v,t) \exp[-P^2/2mk_BT] \quad , \tag{8.69}$$

where $v=p_z/m$ and P is the magnitude of the particle momentum parallel to the surface. We normalize f such that

$$\rho(z,t) = m \int_{-\infty}^{\infty} f(z,v,t)dv \tag{8.70}$$

is the mass density which, integrated over the volume of the gas phase, yields the total mass of gas particles

$$\int_{0}^{\infty} \rho(z,t)dz = N_g m \quad . \tag{8.71}$$

Likewise, the mass current density is given by

$$\begin{aligned} J_m(z,t) &= m \int_{-\infty}^{\infty} v\, f(z,v,t)dv \\ &= \rho(z,t)\, \bar{v}(z,t) \quad , \end{aligned} \tag{8.72}$$

where the second line defines the average velocity field $\bar{v}$. Integrating (8.1) over v, we immediately get the mass balance or continuity equation

$$\frac{\partial \rho}{\partial t} + \frac{\partial(\rho \bar{v})}{\partial z} = 0 \quad . \tag{8.73}$$

To obtain an equation for the mass current, we multiply (8.1) by mv and again integrate over v to obtain the momentum balance equation

$$\frac{\partial(\rho \bar{v})}{\partial t} + \frac{\partial}{\partial z} [\rho\, \bar{v}^2 + P] = \frac{\rho}{m} F(z) - n_0 \rho \bar{v} \quad , \tag{8.74}$$

where we introduced the pressure

$$\begin{aligned} P(z,t) &= m \int_{-\infty}^{\infty} (v - \bar{v})^2\, f(z,v,t)\, dv \\ &= 2\, \rho\, u(z,t) \quad , \end{aligned} \tag{8.75}$$

which is twice the internal energy. With the neglect of two-body interac-tions between the gas particles, u is entirely given by its kinetic energy part. The balance equation for the internal energy is obtained by multiply-ing (8.1) by $m(v-\bar{v})^2/2$ and integrating over v; it reads

$$\frac{\partial(\rho u)}{\partial t} + \frac{\partial}{\partial z} [\rho u \bar{v} + J_q] = -P\, \frac{\partial \bar{v}}{\partial z} - n_0\, P + n_0\, \frac{k_B T}{m}\, \rho \quad . \tag{8.76}$$

The internal energy changes locally because there is a convective inflow
278

$\rho u \tilde{v}$, and because of conduction via a heat current

$$j_q(z,t) = \frac{m}{2} \int_{-\infty}^{\infty} (v - \tilde{v})^3 f(z,v,t)dv \quad . \tag{8.77}$$

The right-hand side of (8.76) represents energy sinks due to the dissipative friction forces. We can obtain a balance equation for the kinetic energy density by multiplying (8.73) by $\tilde{v}$ and using the continuity equation (8.72); it reads

$$\frac{\partial}{\partial t} \left[\tfrac{1}{2} \rho \tilde{v}^2 \right] + \frac{\partial}{\partial z} \left[\tfrac{1}{2} \rho \tilde{v}^2 \tilde{v} + P\tilde{v} \right] = P \frac{\partial \tilde{v}}{\partial z} + \frac{F(z)}{m} \rho \tilde{v} - \eta_0 \rho \tilde{v}^2 \quad . \tag{8.78}$$

For the potential energy density, we get, via the continuity equation (8.73),

$$\frac{\partial}{\partial t} (\rho V_s) + \frac{\partial}{\partial z} (\rho V_s \tilde{v}) = -\rho \tilde{v}F \quad . \tag{8.79}$$

Adding (8.76,78,79), we get for the total energy density,

$$e = u + \tfrac{1}{2} \tilde{v}^2 + \frac{1}{m} V_s$$

$$\frac{\partial(\rho e)}{\partial t} + \frac{\partial}{\partial z} [\rho e \tilde{v} + P\tilde{v} + j_q] = - \eta_0 \left[P + \rho \tilde{v}^2 - \frac{\rho k_B T}{m} \right] \quad . \tag{8.80}$$

A thorough discussion of these balance equations for any macroscopic system, including their derivation from the Liouville equation, has been given by Kreuzer (1981).

Equations (8.73,74,80) have been the starting point for a hydrodynamic theory of thermal desorption developed by Brenig and Schönhammer (1976) and Müller and Brenig (1979). Following Kramers (1940), they study the limits of small and large friction. We assume that the time evolution of mass, momentum, and energy densities varies exponentially with a time constant equal to the desorption time. We deal with large friction if

$$\eta_0 \gg t_d^{-1} \quad . \tag{8.81}$$

In this situation, we can neglect the time derivative in (8.80), and get in the lowest order in η^{-1},

$$P + \rho \tilde{v}^2 \approx \frac{k_B T}{m} \rho - \frac{1}{\eta_0} \frac{\partial}{\partial z} [\rho e \tilde{v} + P\tilde{v} + j_q]$$

$$\approx \frac{k_B T}{m} \rho \quad , \tag{8.82}$$

which, inserted in the momentum balance (8.74), yields, after again dropping the time derivative, the Smoluchowski equation of Brownian motion

$$\frac{k_BT}{m} \frac{\partial \rho}{\partial z} - \rho \frac{F}{m} + n_0 \rho \bar{v} = 0 \quad , \tag{8.83}$$

where we assumed that the gas-solid system is kept in isothermal conditions. This equation can be formally integrated to yield

$$\rho(z) = \left[\rho(a) + \frac{1}{k_BT} \int_z^a n_0(z') \, \rho(z') \bar{v}(z') \, \exp(V_s(z')/k_BT) \, dz' \right]$$

$$* \exp[-V_s(z)/k_BT] \quad . \tag{8.84}$$

Brenig and Schönhammer (1976) argue that the arbitrary point a should be chosen at the outer edge of the surface potential V_s . Moreover, they assume that the friction coefficient n_0 is a function of z, being large only at the bottom of the potential well, and dropping to zero at z=a. Due to the presence of exponential factors, the integral in (8.84) is controlled by the behavior of $n_0(z)$ and $V_s(z)$ around z=a. They consider two specific models:

(i) $n_0(z) = n_{oo} \, e^{-\alpha z}$ and $V_s(z) = -V_0 \, e^{-\alpha z}$ for $z \approx a$ $\qquad$ (8.85)

(ii) $n_0(z) = n_{oo}\Theta(a-z)$ and $V_s(z) = -\frac{1}{2} m\omega_a^2(z - a)^2$ for $z \leq a$.

$$\tag{8.86}$$

In the first case, one finds

$$\rho(z) = \left[\rho(a) + n_{oo} \frac{\rho(a)\bar{v}(a)}{\alpha V_0} \right] \exp[-V_s(z)/k_BT] \tag{8.87}$$

and in the second

$$\rho(z) = \left[\rho(a) + \frac{\rho(a)\bar{v}(a)}{mv_T} \frac{n_{oo}}{2\omega_a} \right] \exp[-V_s(z)/k_BT] \quad , \tag{8.88}$$

where $v_T=(k_BT/2\pi m)^{1/2}$ is the thermal velocity. To calculate the desorption time itself, we look for exponentially decaying solutions of the continuity equation (8.73), and find by integrating it from $-\infty$ to a

$$t_d^{-1} = (\rho(a)\bar{v}(a))/ \int_{-\infty}^a \rho(z)dz$$

$$= \frac{v_0}{v_T} \frac{\bar{v}(a)}{1 + (m\bar{v}(a)/k_BT) \int_{-\infty}^a n_0(z')\exp(V_s(z')/k_BT)dz'} \exp[-V_0/k_BT] \quad , \tag{8.89}$$

where $v_0=\omega_0/2\pi$ is the oscillation frequency at the bottom of the surface potential well. This result for the desorption time is of the form of the

Frenkel-Arrhenius parametrization. It still depends on $\bar{v}(a)$ which must be specified as a boundary condition. The simplest case would be to demand that $\rho(a) = 0$ to account for the fact that, in a desorption experiment, the gas phase is removed by continuous pumping. The continuity equation then implies that $\bar{v}(a) \to \infty$ so that the preexponential factor reads for the two models in (8.86)

(i) $\qquad \nu = \nu_0 \dfrac{2\pi a V_0 v_T}{n_{oo} k_B T}$

(ii) $\qquad \nu = \nu_0 \dfrac{2\omega_a}{n_{oo}} \quad .$ $\hfill (8.90)$

The second case is essentially Kramers' (1940) result in the large friction limit. Brenig and Schönhammer (1976) also consider the small friction limit within the framework of this hydrodynamic theory, essentially recovering the results we reported previously in Sect. 8.2.

Brenig and Schönhammer (1976) and Müller and Brenig (1979), have also set up a macroscopic theory of adsorption along the lines of nonequilibrium thermodynamics. To account for the kinetics of surface processes, one has to consider, for a one component gas, four slowly varying macroscopic quantities, namely the number of adsorbed particles N_a, the number of free gas particles N_g, and the energies E_g and E_a of the gas and the substrate, respectively. Denoting any one of these variables by x_i, they satisfy in the linear regime, a set of rate equations

$$\frac{dx_i}{dt} = \sum_k L_{ik} \frac{\partial S}{\partial x_k} \quad , \hspace{3cm} (8.91)$$

where S is the entropy of the coupled gas-solid system. Müller and Brenig (1979) relate the Onsager coefficients L_{ik} to the sticking and accommodation coefficients, and show that the coupled gas-solid system is characterized by a further kinetic coefficient that couples particle and energy transport.

A continuum model for helium desorption kinetics has recently been presented by Goodstein and Weimer (1983) in which the adsorbate is treated as a film having the thermodynamic properties of bulk liquid helium, and in which the gas phase is treated by elementary kinetic theory. The theory is in particular applied to the phonon reflection experiments by Taborek and Goodstein (1980).

9. Summary and Outlook

9.1 Progress and Problems

Of the two broad fields of adsorption phenomena, chemisorption and physisorption, the latter is, no doubt, the simpler and by now also considerably
better understood.   The wealth of structure and processes in physisorbed
adlayers is immense.   To survey the equilibrium properties of physisorbed
gases, including the wide variety of phase transitions, would require at
least another volume of this size.   They therefore had to be entirely left
out of consideration for this review.   Rather we have endeavoured to present
a unified treatment of the adsorption and desorption kinetics of physisorption.   For the most part we have based the theory on the master equation.

It might be useful for further developments to try to summarize the progress that has been achieved in physisorption kinetics.   Historically, it
might be appropriate to place the beginnings of the field of physisorption
at the Faraday Discussions in 1932.   In the 1930s the concept of the surface
potential was clearly worked out, the possibility of various structures in
the adsorbate was conceived and highly mobile adsorbates had been discovered.   After the pioneering scattering experiments of helium off LiF by
Stern and Estermann, the importance of surface corrugation, previously calculated by Lennard-Jones and his collaborators, was recognized.

In an attempt to understand inelastic scattering off surfaces, Lennard-
Jones proposed the dynamical interaction of physisorbed particles with the
surface via a coupling to the lattice degrees of freedom. His first calculations of desorption times were based on a one-dimensional model in which
desorption was mediated by one-phonon processes.   In the half century since
then, the theory has been improved at the kinetic level by basing it on the
master equation. This led naturally to the inclusion of one-phonon cascades
at the dynamic level.   Further improvements resulted from the inclusion of
multiphonon processes and an extension of the model to a three-dimensional
world.   Although considerably more complex the latter extension proved, cum

grano salis, unnecessary for highly mobile adsorbates. Detailed studies of the role of surface versus bulk phonons now also exist.

With the exception of the mean field approach to physisorption kinetics, all kinetic theories are restricted to vanishingly small coverages. In the future much more work needs to be done to understand physisorption kinetics at nonzero coverage. At a monolayer or so the kinetics of adsorption and desorption can be influenced considerably by (i) a change in the phonon spectrum due to surface loading, in particular for heavy adsorbates; (ii) collective modes will emerge in the adsorbate that can act as energy sinks and sources during adsorption and desorption; (iii) as a multilayer adsorbate desorbs, phase transitions may occur that should show up in the desorption kinetics e.g. via critical slowing down. Even at vanishingly small coverages one should (iv) incorporate surface corrugation and band-structure effects into a kinetic theory.

The theory of sticking and energy accommodation has finally, after forty years of controversy and confusion, seen considerable clarification in the papers by Iche and Nozières and by Brenig. The list of future improvements given above for physisorption kinetics applies here as well.

Our comments have so far been restricted to theoretical efforts. On the experimental side it is surprising how few good data exist on the kinetics of physisorbed gas-solid systems. This is particularly true for desorption kinetics with the noteworthy exception of the work by Goodstein and his collaborators on helium desorption. A gratifying confirmation of the validity and usefulness of the master equation approach was provided by the inelastic scattering experiments by Toennies's group who managed to measure some of the individual bound state-continuum transition rates, confirming earlier theoretical estimates.

9.2 Related Topics I: Photodesorption

This review has been exclusively concerned with kinetic effects in the adsorbate that are triggered and maintained by the thermal energy reservoir of the solid. In this and the next section we want to draw attention to two related fields, namely photodesorption and electron-stimulated desorption of physisorbed species, where a master equation approach has been applied successfully.

A laser beam impinging onto an adsorbate-covered surface of a solid can deposit its energy (i) into the solid directly, heating it up rapidly and leading to laser-induced thermal desorption (Ertl and Neumann 1972; Wedler and Ruhmann 1982; Burgess et al. 1983). (ii) The laser energy can be coupled into the surface bond with which the adsorbed molecule is bound to the surface (Lin and George 1979; Jedrzejek et al. 1981; George et al. 1984). This process is resonant in character in that the laser frequency must match one with which the adsorbed molecule "vibrates" with respect to the surface. If enough energy is deposited into the surface bond, it might rupture leading to photodesoprtion by laser-surface bond coupling. This has not been achieved experimentally yet. (iii) The laser can also be coupled resonantly into some internal degree of freedom of the adsorbed molecule, be it an electronic or vibrational mode. Photodesorption of molecules by resonant laser-molecular vibrational coupling has been realized experimentally for CH_3F on $NaC\ell$ (Heidberg et al. 1980, 1982), for pyridine on $KC\ell$ and Cu (Chuang and Seki,1982; Seki and Chuang 1982), and for NH_3 on Cu (Hussla et al. 1985). The experimental situation has been reviewed by Chuang (1983).

A theory of photodesorption of molecules by resonant laser-molecular vibrational coupling has been proposed by Kreuzer and Lowy (1981) and has been worked out by Gortel, Kreuzer, Piercy and Teshima (1983a,b). The general scheme is depicted in Fig. 9.1. The binding of the molecule to the surface is represented by a surface potential V_S that develops bound states labeled i. The active vibrational mode of the molecule into which the laser is tuned is approximated by a harmonic oscillator of frequency Ω and with quantum number v. Because $\hbar\Omega$ is typically much larger than the spacing between the energies $(E_{i+1}-E_i)$ of bound states in the surface potential, we can, as a first approximation, assume that the internal degrees of freedom of the molecule are decoupled from its interaction with the surface. Thus we can write the energy of an adsorbed molecule as $E_i^v \approx E_i + (v+1/2)\hbar\Omega$. In a photodesorption experiment a molecule, initially thermalized in the surface, i.e., mostly in the state (i,v)=(0,0), will be pumped by the laser through a sequence $(0,0) \rightarrow (0,1) \rightarrow (0,2) \rightarrow \dots$. The state (0,1) is still below the continuum limit and can decay via phonon emission into a state $(i_0,0)$ from where the molecule will cascade down rapidly to (0,0) emitting an avalanche of phonons. This process has been termed resonant heating (Gortel, Kreuzer, Piercy, and Teshima 1983a,b). On metals it is very effectively achieved through the additional coupling to the electronic degrees of freedom (Hussla et al. 1985). Once the laser has pumped the molecule into a state (0,v) that is degenerate with the continuum of v=0, tunneling can occur that leads to desorption.

The theory of Gortel, Kreuzer, Piercy, and Teshima (1983a,b) is based on a master equation for the probability functions n_i^v. They calculate the vari-

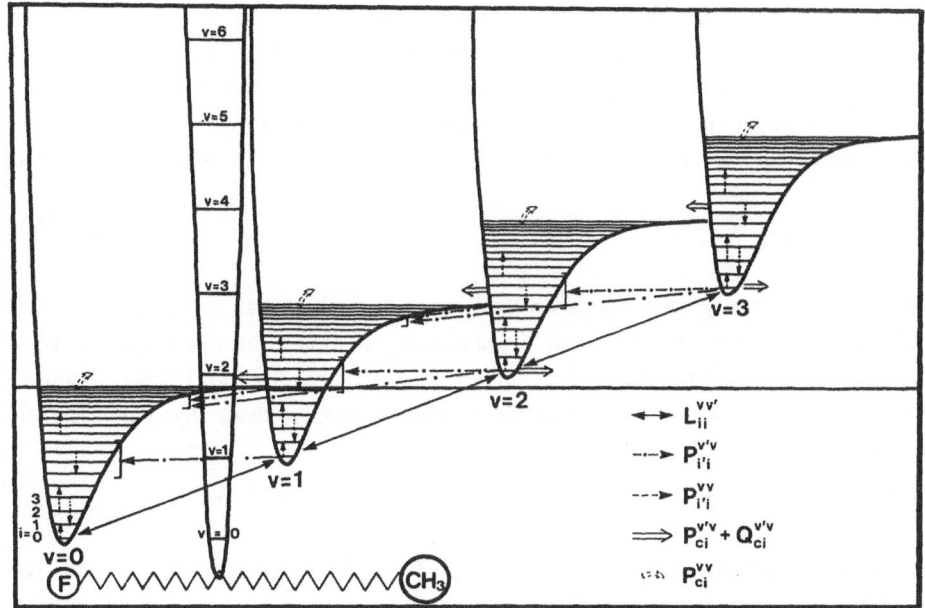

Fig. 9.1. Schematic energy diagram of a molecule adsorbed on a surface with some of the processes included in the theory indicated by arrows with the transition probabilities:
$P_{ii'}^{v'v}$ = phonon mediated bound state–bound state transitions: $(i,v) \rightarrow (i',v')$
$P_{ci}^{v'v}$ = phonon mediated bound state–continuum transitions
$Q_{ci}^{v'v}$ = elastic tunneling into continuum
$L_{ii}^{v'v}$ = laser coupling
(Gortel, Kreuzer, Piercy, and Teshima 1983a)

ous transition probabilities from a Hamiltonian employing Fermi's golden rule. Detailed numerical studies imply that photodesorption by resonant laser-molecular vibrational coupling seems to be possible only for molecules weakly physisorbed on dielectric surfaces, because only the lowest few vibrational levels can be effectively populated by an infrared laser. Also, on metals resonant heating is too effective. For further details we refer to the reviews by Chuang (1983), George et al. (1984), and Gortel et al. (1985).

9.3 Related Topics II: Electron-Stimulated Desorption

Feulner et al. (1981,1983,1984) have reported angular and kinetic energy distributions of neutral molecules desorbed by electron impact from physisorbed adsorbates on metal surfaces. In particular, they report electron-stimulated desorption of neutral N_2O from Ru(001). The data can be understood in the following model: A physisorbed N_2O sitting in the minimum of the surface potential $V_0(z)$ is ionized by electron impact, leading to a sudden transition to a potential curve $V_+(z)$ which differs from $V_0(z)$ by the attractive image potential $V_{im}(z) = -e^2/4(z+z_i)$. The N_2O^+ ion is now accelerated towards the surface implying the propagation of a nonstationary state

$$\psi(t) = \sum_{\tilde{m}} \exp[-i\varepsilon_{\tilde{m}}t] \; |\tilde{m}\rangle\langle\tilde{m}|0\rangle \quad , \tag{9.1}$$

where $|0\rangle$ is the ground state on $V_0(z)$ and $|\tilde{m}\rangle$ are states in $V_+(z)$. Eventually neutralization takes place by capturing a metal electron leading to a transition back onto $V_0(z)$. Desorption as a neutral N_2O will take place if the state so reached has positive energy. The probability for this to happen is

$$P(E,t) = \left|\langle E|\psi(t)\rangle\right|^2 \quad . \tag{9.2}$$

One assumes that neutralization, i.e., electron-ion recombination, is exponential in time with the rate constant given by $r = \alpha n(z)$ where α is the recombination coefficient and $n(z)$ is the electron density a distance z above the surface. Because N_2O and N_2O^+ move more or less classically one can connect the position z and time t via a classical trajectory. The overall probability that N_2O desorbs as a neutral with energy E is then

$$P(E) = \int_0^\infty dt \; e^{-rt} \, P(E,t) \quad . \tag{9.3}$$

Feulner et al. (1984) evaluate these expressions by using wave functions for Morse potentials fitted to $V_0(z)$ and $V_+(z)$. A fit of (9.3) to experimental time of flight data yields good agreement for rate constants around $10^{14} \; s^{-1}$. To avoid complications due to the extended structure of the N_2O molecule one hopes that similar experiments on electron stimulated desorption of neutral rare gas atoms will further clarify the underlying microscopic processes.

9.4 Approaches to Chemisorption Kinetics

In the introductory chapter we briefly discussed the main characteristics that set physisorption apart from chemisorption. The simplifying feature of physisorption was the fact that for a gas particle interacting with a surface, its electronic degrees of freedom can be ignored. To understand the

286

kinetics of adsorption and desorption we then had to identify the sinks and sources of the energy supply. For chemisorbed gas-solid systems the situation is considerably complicated by the fact that substantial rearrangement of the electronic structure occurs in the adsorbed phase. We have, by now, a fair understanding of the resulting equilibrium properties of chemisorption as it emerges from calculations based on (i) cluster models, (ii) functional density theory applied to the jellium model, and (iii) the Anderson-Newns model (for a review see Smith 1980).

The ultimate problem in chemisorption is to unravel the, most often, rather complex reaction pathways that accompany the adsorption of all but the simplest molecules. Let us, as an example, look at the dissociative adsorption of H_2 on a metal. The complete wave function of the coupled H_2-metal system depends on the nuclear coordinates X_1 and X_2 and the electronic coordinates x_1 and x_2 for the two H atoms and on all electronic and ionic degrees of freedom Y of the metal. Fixing the nuclear coordinates X_i we can calculate the energy $E(X,\xi)$ of the system as a function of the center of mass coordinate $X=(X_1+X_2)/2$ and the relative coordinate $\xi=X_1-X_2$ of the H nuclei; a contour map is given in Fig. 9.2. A three-dimensional perspective view of the energy surface is depicted in Fig. 9.3. In this picture, an adsorbing H_2 molecule approaches the surface through the entry valley (arrow in the upper-left corner in Fig. 9.2) first encountering a shallow minimum where it can get trapped in a weakly physisorbed precursor state. Eventually it will try to climb over the saddle point S by stretching its relative separation ξ. If successful, the bond will rip and the two dissociated H atoms will be trapped at the surface in the minimum D. For energetic reasons the system will try to hug the path of steepest ascent and descent; this pathway is commonly parametrized by a reaction coordinate z; its energetics is illustrated in Fig. 9.4.

Due to their large mass (as compared to the mass of the electrons) one can safely treat the motion of the reacting atoms by classical mechanics, thus associating with the reaction cordinate z a conjugate momentum p. The distribution function of the coupled gas-solid system thus becomes a function of z and p and of all other degrees of freedom, electronic and nuclear, that in some sense are orthogonal to the reaction coordinate. If the "surface reaction" proceeds sufficiently slowly and with only small deviations from the reaction coordinate, then the system has sufficient time to settle into a quasi equilibrium at any instant of the reaction. In such situations it is plausible to factorize the total distribution function into a factor $f(z,p,t)$ that depends only on the reaction coordinate z, the momentum p and

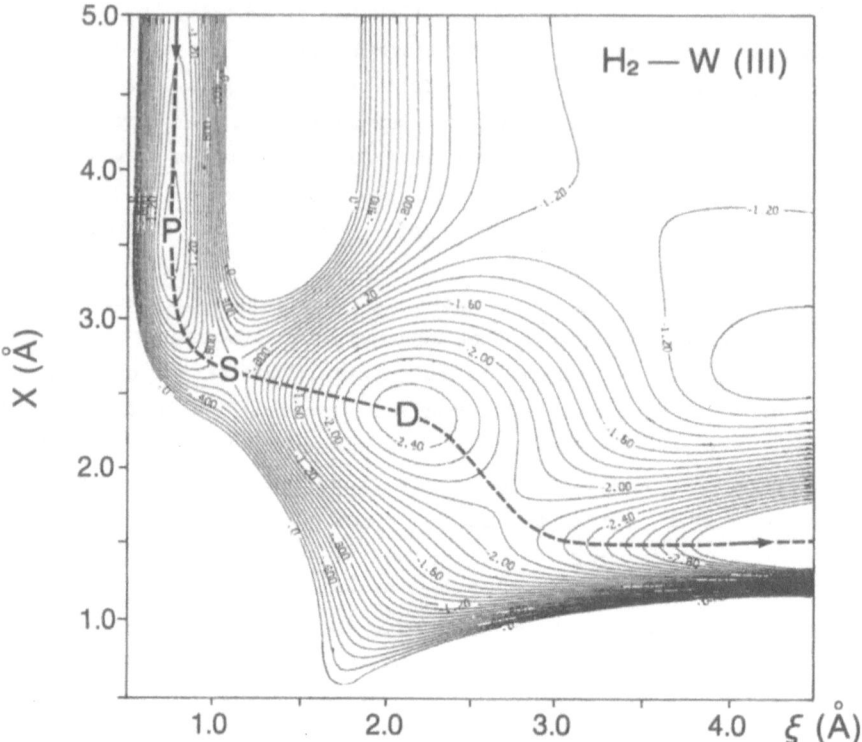

Fig.9.2. Contour plot of the ground state energy surface of H_2 interacting with a W(111) surface. The H_2 molecule is parallel to the surface. Its center of mass coordinate is X and the H≋H bond length is ξ. The reaction path leads from the free molecule (upper-left corner) via a physisorbed precursor P and over a saddle point S to a dissociative state D.

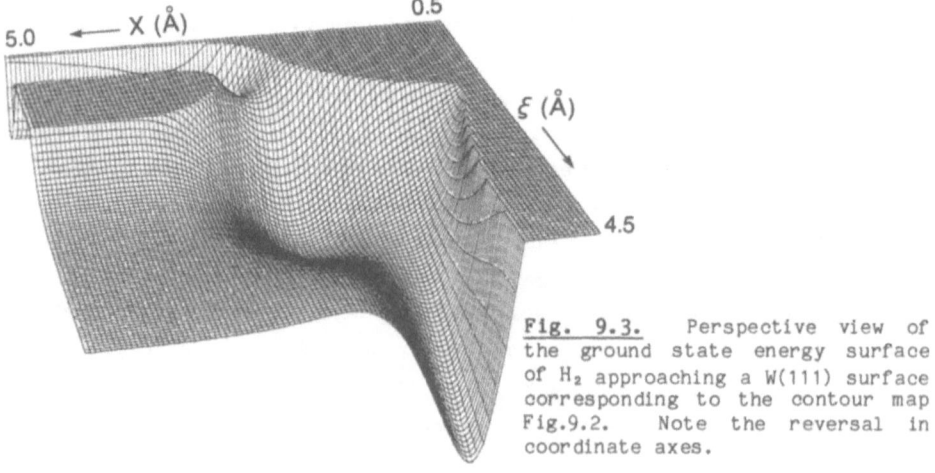

Fig. 9.3. Perspective view of the ground state energy surface of H_2 approaching a W(111) surface corresponding to the contour map Fig.9.2. Note the reversal in coordinate axes.

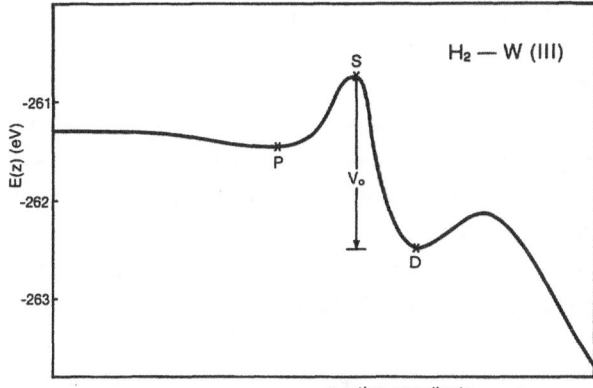

Fig.9.4. Ground state energy of H_2 interacting with a metal along the reaction coordinate z in Fig. 9.2.

on time, and a Boltzmann factor that incorporates all other degrees of freedom. If the momentum and energy exchange along the reaction path is small, one can employ a Brownian motion model. The transport equation for $f(z,p,t)$ is then of Fokker-Planck form; for the present discussion we will adopt the Kramers equation (8.1)

$$\frac{\partial f}{\partial t} + \frac{p}{m} \frac{\partial f}{\partial z} - \frac{\partial E}{\partial z} \frac{\partial f}{\partial p} = \eta \frac{\partial}{\partial p} \left[pf + mk_BT \frac{\partial f}{\partial p} \right] \tag{9.4}$$

Kramers has distinguished three regimes of the friction coefficient η for which the reaction rate constant r is approximately given by

$$r \approx \eta \frac{V_0}{k_BT} e^{-V_0/k_BT} \quad \text{if} \quad \eta < \omega_P \frac{k_BT}{V_0} \quad , \tag{9.5}$$

$$r \approx \omega_S e^{-V_0/k_BT} \quad \text{if} \quad \omega_P \frac{k_BT}{V_0} < \eta < \omega_S \quad , \tag{9.6}$$

$$r \approx \frac{\omega_P \omega_S}{\eta} e^{-V_0/k_BT} \quad \text{if} \quad \eta > \omega_S \quad , \quad \text{where} \tag{9.7}$$

$$(\omega_{P,S})^2 = \frac{2}{M} \partial^2 V/\partial z^2 \big|_{P,S} \quad . \tag{9.8}$$

To understand these regimes we observe that on the grounds of kinematics alone, one would suggest that the rate constant is proportional to the number of particles in the tail of the Maxwell distribution that have sufficient energy to surmount the activation barrier. This argument assumes that for every particle lost over the barrier the Maxwell distribution is instantly replenished. This will, indeed, happen if the thermal coupling to the heat bath of the solid is just right. In this case, Eyring's transition state theory, reviewed briefly in Chap. 1, is applicable yielding a rate constant that is independent of the dynamical coupling as expressed by η. If the energy supply is not sufficient for small η, one expects a decrease of r proportional to η. This situation we encountered in physisorption. In the other extreme, when η becomes large, the heat bath acts as a frictional energy sink inhibiting the motion of the reacting particles to the extend that the rate decreases inversely proportional to η.

D'Agliano et al. (1975) have derived expressions for the friction coefficient η stressing the importance of electron-hole pair excitations for the chemisorption kinetics on metals. This point has since been discussed within the framework of the functional density approach to chemisorption and of the Anderson-Newns model. Let us, for simplicity, follow an H atom approaching a metallic surface. We describe its nuclear motion classically:

$$m \frac{d^2z(t)}{dt^2} = - \frac{dE(z)}{dz} , \qquad (9.9)$$

where $E(z)$ is the ground-state energy of the adsorbing atom at a distance z from the surface. Looking at the electron, we note that as z decreases, the affinity level broadens into a resonance due to the interaction with the metal, cf. Fig.1.1. Within the Anderson-Newns model this is described by a Hamiltonian

$$H = \sum_k \varepsilon_k n_k + \varepsilon_a(z(t))n_a + \sum_k V_{ka}(z(t)) \, \psi_a^\dagger \psi_k + h.c. \qquad (9.10)$$

Here k labels the electronic states of the substrate and ε_a is the affinity level of the adsorbing atom. Because of the finite velocity of the atom there is a nonzero probability that some of its kinetic energy will be transferred to electronic excitations in a nonadiabatic process. This excitation probability can be calculated, once the (time-dependent) interaction V_{ka} has been specified. Note that the model is not quite consistent in that the nuclear motion is calculated on the ground-state energy surface, i.e., assuming that no energy is lost in the approach to the surface. Calculations performed to date are mostly of a qualitative nature discussing the sticking probability (Nørskov and Lundquist 1979; Schönhammer and Gunnarsson 1980; Brako and Newns 1981; Brako 1982,1983; for a recent review see Yoshimori and Tsukada 1985). We note once more that transition state theory ignores such nonadiabatic effects completely; one rather hopes in that approach that the order of magnitude of the reaction rate is determined by the activation energy E_d (known from the ground-state energy surface) and a preexponential factor that is given solely by partition functions, i.e. phase space arguments.

290

Most theoretical approaches to the kinetics of complex chemisorbed systems as they occur in heterogeneous catalysis are phenomenological in nature. One writes down a set of rate equations, and hopes that a small number of rate constants can be fitted to experimental data. Because neither the set of rate equations nor the numerical fit are usually unique, such an approach is more valuable in eliminating unlikely reaction pathways than in identifying the dominant ones. A recent review has been presented by Boudard and Djèga-Mariadassou (1984).

References

Abramovitz, M., Stegun, I.A. (1972): Handbook of Mathematical Functions with Formulae, Graphs, and Mathematical Tables (National Bureau of Standards) (Appl. Math. Series 55, 10th edn with corrections)

Adams, J.E., Doll, J.D. (1982): Thermal desorption of argon and neon from solid xenon. I. Transition state theory rate constants. J. Chem. Phys. **77**, 2964–2967.

Adelman, S.A. (1980): Generalized Langevin equations and many-body problems in chemical dynamics. Adv. Chem. Phys. **44**, 143–253.

d'Agliano, E.G., Kumar, P., Schaich, W., Suhl, H. (1975): Brownian motion model of the interactions between chemical species and metallic electrons: bootstrap derivation and parameter evaluation. Phys. Rev. B **11**, 2122–2143.

Algie, S.H. (1978): Kinetic theories of evaporation. J. Chem. Phys. **69**, 538–543.

Alldredge, G.P., Allen, R.E., de Wette, F.W. (1971): Studies of vibrational surface modes. III. Effect of an adsorbed layer. Phys. Rev. B **4**, 1682–1697.

Allen, R.E., Alldredge, G.P., de Wette, F.W. (1971a): Studies of vibrational surface modes. I. General formulation. Phys. Rev. B **4**, 1648–1660.

–– (1971b): Studies of vibrational surface modes. II. Monoatomic fcc crystals. Phys. Rev. B **4**, 1661–1681.

Allen, R.T., Feuer, P. (1967): Quantum theory of the thermal accommodation coefficient and the effect of two-quantum transitions. In Proceedings of 5th Int. Symposium, Rarified Gas Dynamics, ed. by C.L. Brundin (Academic, New York) pp.109–119.

Anderson, A.C. (1981): The Kapitza thermal boundary resistance between two solids. In Nonequilibrium Superconductivity, Phonons and Kapitza Boundaries, ed. by K. Gray (Plenum, New York) pp.1–30.

Apell, P., Holmberg, C. (1984): Multipole contributions to atom-surface scattering potentials. Solid State Commun. **49**, 1059–1063.

Armand, G. (1968): Collision d'un atome sur réseau métallique en modèle unidemensionnel: coefficients d'accommodation et de capture, vitesse du désorption. Surf. Sci. **9**, 145–164.

–– (1977): Classical theory of desorption rate velocity distribution of desorbed atoms; possibility of a compensation effect. Surf. Sci. **66**, 321–345.

Armand, G., Manson, J.R. (1982): Scattering of neutral particles by an exponential corrugated potential: A solution for high corrugation amplitude. Phys. Rev. B **25**, 6195–6207.

Avouris, Ph., Schmeisser, D., Demuth, J.E. (1982): Observation of rotational excitations of H_2 adsorbed on Ag surfaces. Phys. Rev. Lett. **48**, 199–202.

Bardeen, J. (1940): The image and van der Waals forces at a metallic surface. Phys. Rev. **58**, 727–736.

Baret, J.F. (1969): Theoretical model for an interface allowing a kinetic study of adsorption. J. Coll. Interface Sci. **30**, 1–12.

Barker, J.A., Dion, D.R., Merrill, R.P. (1980): Classical surface scattering computations; Rainbows and energy exchange. Surf. Sci. **95**, 15–52.

Bateman Manuscript Project, Higher Transcendental Functions (1953): ed. by A. Erdelyi (McGraw-Hill, New York)

Batra, I.P., Bagus, P.S., Barker, J.A. (1985): Hartree–Fock studies of helium–surface interaction potentials. Phys. Rev. B **31**, 1737–1743.

Bauer, E., Bonczek, F., Poppa, H., Todd, G. (1975): Thermal desorption of metals from tungsten single crystal surfaces. Surf. Sci. **53**, 87–109.

Baule, B. (1914): Theoretische Behandlung der Erscheinungen in verdünnten Gasen. Ann. Physik **44**, 145–176.

Beeby, J.L., Dobrzynski, L. (1971): The scattering of atoms from surfaces: a model. J. Phys. C **4**, 1269–1278.

Beeby, J.L., Agarwal, B.K. (1982): Theory of the sticking coefficient for atom–surface scattering. Surf. Sci. **122**, 447–458.

Bendow, B., Ying, S.C. (1972): Quantum theory of desorption of adatoms from solid surfaces. J. of Vacuum Sci. & Technol. **9**, 804–807.

— (1973a): Phonon-induced desorption of adatoms from crystal surfaces. I. Formal theory. Phys. Rev. B **7**, 622–636.

— (1973b): Phonon–induced desorption of adatoms from crystal surfaces. II. Numerical computations for a model system. Phys. Rev. B **7**, 637–644.

Benedek, G. (1973): Surface lattice dynamics of ionic crystals by the Green function method. Phys. Status Solidi (b) **58**, 661–671.

— (1976): The Green function approach to the surface lattice dynamics of ionic crystals. Surf. Sci. **61**, 603–634.

— (1982): Surface phonons in ionic crystals. Benedek and Valbusa (1982) pp. 227–255.

— (1983): The spectroscopy of surface vibrations in ionic crystals by inelastic scattering of atoms. Surf. Sci. **126**, 624–640.

Benedek, G., Valbusa, U. (eds.) (1982): <u>Dynamics of Gas-Surface Interaction.</u> Springer Ser. Chem. Phys., Vol. 21 (Springer, Berlin, Heidelberg)

Benedek, G., Brivio, G.P., Miglio, L., Velasco, V.R. (1982): Dispersion relations of surface phonons in LiF(001) and NaF(001). Phys. Rev. B **26.** 497–506.

Benedek, G., Brusdeylins, G., Doak, B.R., Toennies, J.P. (1981): The spectroscopy of surface phonons by inelastic atom scattering. J. Phys. (Paris) **42**, 793–800.

Benedek, G., Brusdeylins, G., Doak, B.R., Skofronick, J.G., Toennies, J.P. (1983): Measurement of the Rayleigh surface–phonon dispersion curve for NaCl (001) from high-resolution He time-of-flight spectroscopy and from kinematical focusing angles. Phys. Rev. B **28**, 2104–2113.

Benton, A.F., White, T.A. (1930): Adsorption of hydrogen by nickel at low temperatures. J. Amer. Chem. Soc. **52**, 2325–2336.

— (1931): Discontinuities in adsorption isotherms. J. Am. Chem. Soc. **53**, 3301–3314.

Bergmann, E., Antonini, J.F. (1975): Application of response theory to adsorption/desorption kinetics. J. Phys. Chem. **79**, 123–126.

Bertel, E., Netzer, F.P. (1980): Adsorption of bromine on the reconstructed Au(100) surface: LEED, thermal desorption and work function measurements. Surf. Sci. **97**, 409–424.

Bienfait, M. (1982): Phase Transitions in Surface Films. Benedek and Valbusa (1982) pp. 94–110.

Bienfait, M., Venables, J.A. (1977): Kinetics of adsorption and desorption using Auger electron spectroscopy: Application to xenon covered (0001) graphite. Surf. Sci. **64**, 425–436.

Bilkadi, L., Parsons, J.D., Mann Jr., J.A. (1980): Brownian motion of particles near a free surface: desorption from a monolayer. J. Chem. Phys. **72**, 960–964.

Black, J.E., Campbell, D.A., Wallis, R.F. (1982): Surface vibrations on body centered cubic and face centered cubic metal surfaces: The (100) Surface. Surf. Sci. 115, 161-182.

Black, J.E., Rahman, T.S., Mills, D.L. (1983): Spectral densities in surface lattice dynamics at large wave vector. Phys. Rev. B 27, 4072-4084.

Black, J.E., Shanes, F.C., Wallis, R.F. (1983): Surface vibrations on face centered cubic metal surfaces: The (111) surfaces. Surf. Sci. 133, 199-215.

Block, J.H. (1982): Field desorption and photon-induced field desorption. In Chemistry and Physics of Solid Surfaces IV, ed. by R. Vanselow, R. Howe, Springer Ser. Chem. Phys. Vol. 20 (Springer, Berlin, Heidelberg) pp.407-434.

Boato, G., Cantini, P., Guidi, C., Tatarek, R., Felcher, G.P. (1979): Bound-state resonances and interaction potential of helium scattered by graphite (0001). Phys. Rev. B 20, 3957-3969.

Böheim, J. (1984): Sticking and accommodation of He on metal surfaces at low temperature. Surf. Sci. 148, 463-477.

Böheim, J., Brenig, W. (1981): Theory of inelastic atom-surface scattering: Examples of energy distributions. Z. Phys. B 41, 243-250.

Böheim, J., Brenig, W., Stutzki, J. (1982): On the low energy limit of reflection and sticking coefficients in atom surface scattering: II. Long range forces. Z. Phys. B 48, 43-49.

Bone, W.A. (1921-22): Discussion. Trans. Faraday Soc. 17, 658-661.

Bone, W.A., Wheeler, R.V. (1906): The combination of hydrogen and oxygen in contact with hot surfaces. Philos. Trans. R. Soc. London, Ser. A 206, 1-67.

Borisov, S.F., Semenov, Yu.G., Suetin, P.E. (1982): Inert-gas energy-accommodation coefficients. Inzhenerno-Fizicheskii Zhurnal 43, 983-988.

Boudard, M. and Djèga-Mariadassou, G. (1984): Kinetics of Heterogeneous Catalytic Reactions (Princeton University Press, Princeton, N.J.)

Bradley, T.L., Dabiri, A.E., Stickney, R.E. (1972): Measurements of the spatial distribution of H_2 desorbed from Ni surfaces: Effects of surface composition and crystal orientation. Surf. Sci. 29, 590-602.

Brako, R. (1982): Energy and momentum transfer in scattering of low-energy atoms on metal surfaces. Surf. Sci. 123 439-455.

-- (1983): Trapping and energy accommodation of atoms on metal surfaces. Solid State Commun. 47, 559-561.

Brako, R., Newns, D.M. (1981): Charge exchange in atom-surface scattering: Thermal versus quantum mechanical non-adiabaticity. Surf. Sci. 108, 253-270.

Brenig, W. (1979): Theory of inelastic atom-surface scattering: Average energy loss and energy distribution. Z. Phys. B 36, 81-87.

-- (1980): On the low energy limit of reflection and sticking coefficients in atom surface scattering. I. Short range forces. Z. Phys. B 36, 227-233.

-- (1982): Microscopic theory of gas-surface interaction. Z. Phys. B 48, 127-136.

Brenig, W., Schönhammer, K. (1976): A hydrodynamic theory of surface reaction rates. Z. Phys. B 24, 91-97.

-- (1979): Spurious phonon instabilities by adsorbed particles, Z. Phys. B 34, 283-285.

Brivio, G.P. (1983): Gas-surface interaction on metals as a localized dynamic perturbation. J. Phys. C 16, L131-L135.

Brivio, G.P., Grimley, T.B. (1983): A model for non-adiabatic coupling on metals: the sticking problem. Surf. Sci. 131, 475-490.

Bruch, L.W. (1983): Theory of physisorption interactions. Surf. Sci. 125, 194-217.

Bruch, L.W., Phillips, J.M., Ni, X.-Z. (1984): Model calculations of adsorbed solids of neon on graphite. Surf. Sci. 136, 361-380.

Brueckner, K.A., Frohberg, J. (1965): The theory of correlated crystals. Suppl. Prog. Theor. Phys., 383#399.

Brueckner, K.A., Gammel, J.L., Weitzner, H. (1958): Theory of finite nuclei. Phys. Rev. 110, 431~445.

Brusdeylins, G., Doak, R.B., Toennies, J.P. (1980): Observation of surface phonons in inelastic scattering of He atoms from LiF(001) crystal surfaces. Phys. Rev. Lett. 44, 1417-1420.

~~ (1981): Observation of selective desorption of one~phonon inelastically scattered He atoms from a LiF crystal surface. J. Chem. Phys. 75, 1784~1793.

~~ (1981): Measurement of the dispersion relation for Rayleigh surface phonons of LiF(001) by inelastic scattering of He atoms. Phys. Rev. Lett. 46, 437-439.

Burgess Jr., D.R., Vishnawathan, R., Hussla, I., Stair, P.C., Weitz, E. (1983): Pulsed Laser induced thermal desorption of CO from copper surfaces. J. Chem. Phys. 79, 5200-5202.

Burgess Jr., D.R., Hussla, I., Stair, P.C., Vishnawathan, R., Weitz, E. 1984. Pulsed laser~induced thermal desorption from surfaces: Instrumentation and procedures. Rev. Sci. Instrum. 55, 1771-1776.

Cabrera, N. (1959): The structure of crystal surfaces. Discuss. Faraday Soc. 28, 16-22.

Cabrera, N., Celli, V., Goodman, F.O., Manson, R. (1970): Scattering of atoms by solid surfaces I. Surf. Sci. 19, 67~92.

Callaway, J. (1974): Quantum Theory of the Solid State, Part A and B (Academic, New York and London)

Cardillo, M.J., Balooch, M., Stickney, R.E. (1975): Detailed balancing and quasi~equilibrium in the adsorption of hydrogen on copper. Surf. Sci. 50,263-278.

Carlos, W.E., Cole, M.W. (1978): Interaction and band structure effects for He on graphite derived from scattering data. Surf. Sci. 77, L173~L176.

~~ (1979): Anistropic He-C pair interaction for a He atom near a graphite surface. Phys. Rev. Lett. 43, 697~700.

~~ (1980a): Band structure and thermodynamic properties of He atoms near a graphite surface. Phys. Rev. B 21, 3713~3720.

~~ (1980b): Interaction between a He atom and a graphite surface. Surf. Sci. 91, 339~357.

Caroli, C., Roulet, B., Saint-James, D. (1978): Effect of adatom-phonon coupling on desorption kinetics in the heavy~adatom limit. Phys. Rev. B 18, 545-558.

Carter, G. (1962): Thermal resolution of desorption energy spectra. Vacuum 12, 245-254.

Casimir, H.B.G., Polder, D. (1948): The influence of retardation on the London~van der Waals forces. Phys. Rev. 73, 360~372.

Cassuto, A., King, D.A. (1981): Rate expressions for adsorption and desorption kinetics with precursor states and lateral interactions. Surf. Sci. 102, 388-404.

Celli, V., Evans, D. (1982): Theory of Atom-surface Scattering. Benedek and Valbusa (1982) pp. 2~39.

Cercignani, C. (1975): Theory and Application of the Boltzmann Equation (Scottish Academic, Edinburgh).

Chambers, C.M., Kinzer, E.T. (1963): Collisions of atoms with crystal surfaces. Bull. Am. Phys. Soc. Ser.II, 8, 552.

~~ (1966): Higher dimensional crystal models: A theory of thermal accommodation coefficients. Surf. Sci. 4, 33~47.

Chan, C.-M, Aris, R., Weinberg, W.H. (1978): An analysis of thermal desorption mass spectra. I. Appl. Surf. Sci. 1, 360~376.

Chandrasekhar, S. (1943): Stochastic problems in physics and astronomy. Rev. Mod. Phys. 15, 1-89.

Chen, S.H.P., Saxena, S.C. (1976): Thermal accommodation coefficients for the gas-covered tungsten-argon system at high temperatures (600-2500K). High Temperature Science 8, 1-9.

Chen, T.S., Alldredge, G.P., de Wette, F.W., Allen, R.E. (1971): Surface and pseudosurface modes in ionic crystals. Phys. Rev. Lett. 26, 1543-1546.

Chen, T.S., de Wette, F.W., Alldredge, G.P. (1977): Studies of vibrational surface modes in ionic crystals. I. Detailed shell model studies for the unrelaxed (001) face of seven crystals having the rocksalt structure. Phys. Rev. B 15, 1167-1186.

Christmann, K. and Demuth, J.E. (1982a): Interaction of inert gases with a nickel (100) surface. I. Adsorption of xenon. Surf. Sci. 120, 291-318.

-- (1982b): The adsorption and reaction of methanol on Pd (100): I. Chemisorption and condensation. J. Chem. Phys. 76, 6308-6317.

-- (1982c): The adsorption and reaction of methanol on Pd (100): II. Thermal desorption and decomposition. J. Chem. Phys. 76, 6318-6327.

Chuang, T.J. (1983): Laser-induced surface interactions. Surf. Sci. Rep. 3, 1-105.

Chuang, T.J., Seki, H. (1982): Resonantly stimulated desorption of pyridine from silver surfaces by polarized infrared laser radiation. Phys. Rev. Lett. 49, 382-386.

Chubb, J.N., Pollard, I.E. (1969): Sticking coefficient and desorption rate variations during growth of condensed gas layers on liquid helium cooled surfaces. VIDE (France) 24, 64-67.

Chung, S., Cole, M.W. (1984): The asymptotic interaction between H, H_2 or a noble gas atom and the surface of NaCℓ, KCℓ, or MgO. Surf. Sci. 145, 269-280.

Clark, A. (1970): The Theory of Adsorption and Catalysis (Academic, New York).

Clausing, P. (1930): Über das Kosinusgesetz der Zurückwerfung als Folge des zweiten Hauptsatzes der Thermodynamik. Ann. Phys. (Leipzig) 4, 533-566.

Cohen, P.I., Unguris, J., Webb, M.B. (1976): Xe monolayer adsorption on Ag (111): I. Structural properties. Surf. Sci. 58, 429-456.

Cohen, S.A., King, J.G. (1973): Measurement of lifetimes and binding energies of atoms adsorbed on surfaces at low temperatures by a rapid-flash desorption technique. Phys. Rev. Lett. 31, 703-710.

Cole, M.W., and Toigo, F. (1982): Atom-phonon interaction at a surface. Surf. Sci. 119, L346-L352.

-- (1982): Kinetics of elementary processes at surfaces. In Interfacial Aspects of Phase Transitions, ed. by B. Mutaftschiev (Reidel, Dordrecht, Holland) p.223.

-- (1985): Energy of immersing a He, Ne, or Ar atom or H_2 molecule into a low-density electron gas. Phys. Rev. B 31, 727-729.

Cole, M.W., Tsong, T.T. (1977): Bound state vibrational spectrum of the 3-9 atom-surface interaction. Surf. Sci. 69, 325-335.

Cole, M.W., Frankl, D.R., Goodstein, D.L. (1981): Probing the helium-graphite interaction. Rev. Mod. Phys. 53, 199-210.

Comsa, G., David, R. (1982): The purely "fast" distribution of H_2 and D_2 molecules desorbing from Cu (100) and Cu (111) surfaces. Surf. Sci. 117, 77-84.

Comsa, G., David, R., Rendulic, K.D. (1977): Velocity distribution of H_2, HD, and D_2 molecules desorbing from polycrystalline Nickel surfaces. Phys. Rev. Lett. 38, 775-778.

Comsa, G., David, R., Schumacher, B.J. (1979): The angular dependence of flux, mean energy and speed ratio for D_2 molecules desorbing from a Ni(111) surface. Surf. Sci. 85, 45-68.

-- (1980): Fast deuterium molecules desorbing from metals. Surf. Sci. 95, L210-L216.

Constabaris, G., Sams Jr., J.R., Halsey, G.D., Jr. (1961): The interaction of H_2, D_2, CH_4 and CD_4 with graphitized carbon black. J. Phys. Chem. **65**, 367-69.

Cowin, J.P. (1985): Comment on "Critical cone in phonon-induced desorption of helium". Phys. Rev. Lett. **54**, 368.

Cowin, J.P., Auerbach, D.J., Becker, C., Wharton, L. (1978): Measurement of fast desorption kinetics of D_2 from tungsten by laser induced thermal desorption. Surf. Sci. **78**, 545-564.

Crljen, Z., Gumhalter, B. (1981): Model Hamiltonian for adatom-substrate polarization interaction. Phys. Lett. A **85**, 48-50.

-- (1982): Quantum model for kinetics of helium adsorption on free-electron metals. I. Polarisation induced dissipation. Surf. Sci. **117**, 116-123.

Czanderna, A.W. (ed.) (1975): Methods of Surface Analysis Vols. 1 and 2 (Elsevier, Amsterdam).

Czarniecki, J., Jaroniec, M. (1983): Studies of adsorption kinetics by means of the stochastic numerical simulation. Surf. Sci. Rep. **3**, 301-354.

Dabiri, A.E., Lee, T.J., Stickney, R.E. (1971): Spatial and speed distributions of H_2 and D_2 desorbed from a polycrystalline nickel surface. Surf. Sci. **26**, 522-544.

Danielson, L.R., Dresser, M.J., Donaldson, E.E., Dickinson, J.T. (1978): Adsorption and desorption of ammonia, hydrogen and nitrogen on ruthenium(0001). Surf. Sci. **71**, 599-614.

Dash, J.G. (1975): Films on Solid Surfaces (Academic, New York).

De, G.S., Landman, U., Rasolt, M. (1980): Microscopic theory of thermal desorption and dissociation processes catalysed by a solid surface. Phys. Rev. B **21**, 3256-3268.

de Boer, J.H. (1968): The Dynamical Character of Adsorption (Clarendon, Oxford).

Degras, D.A. (1974): Recent aspects of adsorption-desorption phenomena. Electron. and Fis. Ap. (Spain), **17**, 169-177.

Dekker, H. (1985): Exact Classical and Quantum mechanics of a particle coupled to a membrane. Physica A **129**, 503-513.

Demuth, J.E., Schmeisser, D., Avouris, Ph. (1981): Resonance scattering of electrons from N_2, CO, O_2, and H_2 adsorbed on a silver surface. Phys. Rev. Lett. **47**, 1166-1169.

Derry, G., Wesner, D., Carlos, W., Frankl, D.R. (1979): Selective adsorption of ^{3}He and ^{4}He on the basal plane surface of graphite. Surf. Sci. **87**, 629-642.

Derry, G., Wesner, D., Krishnaswamy, S.V., Frankl, D.R. (1978): Selective adsorption of ^{3}He and ^{4}He on clean surfaces of NaF and LiF. Surf. Sci. **74**, 245-258.

Devonshire, A.F. (1936): The interaction of atoms and molecules with solid surfaces. V - The diffraction and reflection of molecular rays. Proc. R. Soc. London, Ser. A **156**, 37-44.

-- (1937a): The interaction of atoms and molecules with solid surfaces. VIII - The exchange of energy between a gas and a solid. Proc. R. Soc. London, Ser. A **158**, 269-279.

-- (1937b): The interaction of atoms and molecules with solid surfaces. XII - Critical phenomena in a two-dimensional gas. Proc. R. Soc. London, Ser. A **163**, 132-138.

Diebold, A.C., Adelman, S.A., Mou, C.Y. (1979): Generalized Langevin theory for gas-solid processes: Continuum elastic treatment of surface lattice dynamics. J. Chem. Phys. **71**, 3236-3251.

Djafari-Rouhani, B., Dobrzynski, L., Velasco, V.R., Garcia-Moliner, F. (1981): Dynamics of surfaces with overlayers. Surf. Sci. **110**, 129-150.

Doak, R.B. (1981): Ph.D. Thesis, Massachusetts Institute of Technology.

Doak, R.B., Toennies, J.P. (1982): Inelastic molecular beam scattering from solid surfaces. Surf. Sci. **117**, 1-12.

Doll, J.D. (1980): Classical and statistical theories of gas-surface energy transfer. In Aerosol Microphysics I., ed. by W.H. Marlow, Topics Curr. Phys., Vol. 16 (Springer, Berlin, Heidelberg) pp. 61-88.

Doyen, G. (1979): A model for sticking at metal surfaces. Surf. Sci. **89**, 238-250.

-- (1980): Model for handling the transmission problem in sticking at cold solid surfaces. Phys. Rev. B 22, 497-506.

Doyen, G., Grimley, T.B. (1980): A model hamiltonian for phonon energy transfer in gas-solid collisions. Surf. Sci. **91**, 51-72.

Dresser, M.J., Madey, T.E., Yates, J.T. Jr. (1974): The adsorption of xenon by W(111), and its interaction with preadsorbed oxygen. Surf. Sci. **42**, 533-51.

Drolshagen, G., Heller, E.J. (1983): A time dependent wave packet approach to three-dimensional gas-surface scattering. J. Chem. Phys. **79**, 2072-2082.

Dzyaloshinskii, I.E., Lifshitz, E.M., Pitaevskii, L.P. (1961): The general theory of Van der Waals forces. Adv. Phys. **10**, 165-209.

Efrima, S., Freed, K.F., Jedrzejek, C., Metiu, H. (1980): A one-dimensional microscopic model for thermal desorption of an atom. Applications to the case of weak binding. Chem. Phys. Lett. **74**, 43-48.

Efrima, S., Jedrzejek, C., Freed, K.F., Hood, E., Metiu, H. (1983): A one-dimensional model for phonon-induced desorption. J. Chem. Phys. **79**, 2436-2453.

Ehrlich, G. (1970): Surface structure and gas-solid interactions. J. Vac. Sci. Technol. **7**, 52.

Eichenauer, D., Toennies, J.P. (1984): Theory of one-phonon assisted adsorption and desorption of He atoms from a LiF(001) single crystal surface. In Proceedings of the 17th Jerusalem Symposium on Dynamics of Molecule-Surface Interaction, ed. by B. Pullman, J. Jortner (Reidel, Dordrecht).

Einstein, T.L. (1975): Changes in density of states caused by chemisorption. Phys. Rev. B 12, 1262-1274.

Elgin, R.L., Goodstein, D.L. (1974a): Thermodynamic study of the ⁴He monolayer adsorbed on Grafoil. Phys. Rev. A **9**, 2657-2675.

-- (1974b): Thermodynamic functions for ⁴He submonolayers. Proceedings of the 13-th International Conference on Low Temperature Physics, ed. by K.D. Timmerhaus, W.J. O'Sullivan, E.F. Hammel (Plenum, New York).

Elgin, R.L., Greif, J.M., Goodstein, D.L. (1978): Ground state of the helium atom-graphite surface system. Phys. Rev. Lett. **41**, 1723-1725

Ertl, G., Küppers, J. (1974): Low Energy Electrons and Surface Chemistry (Verlag Chemie, Weinheim)

Ertl, G., Neumann, M. (1972): Laser-induzierte schnelle thermische Desorption von Festkörper-Oberflächen. Z. Naturforsch. Teil A **27**, 1607-1610.

Esser, P., Göpel, W. (1980): "Physical" adsorption on single crystal zinc oxide. Surf. Sci. **97**, 309-318.

Estermann, I., Stern, O. (1930): Beugung von Molekularstrahlen. Z. Phys. **61**, 95-125.

Eucken, A. (1914): Zur Theorie der Adsorption. Verh. Dtsch. Phys. Ges. **16**, 345.

Evans, J., Nord, R.S. (1985): Competitive irreversible random one-, two-, three-, ... point adsorption on two-dimensional lattices. Phys. Rev. B **31**, 1759-1769.

Eyring, H. (1935): Activated complex in chemical reactions. J. Chem. Phys. **3**, 107-115.

Eyring, H. (1938): Theory of absolute rates. Trans. Faraday Soc. **34**, 41-48.

Eyring, H. (1963): Modern Chemical Kinetics (Reinhold, New York).

Ezawa, H. (1971): Phonons in a half space. Ann. Phys. (N.Y.) **67**, 438-460.

Feuchtwang, T.E. (1967a): Dynamics of a semi-infinite crystal lattice in a quasiharmonic approximation. I. The static equilibrium configuration of a semi-infinite lattice. Phys. Rev. 155, 715-730.
— (1967b): Dynamics of a semi-infinite crystal lattice in a quasiharmonic approximation. II. The normal-mode analysis of a semi-infinite lattice. Phys. Rev. 155, 731-744.
Feulner, P., Treichler, R., Menzel, D. (1981): Thresholds and mechanisms in electron-stimulated desorption of ions and neutrals from covalent adsorbates on metals. Phys. Rev. B 24, 7427-7430.
Feulner, P., Riedl, W., Menzel, D. (1983): Angular distributions of neutrals desorbed by electron impact from chemisorbed and physisorbed layers on metal surfaces. Phys. Rev. Lett. 50, 986-989.
Feulner, P., Menzel, D., Kreuzer, H.J., Gortel, Z.W. (1984): Kinetic energy distributions of neutrals desorbed by electron impact from adsorbates on metal surfaces. Phys. Rev. Lett. 53, 671-674.
Fisher, G.B., Erickson, N.E., Madey, T.E. (1977): X-Ray photoemission study of physically adsorbed SF_6. Surf. Sci. 65, 210-228.
Frank, H., Hoinkes, H., Wilsch, H. (1977): Interaction of neutral hydrogen atoms with KCℓ(001): Surf. Sci. 63, 121-142.
Franklin, J.L., Harland, P.W. (1974): Gaseous negative ions. Ann. Rev. Phys. Chem. 25, 485-526.
Friedlander, S.K. (1983): Dynamics of aerosol formation by chemical reaction. Ann. N.Y. Acad. Sci. 404, 354-364.
Füstöss, L. (1982): Role of multiple collisions in the classical models of the energy accommodation coefficient. Surf. Sci. 117, 109-115.

Gadzuk, J.W. (1974): Surface molecules and chemisorption. I. Adatom density of states. Surf. Sci. 43, 44-60.
Gaede, W. (1913): Die äussere Reibung der Gase. Ann. Phys. (Leipzig) 41, 289-336.
Garcia, N., Ibáñez, J. (1976): The effect of the long range part of the gas-solid interaction potential on the adsorption processes. J. Chem. Phys. 64, 4803-4804.
Garcia, N., Celli, V., Goodman, F.O. (1979): Model of the interaction potential of an atom with a hard corrugated surface incorporating a stationary potential well. Surf. Sci. 85, 317-325.
Garcia, N., Celli, V., Manson, J.R. (1980): Recent analysis of accommodation coefficients of light atoms. J. Chem. Phys. 72, 3436-3437.
Garrison, B.J., Adelman, S.A. (1977): Lattice influence on gas-solid desorption: RRKM model. J. Chem. Phys. 67, 2379-2380.
Garrison, B.J., Diestler, D.J., Adelman, S.A. (1977): Quantum-dynamical model for thermal desorption of gases from solid surfaces. J. Chem. Phys. 67, 4317-4320.
Garrison, B.J., Adelman, S.A. (1977a): Generalized Langevin theory for gas-solid processes: inelastic scattering studies. Surf. Sci. 66, 253-271.
— (1977b): Lattice influence on gas-solid desorption: RRKM model. J. Chem. Phys. 67, 2379-2380.
George, T.F., Lin, J., Beri, A.C., Murphy, W.C. (1984): Theory of Laser-stimulated surface reactions. Progr. Surf. Sci. 16, 139-274.
Gilbey, D.M. (1961): A re-examination of thermal accommodation coefficient theory. J. Phys. Chem. Solids 23, 1453-1461.
Glachant, A., Bardi, U. (1979): Thermodynamics and kinetics on Xe monolayer adsorption on Cu(100) by LEED and AES. Surf. Sci. 87, 187-202.
Gland, J.L., Kollin, E.B. (1981a): Ammonia adsorption of the Pt(111) and Pt(S) - 6(111) x (111) surfaces. J. Vac. Sci. Tech. 18, 604-606.
— (1981b): Ammonia adsorption on the Pt (111) and Pt(S) - 6 (111) x (111) Surfaces. Surf. Sci. 104, 478-90.
Goldys, E., Gortel, Z.W., Kreuzer, H.J. (1981): Surface Debye temperature in desorption kinetics. Solid State Commun. 40, 963-965.

-- (1982): Desorption kinetics mediated by surface phonon modes. Surf. Sci. **116**, 33-65.

Goodman, F.O. (1962): The dynamics of simple cubic lattices. I. Applications to the theory of thermal accommodation coefficients. J. Phys. Chem. Sol. **23**, 1269-1290.

-- (1965): Response functions and thermal motions of a simple n-dimensional lattice model. Surf. Sci. **3**, 386-414.

-- (1966): One-dimensional theory of desorption. Surf. Sci. **5**, 283-308.

-- (1970): Quantum mechanical basis for the cubes models in gas-surface scattering theory, and an experimental test. J. Chem. Phys. **53**, 2281-2283.

-- (1971): Scattering of atoms by solid surfaces. Critical review of the Lennard-Jones, Devonshire and Strachan theory of inelastic gas-surface interactions, and some improvements. Surf. Sci. **24**, 667-699.

-- (1972a): Quantum mechanical treatment of the one-phonon inelastic scattering of gas atoms in three dimensions by a simplified continuum model of a solid. Surf. Sci. **30**,1-42

-- (1972b): One-phonon inelastic scattering of gas atoms in three dimensions by a simplified continuum model of a solid: Calculation of energy accommocation coefficients. J. Chem. Phys. **56**, 6082-6088.

-- (1975): Thermal accommodation. Progr. Surf. Sci. **5**, 261-375.

-- (1976): Application of a one-dimensional classical model of atom-oscillator scattering to atom-surface accommodation coefficient theory. Surf. Sci. **60**, 45-65.

-- (1980): Accommodation and adsorption of gases on solid surfaces at low temperatures. Surf. Sci. **92**, 185-190.

-- (1981): On the theory of accommodation and adsorption of atoms on solid surfaces at low energies. Surf. Sci. **111**, 279-299.

-- (1983a): Nonequilibrium desorption of helium. Phys. Rev. B **27**, 6478-6480.

-- (1983b): Classical and quantum-mechanical model of thermal and flash desorption of atoms from surfaces. J. Chem. Phys. **78**, 1582-1587.

Goodman, F.O., Garcia, N. (1979): Sticking coefficients of atoms on solid surfaces at low temperatures. Phys. Rev. B **20**, 813-814.

-- (1982): Nonequilibrium desorption of atoms and molecules from surfaces. Surf. Sci. **120**, 251-261.

Goodman, F.O., Gillerlain, J.D. (1971): Correction to the Devonshire theory of accommodation coefficients for bound-state transitions in gas-surface interactions. J. Chem. Phys. **54**, 3077-3083.

Goodman, F.O., Romero, I. (1978): One-phonon scattering of atoms in three dimensions by a simplified continumm model of a surface: thermal desorption. J. Chem. Phys. **69**, 1086-1091.

Goodman, F.O., Wachman, Y. (1976): Dynamics of Gas-Surface Scattering (Academic, New York).

Goodstein, D.L. (1984): The adsorption and desorption of helium films. In Many Body Phenomena at Surfaces, ed. by. D. Langreth, H. Suhl (Academic Press, San Francisco) p.277.

Goodstein, D.L., Weimer, M. (1983): A simple model of helium desorption kinetics. Surf. Sci. **125**, 227-252.

Goodstein, D.L., Hamilton, J.J., Lysek, M.J., Vidali, G. (1984): Phase diagrams of multilayer adsorbed methane. Surf. Sci. **148**, 187-199.

Goodstein, D.L., Maboudian, R., Scaramuzzi, F., Sinvani, M., Vidali, G. (1985): Experiments on quantum and thermal desorption from ⁴He films. Phys. Rev. Lett. **54**, 2034-2037.

Gortel, Z.W., Kreuzer, H.J. (1979): Desorption times in flash desorption and isothermal desorption experiments. Chem. Phys. Lett. **67**, 197-201.

-- (1980): Quantum statistical theory of localized physisorption. Int. J. of Quantum Chem. Quantum Chemistry Symposium **14**, 617-628.

-- (1983a): Negligible adsorbent cooling in thermal desorption. Surf. Sci. 131, L359-L366.

-- (1983b): Flash desorption and thermalization. Surf. Sci. 133, 484-498.

-- (1984): Time of flight spectra in photodesorption via laser-adsorbate coupling. Phys. Rev. B 29, 6926-6931.

-- (1985): Forward peaking and thermalization in the desorption of helium. Phys. Rev. B 31, 3330-3337.

Gortel, Z.W., Kreuzer, H.J., Piercy, P. (1985): Photodesorption by infrared laser-adsorbate coupling: A review of the theoretical approaches. In Adv. in Multi-photon Processes and Spectroscoy, ed. by S.H. Lin (World Scientific Publishing, Singapore).

Gortel, Z.W., Kreuzer, H.J., Spaner, D. (1980): Quantum statistical theory of flash desorption. J. Chem. Phys. 72, 234-246.

Gortel, Z.W., Kreuzer, H.J., Teshima, R. (1980a): Calculation of desorption times using local or nonlocal surface potentials. Can. J. Phys. 58, 376-383.

-- (1980b): Multiphonon processes in a quantum-statistical theory of desorption. Phys. Rev. B 22, 512-530.

-- (1980c): Desorption by phonon cascades for gas-solid systems with many physisorbed surface bound states. Phys. Rev. B 22, 5655-5670.

Gortel, Z.W., Kreuzer, H.J., Teshima, R., Turski, L.A. (1981): Kinetic equations for desorption. Phys. Rev. B 24, 4456-4469.

Gortel, Z.W., Kreuzer, H.J., Piercy, P., Teshima, R. (1983a): Theory of photodesorption of molecules by resonant laser-molecular vibrational coupling. Phys. Rev. B 27, 5066-5083.

-- (1983b): Resonant heating in photodesorption via laser-adsorbate coupling. Phys. Rev. B 28, 2119-2124.

Gortel, Z.W., Kreuzer, H.J., Schäff, M., Wedler, G. (1983): Time of flight spectra in thermal desorption of helium. Surf. Sci. 134, 577-600.

Gortel, Z.W., Kreuzer, H.J., Sommer, E. (1983): Desorption kinetics: corrections to the rigid wall approximation. Surf. Sci. 133, L481-L485.

Gortel, Z.W., Kreuzer, H.J., Wedler, G., Schäff, M. (1984): Time of flight spectra in chemisorption. Surf. Sci. 143, 287-302.

Govers, T.R., Mattera, L., Scoles, G. (1980): Molecular beam experiments on the sticking and accommodation of molecular hydrogen on a low-temperature substrate. J. Chem. Phys. 72, 5446-5455.

Gradshteyn, I.S., Ryzhik, I.M. (1965): Table of Integrals, Series, and Products. Academic, N.Y.

Grimmelmann, E.K., Tully, J.C., Cardillo, M.J. (1980): Hard cube model analysis of gas-surface energy accommocation. J. Chem. Phys. 72, 1039-1043.

Grimmelmann, E.K., Tully, J.C., Helfand, E. (1981): Molecular dynamics of infrequent events: thermal desorption of xenon from a platinum surface. J. Chem. Phys. 74, 5300-5310.

Grunze, M. (1984): Thermodynamic and kinetic phenomena in adsorbed layers. In Proc. 2nd Israel Materials Engineering Conf. Beer Sheva, Israel.

Gumhalter, B., Crljen, Z. (1983): Quantum model for kinetics of helium atoms near surfaces of free-electron metals. II. Overlap-induced dissipation. Surf. Sci. 126, 666-674.

-- (1984): The effect of the electronic surface response on sticking of He atoms on metallic substrates. Surf. Sci. 139, 231-238.

Gurney, R.W. (1935): Theory of electrical double layers in adsorbed films. Phys. Rev. 47, 479-482.

Habenschaden, E., Küppers, J. (1984): Evaluation of flash desorption spectra. Surf. Sci. 138, L147-L150.

Haber, F. (1914): Modern Chemical Industry. (Hurter Memorial Lecture): J. Ind. Eng. Chem. 6, 325-331.

-- (1914): Discussion of R. Marc, Über die Kinetik der Adsorption. Z. Elektrochem. 20, 521.

Hamilton, J.J., Goodstein, D.L. (1983): Thermodynamic study of methane multilayers adsorbed on graphite. Phys. Rev. B 28, 3838-3848.

Hayward, D.O., King, D.A., Tompkins, F.C. (1967): Variation of sticking probabilities with temperature and coverage, and desorption spectra for nitrogen on tungsten films. Proc. R. Soc. London, Ser. A 297, 321-335.

Heidberg, J., Stein, H., Riehl, E., Nestmann, A. (1980): Evaporation and desorption by resonant excitation of molecular normal vibrations with laser infrared. Z. Phys. Chem. (N.F.) 121, 145-164.

Heidberg, J., Stein, H., Riehl, E. (1982): Resonance, rate, and quantum yield of infrared-laser-induced desorption by multiquantum vibrational excitation of the adsorbate CH_3F on NaCℓ. Phys. Rev. Lett. 49, 666-669.

Hill, T.L. (1952): Theory of physical adsorption. Adv. Catal. 4, 212-258.

von Himbergen, J.E., Silbey, R. (1979): Applications of the Hellman-Feynman theorem in surface physics. Phys. Rev. B 20, 567-575.

Hobson, J.P. (1974): Physical adsorption. CRC Crit. Rev. Solid State Sci. 4, 221-245.

Hoinkes, H. (1980): The physical interaction potential of gas atoms with single-crystal surfaces, determined from gas-surface diffraction experiments. Rev. Mod. Phys. 52, 933-970.

Holloway, S., Beeby, J.L. (1975): The theory of atomic desorption within the harmonic approximation. J. Phys. C 8, 3531-3540.

Holloway, S., Jewsbury, P. (1976): The theory of desorption from a general potential. J. Phys. C 9, 1907-1918.

Holloway, S., Jewsbury, P., Beeby, J.L. (1977): Theoretical analysis of desorption from surfaces. Surf. Sci. 63, 339-347.

Holmberg, C., Apell, P. (1984): Van der Waals interaction in atom-surface scattering. Phys. Rev. B 30, 5721-5733.

-- (1984): Improved description of the Van der Waals interaction in Physisorption. Solid State Commun. 49, 513-517.

Hölzl, J., Schulte, F.K. (1979): Work Function of Metals, in Springer Tracts Mod. Phys., Vol. 85 (Springer, Berlin, Heidelberg).

Honig, J.M. (1979): Systematization of the thermodynamics of gas adsorption phenomena. J. Colloid Interface Sci. 70, 83-89.

Hood, E., Jedrzejek, C., Freed, K.F., Metiu, H. (1984): A one-dimensional model for phonon-induced desorption. II. Numerical analysis of the desorption of noble gas atoms (argon, krypton, and xenon) from tungsten and carbon monoxide from copper. J. Chem. Phys. 81, 3277-3293.

Hopkins, B.J., Williams, C.B., Willmer, P.C. (1971): Chemical and physical adsorption of oxygen on the (110) plane of tungsten. Surf. Sci. 25, 633-642.

Horton, D.R., Masel, R.I. (1982): Angular resolved flash desorption of hydrogen from recrystalized nickel. Surf. Sci. 116, 13-21.

Horton, D.R., Banholzer, W.F., Masel, R.I. (1982): Directed desorption as a probe of the structure of the desorption site. Surf. Sci. 116, 22-32.

Hughes, F.L. (1959): Mean adsorption lifetime of Rb on etched tungsten single crystals: Neutrals. Phys. Rev. 113, 1036-1038.

Hurst, J.E., Becker, C.A., Cowin, J.P., Janda, K.C., Wharton, L., Auerbach, D.J. (1979): Observation of direct inelastic scattering in the presence of trapping - desorption scattering: Xe on Pt(111). Phys. Rev. Lett. 43, 1175-1177.

Hurst, J.E., Wharton, L., Janda, K.C., Auerbach, D.J. (1983): Direct inelastic scattering Ar from Pt(111). J. Chem. Phys. 78, 1559-1581.

Hussla, I., Seki, H., Chuang, T.J., Gortel, Z.W., Kreuzer, H.J., Piercy, P. (1985): Infrared laser-induced photodesorption of NH_3 and ND_3 adsorbed on Cu(100) and Ag(films). Phys. Rev. B 32, 3489-3501.

Iannotta, S., Valbusa, V. (1980): High resolution H atom scattering from NaCl (001). Surf. Sci. **100**, 28↔34.

Ibach, H., Bruchmann, L. (1980): Observation of surface phonons on Ni(111) by electron-loss spectroscopy. Phys. Rev. Lett. **44**, 36↔39.

Ibach, H., Erley, W., Wagner, H. (1980): The preexponential factor in desorption → CO on Ni (111). Surf. Sci. **92**, 29↔42.

Iche, G., Nozières, P. (1976): A simple stochastic description of desorption rates. J. Phys. (Paris) **37**, 1313-1323.

Iosilevski, Ya.A. (1971): On the dynamics of surface atoms. Phys. Stat. Solidi (b) **46**, 125↔135.

Jack, D.B., Kreuzer, H.J. (1982): Derivation of Kramers' equation, friction coefficient and macroscopic laws for physisorption. Phys. Rev. B **26**, 6516-6529.

-- (1985): On the Kramers' equation for physisorption. Phys. Rev. B **31**, 2514-2516.

Janda, K.C., Hurst, J.E., Becker, C.A., Cowin, J.P., Auerbach, D.J., Warton, L. (1980): Direct measurement of velocity distributions in argon beam-tungsten surface scattering. J. Chem. Phys. **72**, 2403-2410.

Jedrzejek, C. (1984): Quantum Dynamical Model of Desorption Processes of Atom from Surface (Jagiellonian University Press, Krakow).

Jedrzejek, C., Freed, K.F., Efrima, S., Metiu, H. (1981): A one-dimensional microscopic quantum mechanical theory of light enhanced desorption. Surf. Sci. **109**, 191↔206.

-- (1981): A one-dimensional microscopic model for the rate of thermal desorption of an atom. The role of multiphonon processes. Chem. Phys. Lett. **79**, 227-232.

Jedrzejek, C., Gijzeman, O.L.J., Freed, K.F. (1981): Theoretical test of none-quilibrium experimental method for measuring heats of adsorption. Surf. Sci. **107**, 43-50.

Jewsbury, P. (1977a): The interaction of adsorbed molecules with surfaces: I. Desorption. J. Phys. C **10**, 671↔680.

-- (1977b): The interaction of adsorbed molecules with surfaces: II Dissociation. J. Phys. C **10**, 681↔687.

Jewsbury, P., Beeby, J.L. (1975): The theory of atomic desorption for general potentials. J. Phys. C **8**, 3541-3548.

Jiang, X.P., Toigo, F., Cole, M.W. (1984a): The dispersion force of physical adsorption. I. Local theory. Surf. Sci. **145**, 281↔293.

-- (1984b): The dispersion force of physical adsorption. II. Nonlocal theory. Surf. Sci. **148**, 21↔36.

Jody, B.J., Fain, P.C., Saxena, S.C. (1977): Thermal accommodation coefficients of neon and krypton on gas-covered platinum. Chem. Phys. Lett. **48**, 545-549.

Kac, M., Logan, J. (1979): Fluctuations. In Studies in Statistical Mechanics, ed. by E.W. Montroll, J.L. Leibovitz (North-Holland, Amsterdam) pp. 3↔60.

van Kampen, N.G. (1961). A power series expansion of the master equation. Can. J. Phys. **39**, 551-567.

-- (1981): Stochastic Processes in Physics and Chemistry (North-Holland, Amsterdam).

Kayser, H. (1881): Über die Verdichtung von Gasen an Oberflächen in ihrer Abhängigkeit von Druck und Temperatur. Ann. Physik und Chemie 12, 450-468 and 526-537.

Keck, J.C. (1967): Generalized diffusion theory of nonequilibrium reaction rates. J. Chem. Phys. **46**, 4211-4213.

Keesom, P.H., Pearlman, N. (1955): Atomic heat of graphite between 1 and 20 K. Phys. Rev. **99**, 1119↔1124.

Kidnay, A.J., Hiza, M.J. (1970): Physical adsorption in cryogenic engineering. Cryogenics 10 271-274.

Kim, S.K. (1958): Mean first passage time for a random walker and its application to chemical kinetics. J. Chem. Phys. 28, 1057-1067.

King, D.A. (1968): Nitrogen adsorption on nickel and palladium films. Surf. Sci. 9, 375-395.

-- (1977): The influence of weakly bound intermediate states on thermal desorption kinetics. Surf. Sci. 64, 43-51.

-- (1978): Kinetics of adsorption, desorption, and migration at single-crystal metal surfaces. CRC Crit. Rev. Sol. State and Mat. Sci. 7, 167-208.

King, D.A., Tompkins, F.C. (1968): Sticking probabilities, redistribution phenomena and desorption spectra for nitrogen on molybdenum and titanium films. Trans. Faraday Soc. 64, 496-506.

Kisliuk, P. (1957): The sticking probabilities of gases chemisorbed on the surfaces of solids. J. Phys. Chem. Sol. 3, 95-101.

-- (1958): The sticking probabilities of gases chemisorbed on the surfaces of solids II. J. Phys. Chem. Sol. 5, 78-84.

-- (1975): Thermal desorption from metal surfaces: A review. Surf. Sci. 47, 384-402.

Kleiman, G.G., Landman, U. (1973): Theory of physisorption: He on Metals. Phys. Rev. B 8, 5484-5495.

-- (1973): Prediction of physisorption energies: He on metals. Phys. Rev. Lett. 31, 707-710.

-- (1974): Effect of spatial dispersion upon physisorption energies: He on metals. Phys. Rev. Lett. 33, 524-527.

-- (1976): The interaction of rare gases with metals. Sol. State Commun. 18, 819-822.

Klein, J.R., Chan, M.H.W., Cole, M.W. (1984): Liquid-vapor critical point of physisorbed films. Surf. Sci. 148, 200-211.

Knowles, T.R., Suhl, H. (1977): Sticking coefficient of atoms on solid surfaces at low temperatures. Phys. Rev. Lett. 39, 1417-1420.

Knudsen, M. (1909): Die Molekularströmung der Gase durch Öffnungen und die Effusion. Ann. Phys. (Leipzig) 28, 999-1016.

-- (1911): Molekularströmung des Wasserstoffs durch Röhren und das Hitzdrahtmanometer. Ann. Phys. (Leipzig) 35, 389-396.

-- (1915): Das Cosinusgesetz in der kinetischen Gastheorie. Ann. Phys. (Leipzig) 48, 1113-1121.

Kohrt, C., Gomer, R. (1970): Adsorption of oxygen on the (110) plane of tungsten. J. Chem. Phys. 52, 3283-3294.

Kornelsen, E.V. (1969): Ionic Entrapment in tungsten monocrystals; a survey of effects observed in thermal desorption. J. Vac. Sci. Technol. 6, 172-174.

Koster, G.P. (1957): Space groups and their representations. Solid State Physics 5, 174-256.

Kouptsidis, J., Menzel, D. (1967): Akkommodation der Edelgase an reinen Wolfram- und Molybdänoberflächen bei Zimmertemperatur. Ber. Bunsenges. physik. Chem. 71, 720-730

-- (1969): Thermal accommodation of the helium isotopes on clean tungsten surfaces. Z. Naturforschung - A 24, 479-480.

-- (1970): Accommodation of the rare gases on clean tungsten surfaces between 77 and 380K. Ber. Bunsenges. physik. Chem. 74, 512-520.

Kramers, H.A. (1940): Brownian Motion in a field of force and the diffusion model of chemical reactions. Physica 7, 284-304.

Kreuzer, H.J. (1980): Quantum statistical theory of adsorption and desorption of a gas at a solid surface. Surf. Sci. 100, 178-198.

-- (1981): Nonequilibrium Thermodynamics and its Statistical Foundations (Oxford University Press, Oxford) [Paperback edition (1983)]

Kreuzer, H.J., Gortel, Z.W. (1980): Limitations of the relaxation time approach to desorption. Chem. Phys. Lett. **73**, 365–369.

Kreuzer, H.J., Lowy, D.N. (1981): Photodesorption of diatomic molecules by laser–molecular vibrational coupling. Chem. Phys. Lett. **78**, 50–53.

Kreuzer, H.J., Summerside, P. (1981): Physisorption kinetics and equilibrium properties of gas-solid systems approaching monolayer coverage. Surf. Sci. **111**, 102–118.

Kreuzer, H.J., Teshima, R. (1981): Desorption times from rate equations, the master equation, and the Fokker-Planck equation. Phys. Rev. B **24**, 4470–4483.

Kromhout, R.A., Linder, B. (1979): On the dipole moments of physisorbed rare gas atoms. Chem. Phys. Lett. **61**, 283–287.

Küppers, J., Nitschke, F., Wandelt, K., Ertl, G. (1979): The adsorption of Xe on Pd(110). Surf. Sci. **87**, 295–314.

Kusunoki, I. (1974): Sticking coefficient curves expected for multilayer adsorption. J. Phys. Chem. **78**, 748–751.

Landau, L.D. (1935): On the theory of the accommodation coefficient. Physik. Z. Sowjet. **8**, 489–497.

Landman, U., Kleiman, G.G. (1975): Local and nonlocal effects in the theory of physisorption. J. Vac. Sci. Technol. 12, 206–209.

—— (1977): Microscopic approaches to physisorption: Theoretical and experimental aspects. In Surface and Defect Properties of Solids, Vol. 6. Specialist Periodical Reports. The Chemical Society, London, pp. 1–105.

Lang, N.D. (1981): Interaction between closed-shell systems and metal surfaces. Phys. Rev. Lett. **46**, 842–845.

Lang, N.D., Nörskov, J.K. (1983): Interaction of helium with a metal surface. Phys. Rev. B **27**, 4612–4616.

Lang, N.D., Williams, A.R. (1978): Theory of atomic chemisorption on simple metals. Phys. Rev. B **18**, 616–636.

Langbein, D. (1974): Theory of Van der Waals Attraction, in Springer Tracts Mod. Phys., Vol. 72 (Springer, Berlin, Heidelberg)..

Langmuir, I. (1916): The constitution and fundamental properties of solids and liquids. J. Am. Chem. Soc. 38, 2221-2295.

Lax, M. (1960): Cascade capture of electrons in solids. Phys. Rev. **119**, 1502–1523.

Le Lay, G., Manneville, M., Kern, R. (1977): Isothermal desorption spectroscopy for the study of two–dimensional condensed phases. Investigation of the Au (deposit)/Si (111) (substrate) system; application to the Xe/(0001) graphite system. Surf. Sci. **65**, 261–276.

Le Roy, R.J. (1978): Determining potential energy constants for atom- and molecule–surface interactions. Surf. Sci. **59**, 541–553.

Lee, T.J. (1974): The physical adsorption of helium on solid argon, krypton and xenon. Surf. Sci. **44**, 389–400.

Lennard–Jones, J.E. (1932): Processes of adsorption and diffusion on solid surfaces. Trans. Faraday Soc. **28**, 333–359.

—— (1937): The interaction of atoms and molecules with solid surfaces. XI – The dispersal of energy from an activated link. Proc. R. Soc. London, Ser. A 163, 127–131.

Lennard–Jones, J.E., Dent, B.M. (1928): Cohesion at a crystal surface. Trans. Faraday Soc. **24**, 92–108.

Lennard–Jones, J.E., Devonshire, A.F. (1936a): The interaction of atoms and molecules with solid surfaces. III – The condensation and evaporation of atoms and molecules. Proc. R. Soc. London, Ser. A **156**, 6-28.

—— (1936b): The interaction of atoms and molecules with solid surfaces. IV – The condensation and evaporation of atoms and molecules. Proc. R. Soc. London, Ser. A **156**, 29–36.

--- (1937a): The interaction of atoms and molecules with solid surfaces. VI – The behavior of adsorbed helium at low temperatures. Proc. R. Soc. London, Ser. A 158, 242-252.

--- (1937b): The interaction of atoms and molecules with solid surfaces. VII – The diffraction of atoms by a surface. Proc. R. Soc. London, Ser. A 158, 253-268.

Lennard-Jones, J.E., Goodwin, E.T. (1937): The interaction of atoms and molecules with solid surfaces. X – The activation of adsorbed atoms by metallic electrons. Proc. R. Soc. London, Ser. A 163, 101-127.

Lennard-Jones, J.E., Strachan, C. (1935): The interaction of atoms and molecules with solid surfaces. I – The activation of adsorbed atoms to higher vibrational states. Proc. R. Soc. London, Ser. A 150, 442-455.

Leuthäuser, U. (1980): Generalized quasi-chemical approximation for a lattice gas: Application to CO on Ru. Z. Phys. B 37, 65-67.

--- (1981): Kinetic theory of adsorption and desorption. Z. Phys. B 44, 101-108.

--- (1983): Kinetic theory of desorption: Energy and angular distributions. Z. Phys. B 50, 65-69.

--- (1984): Kinetic theory of gas surface scatterings: Energy and angular distributions of the fast part. Surf. Sci. 145, 48-61.

Liebsch, A., Harris, J., Meinert, M. (1984): Interaction of helium with a graphite surface. Surf. Sci. 145, 207-222.

Lilienkamp, G., Toennies, P.J. (1982): Observation of resonant energy transfer between a selectively adsorbed He atom and surface Rayleigh phonons of the LiF(001) single-crystal substrate. Phys. Rev. B 26, 4752-4755.

--- (1983): The observation of one-phonon assisted selective desorption and adsorption of He atoms in defined vibrational levels on a LiF (001) single crystal plane. J. Chem. Phys. 78, 5210-5224.

Lin, J., George, T.F. (1979): Kinetic model of laser-controlled heterogeneous processes. Chem. Phys. Lett. 66, 5-8.

Lin, Y.W., Wolken Jr. G. (1976a): Theoretical study of gas-solid energy transfer: An isotropic Debye model. J. Chem. Phys. 65, 2634-2649.

--- (1976b): Theoretical study of energy transfer in molecule-solid collisions: H_2, D_2 + Ag(111). J. Chem. Phys. 65, 3729-3734.

Lin, Y.W., Adelman, S.A.; Wolken Jr. G. (1977): A test of the first order distorted wave Born approximation in gas-solid energy transfer. Surf. Sci. 66, 376-379.

Logan, R.M. (1969): Calculation of the energy accommodation coefficient using the soft-cube model. Surf. Sci. 15, 387-402.

Logan, R.M., Keck, J.C. (1968): Classical theory for the interaction of gas atoms with solid surfaces. J. Chem. Phys. 49, 860-876.

Logan, R.M., Stickney, R.E. (1966): Simple classical model for the scattering of gas atoms from a solid surface. J. Chem. Phys. 44, 195-201.

London, F. (1930): Über einige Eigenschaften und Anwendungen der Molekularkräfte. Z. f. Phys. Chem. B 11, 222-251.

Lundqvist, B.I., Gunnarsson, O., Hjelmberg, H., Nórskov, J. K. (1979): Theoretical description of molecule-metal interaction and surface reactions. Surf. Sci. 89, 196-225.

Madey, T.E. (1972): Adsorption and displacement processes on W(111) involving CH_4, H_2, and O_2. Surf. Sci. 29, 571-89.

Madey, T.E., Yates Jr., J.T. (1977): Desorption methods as probes of kinetics and bonding at surfaces. Surf. Sci. 63, 203-231.

Madix, R.J. (1980): The kinetics of elementary reactions on single-crystal surfaces. AIP Conf. Proc. (USA) 61, 39-56.

Madix, R.J., Benziger, J. (1978): Kinetic processes on metal single-crystal surfaces. Ann. Rev. Phys. Chem. 29, 285-306.

Mahanty, J., Ninham, B.W. (1974): Dispersion Forces (Academic, New York).
Manson, J.R. (1972): Simple model for the energy accommodation coefficient. J. Chem. Phys. 56, 3451-3455.
Manson, J.R., Tompkins, J. (1976): Two phonon contributions to the energy accommodation coefficient. In Proceedings of Int. Symposium Rarefied Gas Dynamics, ed. by J.L.Potter. Progr. in Astronautics and Aeronautics 51, 603-619.
Maradudin, A.A., Wallis, R.F., Dobrzynski, L. (1980): Surface phonons and polaritons. In Handbook of Surfaces and Interfaces III, ed. by L. Dobrzynski (Garland STPM Press, New York).
Maradudin, A.A., Montroll, E.W., Weiss, G.H., Ipatova, I.P. (1971): Theory of Lattice Dynamics in the Harmonic Approximation (Academic, New York).
Margenau, H., Pollard, W.G. (1941): The forces between neutral molecules and metallic surfaces. Phys. Rev. 60, 128-134.
Masel, R.I., Merrill, R.P., Miller, W.H. (1974): Semiclassical trajectory calculation of He scattering from W(112). Surf. Sci. 46, 681-688.
Mavroyannis, C. (1963): The interaction of neutral molecules with dielectric surfaces. Mol. Phys. 6, 593-600.
McCarroll, B. (1969): Analysis of thermal desorption spectra. J. Appl. Phys. 40, 1-9.
McCarroll, B., Ehrlich, G. (1963): Trapping and energy transfer in atomic collisions with a crystal surface. J. Chem. Phys. 38, 523-532.
McElhinney, G., Pritchard, J. (1976): The adsorption of Xe and CO on Au(100). Surf. Sci. 60 397-410.
McElhinney, G., Papp, H., Pritchard, J. (1976): The adsorption of Xe and CO on Ag (111). Surf. Sci. 54 617-634.
McLachlan, A.D. (1964): Van der Waals forces between an atom and a surface. Mol. Phys. 7, 381-388.
Mennicke, S., Wagner, W., Dittmar, W. (1973): Bestimmung der Aktivierungsenergie der Desorption des Stickstoffes von einer Kaliumoberfläche bei tiefen Temperaturen. Surf. Sci. 36, 805-809.
Menzel, D. (1975): Desorption phenomena. In Interactions on Metal Surfaces, ed. by R. Gomer, Topics Appl. Phys., Vol.4 (Springer, Berlin, Heidelberg) pp. 102-143.
-- (1982): Thermal desorption. In Chemistry and Physics of Solid Surfaces IV, ed. by R. Vanselow, R. Howe (Springer-Verlag, Berlin) pp. 389-406.
-- (1982): Recent developments in electron and photon stimulated desorption. J. Vac. Sci. Technol. 20, 538-543.
Messiah, A. (1961): Quantum Mechanics (North-Holland, Amsterdam).
Modak, A.T., Pagni, P.J. (1973): Atom trapping on surfaces. J. Chem. Phys. 59, 2019-2031.
-- (1976): Time and space dependent energy distributions of atoms from surfaces. J. Chem. Phys. 65, 1327-1344.
Montroll, E.W., Shuler, K.E. (1958): The application of the theory of stochastic processes to chemical kinetics. Adv. Chem. Phys. 1,361-399.
Muirhead, R.J., Dash, J.G., Krim, J. (1984): Wetting and nonwetting of molecular films at zero temperature. Phys. Rev. B 29, 5074-5080.
Mülfahrt, P. (1900): Ueber Adsorption von Gasen an Glaspulver. Ann. Phys. (Leipzig) 3, 328-352.
Müller, H., Brenig, W. (1979): Kinetic theory of gas-surface interaction. Z. Phys. B 34, 165-173.
Muscat, J.P., Newns, D.M. (1978): Chemisorption on metals. Progr. Surf. Sci. 9,1-43.
Musser, S.W., Rieder, K.H. (1970): Influence of surface force constant changes on surface mode frequencies. Phys. Rev. B 2, 3034-3039.

Nath, K., Gortel, Z.W., Kreuzer, H.J. (1985): Van der Waals interaction of rare gases and hydrogen on semiconductor surfaces. Surf. Sci. 155, 596-606.

Newns, D.M. (1969): Self-consistent model of hydrogen chemisorption. Phys. Rev. **178**, 1123-1135.

Nieuwenhuys, B.E., Sachtler, W.M.H. (1974): Crystal face specificity of xenon adsorption on iridium field emitters. Surf. Sci. **45**, 513-529.

Nitzan, A., Carmeli, B. (1982): Non-markoffian theory of activated rate processes II. Thermal Desorption. Isr. J. Chem. 22, 360-364.

Nórskov, J.K., Lundquist, B.I. (1979): Correlation between sticking probability and adsorbate-induced electron structure. Surf. Sci. **89**, 251-261.

Novaco, A.D. (1973): Cluster expansion for superlattice physisorption: Helium adsorbed upon graphite. Phys. Rev. A **7**, 1653-1659.

Opila, R., Gomer, R. (1981): Thermal desorption of Xe from the W(110) plane. Surf. Sci. 112, 1-22.

Østgaard, E. (1968): Treatment of the hard core in the two-nucleon potential in nuclear matter calculations. Phys. Rev. **168**, 1139-1144.

Pagni, P.J. (1973): Comparison of diffusion theory adsorption and desorption rate constants with experimental lifetimes. J. Chem. Phys. **58**, 2940-2954.

Pagni, P.J., Keck, J.C. (1973): Diffusion theory for adsorption and desorption of gas atoms at surfaces. J. Chem. Phys. **58**, 1162-1177.

Palmberg, P.W. (1971): Physical adsorption of xenon on Pd(100). Surf. Sci. **25**, 598-608.

Palmer, R.L., Smith Jr., J.N., Saltsburg, H., O'Keefe, D.R. (1970): Measurement of the reflection, adsorption, and desorption of gases from smooth metal surfaces. J. Chem. Phys. **53**, 1666-1676.

Papp, H., Pritchard, J. (1975): The adsorption of Xe and CO on a Cu (311) single crystal surface. Surf. Sci. **53**, 371-82.

Persson, B.N.J., Apell, P. (1983): Sum rules for surface response functions with application to the van der Waals interaction between an atom and a metal. Phys. Rev. B **27**, 6058-6065.

Perrson, B.N.J., Zaremba, E. (1984): Reference-plane position for the atom-surface van der Waals interaction. Phys. Rev. **30**, 5669-5679.

Petermann, L.A. (1971): The interpretation of slow desorption kinetics. 2nd Intern. Symp. on Adsorption-Desorption Phenomena, pp.14-17 April, Florence, Italy.

Pfnür, H., Feulner, P., Engelhardt, H.A., Menzel, D. (1978): An example of "fast" desorption: anomalously high pre-exponentials for CO desorption from Ru (001). Chem. Phys. Lett. **59**, 481-486.

Pfnür, H., Menzel, D. (1983): The influence of adsorbate interactions on kinetics and equilibrium for CO on Ru (001): I. Adsorption kinetics. J. Chem. Phys. **79**, 2400-2410.

Pfnür, H., Feulner, P., Menzel, D. (1983): The influence of adsorbate interactions on kinetics and equilibrium for CO on Ru (001): II. Desorption kinetics and equilibrium. J. Chem. Phys. **79**, 4613-4623.

Pollard, W.G. (1941): Exchange forces between neutral molecules and a metal surface. Phys. Rev. **60**, 578-585.

Polanyi, J.C., Wolf, R.J. (1985): Dynamics of simple gas-surface interaction. II. Rotationally inelastic collisions at rigid and moving surfaces. J. Chem. Phys. **82**, 1555-1566.

Polanyi, M. (1932): Theories of the adsorption of gases. A general survey and some additional remarks. Trans. Faraday Soc. 28, 316-333.

Price, G.L. (1974): Potential energies of adsorbed rare gases on graphite. Surf. Sci. **46**, 697-702.

Prosen, E.J.R., Sachs, R.G. (1942): The interaction between a molecule and a metal surface. Phys. Rev. **61**, 65-73.

Rawlings, K.J., Foulias, S.D., Price, G.G., Hopkins, B.J. (1982): Thermal desorption using AES. Some examples of adsorbate-adsorbate interactions. Surf. Sci. 118, 47-56.

Redhead, P.A. (1962): Thermal desorption of gases. Vacuum 12, 203-211.

Redondo, A., Zeiri, Y., Goddard III, W.A. (1984a): Rates of desorption from solid surfaces: Coverage dependence. Surf. Sci. 136, 41-58.

— (1984b): Classical stochastic diffusion theory for thermal desorption from solid surfaces. J. Vac. Sci. Technol. B 2, 550-560.

Reyes, J., Chavira, E., Romero, I., Goodman, F.O. (1984): Angle resolved thermal and flash desorption: Analytical expressions for the flux, average energy, and speed ratio. Surf. Sci. 148, 155-166.

Richard, A.M., Depristo, A.E. (1983): On the surface temperature dependence of non-equilibrium translational energy accommodation coefficients. Surf. Sci. 124, 241-252.

Riley, M.E., Diestler, D.J. (1984): Energy transfer in the collision of an atom with a cold surface. J. Chem. Phys. 81, 6361-6366.

Roberts, J.K. (1935): The adsorption of hydrogen on tungsten. Proc. Roy. Soc. (London) A 152, 445-480.

— (1939): Some Problems in Adsorption (Cambridge University Press, Cambridge).

Roberts, M.W., McKee, C.S. (1978): Chemistry of the Metal-Gas Interface (Clarendon Press, Oxford).

Roberts, R.H., Pritchard, J. (1976): Monolayer structures of Kr and Xe on Ag(111) and Cu (211). Surf. Sci. 54, 687-691.

Rodersan, T.S., Mills, D. L., Block, J.E. (1983): Sensitivity of electron - energy-loss spectra to adsorption site: Au ordered overlayer on the Ni(111) surface. Phys. Rev. B 27, 4059-4071.

Rogers Jr., J.W., Campbell, C.T., Hance, R.L., White, J.M. (1980): An electron spectroscopy study of ammonia adsorption on clean and oxidized aluminum. Surf. Sci. 97, 425-447.

de Rouffignac, E., Alldredge, G.P., de Wette, F.W. (1981a): Lattice dynamics of graphite slabs. Phys. Rev. B 23, 4208-4219.

— (1981b): Dynamics of xenon - covered graphite slabs. Phys. Rev. B 24, 6050-6059.

Rybolt, T.R., Pierotti, R.A. (1979): Rare gas-graphite interaction potentials. J. Chem. Phys. 70, 4413-4419.

Sams, J.R. (1974): Applications of statistical mechanics to physical adsorption. Proc. Surf. Membr. Sci. 8, 1-48.

Saxena, S.C., Joshi, R.K. (1981): Thermal Accommodation and Adsorption Coefficients of Gases (McGraw-Hill, New York).

Scales, W.W. (1958): Specific heat of LiF and KI at low temperatures. Phys. Rev. 112, 49-54.

Schaich, W.L. (1974): Brownian motion model of surface chemical reactions. Derivation on the large mass limit. J. Chem. Phys. 60, 1087-1093.

Schick, M. (1981): The classification of order-disorder transitions on surfaces. Prog. Surf. Sci. 11, 245-292.

Schmeisser, D., Demuth, J.E., Avouris, Ph. (1982): Electron-energy-loss studies of physisorbed O_2 and N_2 on Ag and Cu surfaces. Phys. Rev. B 26, 4857-4863.

Schmeits, M., Lucas, A.A. (1983): Physical adsorption and surface plasmons. Prog. Surf. Sci. 14, 1-51.

Schmidt, L.D. (1980): Precursor intermediates in adsorption, desorption and reaction. AIP Conf. Proc. (USA) 61, 83-96.

Schmit, J.N. (1976): The effect of long-range three-particle forces on the surface energy and physisorption energy of rare gas crystals. Surf. Sci. 55, 589-600.

Schönhammer, K. (1979): On the Kisliuk model for adsorption and desorption kinetics. Surf. Sci. 83, L633-L636.

Schönhammer, K., Gunnarsson, O. (1980): Sticking probability on metal surfaces: Contribution from electron-hole pair excitations. Phys. Rev. B 22, 1629-1637.

-- (1981): Sticking probability on metal surfaces: Temperature dependence of the electron-hole pair mechanism. Phys. Rev. B 24, 7084-7092.

-- (1982): Sticking and inelasting scattering at metal surfaces: The electron-hole pair mechanism. Surf. Sci. 117, 53-59.

-- (1983a): Electronic friction and covalent chemisorption. Phys. Rev. B 27, 5113-5115.

-- (1983b): Energy dissipation at metal surfaces: Electronic versus vibrational excitations. J. Electron Spectr. and Related Phenom. 29, 91-103.

Schrödinger, E. (1914): Zur Dynamik elastisch gekoppelter Punktsysteme. Ann. Phys. (Leipzig) 44, 916-934.

Schumann, O. (1886): Über die Dicke der adsorbirten Luftschicht auf Glasflächen. Ann. Physik und Chemie 27, 91-94.

Schweber, S.S. (1961): An Introduction to Relativistic Quantum Field Theory (Harper and Row, New York).

Sedlmeir, R., Brenig, W. (1980): Inelastic Atom-Surface Scattering: A Comparison of Classical and Quantum Treatments. Z. Physik B 36, 245-250.

Seki, H., Chuang, T.J. (1982): The detection by SERS of resonantly excited desorption of pyridine from silver island films by IR laser absorption. Solid State Commun. 44, 473-475.

Sexton, B.A., Mitchell, G.E. (1980): Vibrational spectra of ammonia chemisorbed on platinum (111): I. Identification of chemisorbed states. Surf. Sci. 99, 523-538.

Shields, F.D. (1975): An acoustical method for determining the thermal and momentum accommodation of gases on solids. J. Chem. Phys. 62, 1248-1252.

-- (1983): Energy and momentum accommocation coefficients on platinum and silver. J. Chem. Phys. 78, 3329-3333.

Shigeishi, R.A. (1975): Adsorption of methane on a tungsten field emitter. Surf. Sci. 51, 377-395.

Sinvani, M., Taborek, P., Goodstein, D.L. (1982): Direct measurement of desorption kinetics of ^{4}He at low temperatures. Phys. Rev. Lett. 48, 1259-1263.

Sinvani, M., Cole, M.W., Goodstein, D.L. (1983): Sticking probability of ^{4}He on solid surfaces at low temperatures. Phys. Rev. Lett. 51, 188-191.

Sinvani, M., Goodstein, D.L. (1983): Relaxation time of an adsorbing ^{4}He film. Surf. Sci. 125, 291

-- (1983): Direct and thermal desorption of ^{4}He films. Phys. Lett. A 95, 59-62.

Sinvani, M., Goodstein, D.L., Cole, M.W. (1984): Scattering of low-energy helium atoms from a low-temperature solid surface. Phys. Rev. B 29, 3905-3907.

Sinvani, M., Goodstein, D.L., Cole, M.W., Taborek, P. (1984): Desorption of helium atoms from thin films. Phys. Rev. B 30, 1231-1248.

Slater, J.C. (1951): A simplification of the Hartree-Fock method. Phys. Rev. 81, 385-390.

Smith, J.N., Palmer, R.W. (1972): Molecular beam study of oxidation of deuterium on a (111) Platinum surface. J. Chem. Phys. 56, 13-20.

Smith, J.R. (ed.) (1980): Theory of Chemisorption, Topics Curr. Phys., Vol.19 (Springer, Berlin, Heidelberg).

Sols, F., Flores, F., Garcia, N. (1984): Friction and sticking coefficients of rare gases approaching a metal surface. Surf. Sci. 137, 167-180.

Sommer, E., Kreuzer, H.J. (1982a): Mean field theory of multilayer physisorption. II. Thermodynamic functions for ^{3}He and ^{4}He adsorbed on graphite. Phys. Rev. B **26**, 658~668.

~- (1982b): Mean field theory of multilayer physisorption. III. Desorption kinetics. Phys. Rev. B **26**, 4094~4105.

~- (1982c): Physisorption kinetics from mean field theory: compensation effect near monolayer coverage. Phys. Rev. Lett. **49**, 61~64.

~- (1982d): Small prefactors and compensation effect in physisorption kinetics. Surf. Sci. **119**, L331~L338.

Somorjai, G. (1972): <u>Principles of Surface Chemistry</u> (Prentice~Hall, N.J.).

Steele, W.A. (1973a): The physical interaction of gases with crystalline solids. I. Gas-solid energies and properties of isolated adsorbed atoms. Surf. Sci. **36**, 317~352.

~- (1973b): The physical interaction of gases with crystalline solids. II. Two~dimensional second and third virial coefficients. Surf. Sci. **39**, 149~175.

~- (1974): <u>The Interaction of Gases With Solid Surfaces.</u> (Pergamon, New York).

Steinbrüchel, C. (1977): Model calculation of gas-surface energy accomodation coefficients: the importance of multiple collisions and of trapping. Surf. Sci. **66**, 131~144.

~- (1975): On the interpretation of adsorption and desorption kinetics experiments. Surf. Sci. 51, 539~545.

~- (1979): Desorption kinetics of one~and two-step mechanisms. Surf. Sci. 81, L645~L650.

Steinbrüchel, C., Schmidt, L.D. (1973): Condensation of gases on metals: the one~dimensional square well model. J. Phys. Chem. Solids 34, 1379-84.

~- (1975): Energy transfer in adsorption. J. Vac. Sci. Technol. 12, 204-205.

Stoll, A.G., Smith, D.L., Merrill, R.P. (1971): Scattering of the rare gases (He, Ne, Ar, Kr, and Xe) from Platinum (111) surfaces. J. Chem. Phys. **54**, 163-169.

Strachan, C. (1935): The interaction of atoms and molecules with solid surfaces. II ~ The evaporation of adsorbed atoms. Proc. R. Soc. London, Ser. A 150, 456~464.

~- (1937): The interaction of atoms and molecules with solid surfaces. IX-The emission and absorption of energy by a solid. Proc. R. Soc. London, Ser. A 158, 591-605.

Stutzki, J., Brenig, W. (1981): Theory of inelastic atom - surface scattering: a three~dimensional treatment. Z. Phys. B **45**, 49~59.

Sullivan, D.E. (1979): Van der Waals model of adsorption. Phys. Rev. B 20, 3991~4000.

Summerside, P., Sommer, E., Kreuzer, H.J., Teshima, R. (1982): Mean field theory of multilayer physisorption. Adsorbate densities and surface potentials. Phys. Rev. B 25, 6235-6254.

Suzanne, J., Coulomb, J.P., Bienfait, M. (1973): Auger electron spectroscopy and LEED studies of adsorption isotherms: Xenon on (0001) graphite. Surf. Sci. **40**, 414~418.

~- (1974): Transition bidimensionnelle du premier ordre; cas du xe'non adsorbe' sur la face (0001) du graphite. Surf. Sci. **44**, 141~156.

~- (1975): Thermodynamics and kinetics of the first monolayer adsorption of xenon on the (0001) graphite face. J. Cryst. Growth. (Netherlands) 31, 87~91.

Taborek, P. (1982): Critical cone in phonon-induced desorption of helium. Phys. Rev. Lett. **48**, 1737~1741.

Taborek, P., Goodstein, D.L. (1979): Phonon reflection at a sapphire-vacuum interface. J. Phys. C: Solid State Phys. 12, 4737~4751.

~- (1980a): Diffuse reflection of phonons and the anomalous Kapitza resistance. Phys. Rev. B 22, 1550-1563.

-- (1980b): Phonon focussing catastrophes. Solid State Commun. **33**, 1191-1194.

-- (1981): Phonon reflection at noble gas interfaces. Solid State Commun. **38**, 215-218.

Takaishi, T. (1975): Interactions between physically adsorbed molecules. Progr. Surf. Sci. **6**, 43-62.

Tikhonov, A.N., Samarskii,A.A. (1963): Equations of Mathematical Physics (Pergamon, New York).

Thomas, L.B. (1967): Thermal accommodation of gases on solids. In Fundamentals of Gas-Surface Interactions, ed. by H. Saltsburg, J.N. Smith, Jr., M. Rogers (Academic Press, New York) pp. 346-391.

Toennies, J.P. (1982a): Measurement of surface phonon dispersion curves of alkali-halide single crystals by time of flight spectroscopy of He atom beams. Phys. Scripta **T1**, 89-92.

-- (1982b): Phonon interactions in atom scattering from surfaces. (Benedek, Valbusa, 1982) pp. 208-255.

Tolk, N.H., Traum, M.M., Tully, J.C., Madey, T.E. (eds.) (1983): Desorption Induced by Electronic Transitions - DIET I, Springer Ser. Chem. Phys. Vol. 24 (Springer, Berlin, Heidelberg).

Toloukian, Y.S., Powell, R.W., Ho, C.Y., Klemens, P.G. (Eds.) (1970): Thermophysical Properties of Matter (IFI/Plenum, New York)

Tomanek, D., Kreuzer, H.J., Block, J.H. (1985): Tight-binding approach to field desorption: N_2 on Fe(111). Surf. Sci. **157**, L315-L322.

Tompkins, F.C. (1978a): Historical review: chemisorption on metals - retrosepect and prospect. CRC. Crit. Rev. Solid State and Materials Sci. **7**, 81-100.

-- (1978b): Chemisorption of Gases on Metals (Academic, London).

Trilling, L. (1970): The interaction of monoatomic inert gas molecules with a continuous elastic solid. Surf. Sci. **21**, 337-365.

Tully, J.C. (1980a): Theories of the dynamics of inelastic and reactive processes at surfaces. Ann. Rev. Phys. Chem. **31**, 319-343.

-- (1980b): Dynamics of gas-surface interactions: Langevin model applied to fcc and bcc surfaces. J. Chem. Phys. **73**, 1975-1985.

-- (1981): Dynamics of gas-surface interactions: Thermal desorption of Ar and Xe from platinum. Surf. Sci. **111**, 461-478.

Umbach, E., Menzel, D. (1982): Surface Science Index 1956-1977. (Fachinformationszentrum, Karlsruhe).

Unguris, J., Bruch, L.W., Moog, E.R., Webb, M.B. (1979): Xe adsorption on Ag(111): experiment. Surf. Sci. **87**, 415-436.

Venables, J.A., Bienfait, M. (1976): On the reaction order in thermal desorption spectroscopy. Surf. Sci. **61**, 667-672.

Velasco, V.R., Hardouin Duparc, O., Djafari - Rouhaui, B. (1982): Elastic surface waves in hexagonal molecular crystals with overlayers. Surf. Sci. **114**, 574-586.

Verwoerd, W.S. (1983): Diffusion effects on thermal desorption spectra of hydrogen from (111) surfaces of diamond-type crystals. Surf. Sci. **125**, 575-594.

Vidali, G., Cole, M.W. (1981): The interaction between an atom and a surface at large separation. Surf. Sci. **110**, 10-18.

-- (1984): Lateral variation of the physisorption potential for noble gases on graphite. Phys. Rev. B **29**, 6736-6738.

Vidali, G., Cole, M.W., Klein, J.R. (1983): Shape of physical adsorption potentials. Phys. Rev. B **28**, 3064-3073.

Volmer, M. (1932): The migration of adsorbed molecules on surfaces of solids. Trans. Faraday Soc. **28**, 359-363.

Volmer, M., Adhikari, G. (1926): Versuche über Kristallwachstum und Auflösung. Z. Phys. (Leipzig) **35**, 170-176.

Wallace, J.L., Goodstein, D.L. (1970): ⁴He on copper: some effects of pread-
    sorbed noble-gas layers. J. Low Temp. Phys. **3**, 283-300.
Wandelt, K., Hulse, J., Küppers, J. (1981): Site - selective adsorption of
    xenon on a stepped Ru(0001) surface. Surf. Sci. **104**, 212-239.
Wang, C., Gomer, R. (1979): Sticking coefficients of CO, O_2 and Xe on the
    (110) and (100) planes of tungsten. Surf. Sci. **84**, 329-354.
-- (1980): Adsorption and coadsorption with oxygen of xenon on the (110) and
    (100) planes of tungsten. Surf. Sci. **91**, 533-550.
Ward, C.A., Findlay, R.D., Rizk, M. (1982a): Statistical rate theory of inter-
    facial transport. I. Theoretical development. J. Chem. Phys. **76**.
    5599-5605.
-- (1982b): Statistical rate theory of interfacial transport III. Predicted
    rate of nondissociative adsorption. J. Chem. Phys. **76**, 5615-5623.
Weaver, D.L. (1979): Exact time-dependent desorption from a surface. Surf.
    Sci. **90**, 197-200.
Wedler, G., Ruhmann, H. (1982): Laser-induced thermal desorption of carbon
    monoxide from Fe(110) surfaces. Surf. Sci. **121**,464-486.
Weimer, M., Goodstein, D. (1983): A continuum model of helium desorption
    kinetics. Phys. Rev. Lett. **50**, 193-196.
Weinberg, S. (1963): Quasiparticles and the Born series. Phys. Rev. **131**,
    440-460.
Weinberg, W. H., Merrill, R. P. (1971): A simple classical model for trapping
    in gas-surface interactions. J. Vacuum Sci. Tech. **8**, 718-724.
Whitemouse, S.B., Jewsbury, P. (1979): Theoretical analysis of thermal
    desorption. Appl. Phys. **19**, 387-398.
Williams, A.R., Feibelman, P.J., Lang, N.D. (1982): Green's function methods
    for electronic-structure calculations. Phys. Rev. B **26**, 5433-5444.
Williams, R.D., Cole, M.W., Koonin, S.E. (1983): Coupling of phonons to a
    helium atom adsorbed on graphite. Phys. Rev. B **28**, 1076-1080.
van Willigen, W. (1968): Angular distribution of hydrogen molecules desorbed
    from metal surfaces. Phys. Lett. A **28**, 80-81.
Woodruff, D.P. (1981): Surface periodicity, crystallography and structure.
    In The Chemical Physics of Solid Surfaces and Heterogeneous Catalysis
    Vol. 1, ed. by D.A. King, D. P. Woodruff (Elsevier, Amsterdam) pp.
    81-181.

Yates Jr., J.T., Madey, T.E. (1971): The adsorption of methane by tungsten
    (100). Surf. Sci. **28**, 437-459.
Yates Jr., J.T., Erickson, N.E. (1974): X-ray photoelectron spectroscopic
    study of the physical adsorption of xenon and the chemisorption of
    oxygen on tungsten (111). Surf. Sci. **44**, 489-514.
Ying, S.C., Smith, J.R., Kohn, W. (1975): Density-functional theory of chemi-
    sorption on metal surfaces. Phys. Rev. B **11**, 1483-1496.
Yoshimori, A., Odoi, Y. (1984): Theory of coverage dependence of sticking
    probability. Surf. Sci. **143**, 37-45.
Yoshimori, A., Tsukada, M. (eds.) (1985): Dynamical Processes and Ordering on
    Solid Surfaces, Springer Ser. Solid-State Sci., Vol.59 (Springer,
    Berlin, Heidelberg).
Young, D.M., Crowell, A.D. (1962): Physical Adsorption of Gases (Butter-
    worths, London)

Zaremba, E., Kohn, W. (1976): Van der Waals interaction between an atom and
    a solid surface. Phys. Rev. B **13**, 2270-2285.
-- (1977): Theory of helium adsorption on simple and noble-metal surfaces.
    Phys. Rev. B **15**, 1769-1781.
Zatsepin, V.M. (1978): Theory of thermal desorption. Theor. and Exp. Chem.
    (USA) **14**, 404-407.

Zeiri, Y., Redondo, A., Goodard III, W.A. (1983): Classical stochastic diffusion theory for desorption from solid surfaces. Surf. Sci. 131, 221-238.

Zhdanov, V.P. (1981a): Effect of the lateral interaction of adsorbed molecules on preexponential factor of the desorption rate constant. Surf. Sci. 111, L662-L666.

-- (1981b): Lattice-gas model for description of the adsorbed molecules of two kinds. Surf. Sci. 111, 63-79.

Zwanzig, R.W. (1960): Collision of a gas atom with a cold surface. J. Chem. Phys. 32, 1173-1177.

Subject Index

Absolute rate theory see Transition
  state theory
Absorption of phonons 38-41,88,106,
  114,124,178,205,220
Accommodation 11,21,210,223,230,
  251-258,283
Accommodation coefficient 223,240,
  246-248,250,281
  classical theories 256-258
  contribution from bound states
    251-253,255-256
  definition 22,246,251
  experimental results 22,227-228
  high temperature limit 254-255
  low temperature limit 250,254-256
  master equation approach 251-252
  nonequilibrium 255
  prompt 253
  quantum theories 251-256
  relation to sticking coefficient
    240-241,246-250,256
  role of surface phonons 254
  role of two-phonon processes 255
Acoustic mismatch 200,204
Acoustic modes see Phonon dynamics
Action variable 266,268
Activated complex 13-14
Activation barrier 15,186,198,289
Activation energy 13,147
Adjoint equation see Master equation
Adsorbate 1-4,14-15,93,96,98,101,
  153,193-194,214,229,231,282
Adsorption 1-3,10-11,96-98,103,231,
  269,283
  chemical see Chemisorption
  dissociative 287-288
  heat of 3-4,12-13,18,20,43-44,107,
    152,165,182-184,186,198,226
  isotherm 18-19,183
  localized 14,43,93-94,98-99,101,
    107,145,164-173
  mobile 10,14,43,46,94,98,101-102,
    112,128,142,145-146,159,168-169,
    172,197,252-253,277,282-283

physical see Physisorption
  rate 14,109
  site 3,10,12,27-28,107-109
  time 11
Affinity 1-3
Ag 6,24,36,50,131,228,245-246
Al 8,24,34-36
Alkali halide 4,22,32,63,82,175
Anderson-Newns model 287,290
Ar 5,24,37,50,132,149,179-181,183,
  220-223,227-228,240-242,245-246,
  248-249,256,258
Attempt frequency 13,102-103
Au 8,17,24,36,245
Auger electron spectroscopy 17

Backward equation see Master equation
Balance equation 260,261
  for internal energy 278
  for kinetic energy 279
  for mass (continuity equation)
    277-281
  for momentum 277-278
  for potential energy 279
  for total energy 277,279
Band structure 283
  two-dimensional 32-33
Bessel functions 27,30,243
Bolometer 18,199-204,208,210, 223-
  224
Boltzmann distribution 137,212 see
  also Maxwell-Boltzmann statis-
  tics/distribution
Boltzmann equation 132
Born-von Karman boundary conditions
  67
Bose-Einstein statistics 46,71,101,
  107,110,165
Bravais net, two-dimensional 68
Brillouin zone
  three-dimensional 68,74
  two-dimensional 32-33,68,74,79,83,
    85
Brownian motion 259,276,279,289

Brueckner-Hartree-Fock theory   see
  Hartree-Fock theory
Bulk Debye model   see Debye model

C   see Graphite
Cascade model   16,114,165,168,175,
  181,205-207,219-220,230,282
Catalysis   291
CH_4, CD_4 4,8
CH_3F 284
CH_3OH 9
Chapman-Kolmogorov equation   91-92,
  261
Chemical potential   50-51,107,181,
  186,194,199,202,204,252
  ideal gas   18,101,108
Chemisorption   3-4,15-16,43,63,179,
  181,185-186,198,209,226,282
  kinetics   12,259,276,286-291
Classical scaling theory   257
Classical trajectory   131,134,136
CO, CO_2 9,17,179,209,227-228
Coarse graining   161,192
Cole-Toigo corrections   see Dynam-
  ics of atom-solid interaction;
  Desorption time
Collective modes   283
Collision duration   97,134-135
Collisions   191,197,208-209,218
Compensation effect   183,185
Condensation coefficient   102
Conduction electrons   35
Confluent hypergeometric function
  115
Conservation laws   see Balance equa-
  tion
Constantan   5,185-186,198-199,202
Continuity equation   see Balance
  equation
Cooling   16-17,186-190
Cooling rate   199
Correlation function   63,112,122,125
  -128,130,179,277
  damping factor   128
Correlations   105,276
Corrugation   see Surface potential
Cosine law   198,210-211,216
Coverage   10,15,43,51,111,182,184,
  187,189,191,223,225-226,232,283
  definition   12,99
  phenomenological rate equation
    12,96,99,102,107,232
Critical cone   159,206
Critical temperature for trapping
  247
Critical trapping velocity   248
Cs   35
Cu   5,24,36,179,211,284

Current density   208-209,233,252,278

D, D_2 8,40,224-225
Debye frequency   133,152,171,173,
  175-176,179,181,191,220,268
  at surface   77
  infinite solid   70-71
  semi-infinite solid   69,158
Debye model
  infinite elastic continuum (bulk
    Debye model)   68-77,88,112,114,
    117,126,128,157-160,165,170,213,
    216,236,254,257
  semi-infinite elastic continuum
    (surface Debye model)   68-77,88,
    112,114,157-159,161,213,216,
    238-239
Debye temperature   120,133,145,147,
  165,193,254,263
  localized adsorption   172-173
  mobile adsorption   158-159,172
Debye-Waller factor   39,61,126
Density functional theory   24,34,
  287,290
Density of states   3,168
  bulk Debye model   69-70,74
  for adsorbed particles   33,261
  phonons   69,158-159,214
  surface Debye model   71-77,238
Density operator/matrix   89,104-106,
  277
Derivative coupling   see Dynamic
  atom-solid coupling
Desorption   10-11,13,96-98,102-103,
  152-156,186,190,195,205,214,218,220,
  223,229-230,256,259,273-274,283-284
  cooling   see Cooling
  from localized adsorbate   164-173
  one-dimensional model   114,159,
    168,175,179,197
  three-dimensional model   159-163,
    168-169,175
Desorption/Adsorption energy   see
  Desorption rate constant; Adsorption
Desorption flux   see Flux
Desorption rate constant   15,99-100,
  109,146,165,168-169,172,175
  differential rate   164-169,180
  energy parameter   16,144,147,149,
    200-202,221
  Frenkel-Arrhenius parametrization
    12,16,20,102,144-145,147,182,185,
    198-201,281
  in large friction limit   281
  in small friction limit   260,269,
    281

multiphonon contributions 176–181
  see also Multiphonon processes
Pade approximation 156
prefactor 12-14,101,145,149,
  166-168,182-184,186,198-202,281
quasi-equilibrium approximation
  100
Desorption time
  Cole-Toigo corrections 173-176
  from Fokker-Planck/Kramers
    equation 155,260,269
  model system with two bound
    states 143-146
  perturbation theory 100,154-155,
    164,168,182
  surface Debye model 157-159,216
  systems with shallow bound states
    146-150,148-149
Detailed balance 94-96,99-102,117-
  118,128,138,150-151,163,
  211-212,232-233,252
  definition 93
Detector response function 208,210
Dielectric function 24
Differential desorption rate see
  Desorption rate constant
Diffraction 37,240
  into bound states (elastic
    sticking) 240,256
  see also Scattering
Diffusion coefficient 260
Diffusion equation 276
Direct phonon process 203-204
Dispersion force see van der Waals
  interaction
Dispersion relation
  bulk phonons 78
  surface phonons 39,41,58,70,78-86
Displacement vector 55-56,58,62,64,
  126
Dissociation 13,198
Dissociative adsorption see Adsorp-
  tion
Double layer 3
Dressing transformation 123-124
Dynamic atom-solid interaction
  55-64,87,97,125,128,157,170,211,282
  Cole-Toigo corrections 56,60,122,
    173
  derivative coupling 55,60-61,122,
    173,177

Effective medium theory 36
Effusion 208,210-212
Elastic constants 64,239
Elastic continuum see Debye model
Elastic desorption 240

Elastic sticking see Diffraction
  into bound states
Electron energy loss spectroscopy
  18
Electron-hole excitation 290
Electron stimulated desorption 11,
  283,285-286
Electron stimulated field desorption
  11
Electron transfer in chemisorption
  1-4
Emission of phonons 39,87,102,106,
  114,118,124,205,284
Energy accommodation coefficient
  see Accommodation coefficient
Energy current density 187-188,
  252-253
Energy dependent sticking coeffi-
  cient see Sticking coefficient
Energy distribution 236-237
  desorbed particles see Time of
    flight spectra
  scattered particles 131
Energy transfer 236,252
  soft cube model 134-135,137
Equilibrium constant 13
Evaporation 11,100,184-185,198,211,
  223
Ewald diagram 38-39

Fe 9
Fermi-Dirac statistics 46,101,107-
  108,110,182
Fermi level 1-3,35
Field desorption 11
Field emission microscopy 18
Field evaporation 11
Field operator 87
First-order reaction see Reaction
  order
First passage time 96,155,157
Flash adsorption 195-196
Flash desorption
  definition 20,190
  forward peaking 20,113,219
  initial/boundary conditions 113,
    161,191
  nonequilibrium effects 161-163,
    205-206,218
  thermalization 113,162,190-197,
    207,214,218,220
  time of flight spectra 198,
    218-219
Fluctuations 267,270-275
Flux 12,223,229-230,233-234,251-253
  desorbed particles 20,113-114,
    198,201,207-212,215

Fokker-Planck equation 89,155,259-
261,276-277,289
Forward peaking see Time of flight
spectra; Flash desorption
Fourier law of heat transfer 16,
187
Frenkel-Arrhenius parametrization
see Desorption rate constant
Friction coefficient 255,260-261,
265-266,268,270,276-277,280,289-290
Friction function 267-268
Friction law for adsorbed particles
269-275
Friction tensor 276
Full sticking coefficient see Stick-
ing coefficient

Gas-solid interaction 1,86-88,211
see also Surface potential; Dynamic
atom-solid interaction
Gas-surface scattering see Scattering
Ge 24
Grafoil 44
Graphite 4-5,8,10,17-20,25,27-31,
33,40,44,50-51,86,101,119,146-147,
149,164-167,169,174,180,182-183,187-
189,192-193,195-196,202,209,
214-217
Green's function 82,123,125,187

H, H_2 8,24,37,40,146,149,211,223-225,
227,238,257,277,287
Hard cube model 244,258
Harmonic chain see Soft cube model;
Zwanzig model
Hartree-Fock theory 35,107,182
Brueckner-Hartree-Fock theory
45-54
temperature dependent 45,110,141
He 5,10-11,17-20,24-25,27-31,33-37,
39-40,44-45,48-54,101,113,118-120,
131-132,142,146-150,161-163,169,
174,176,182-189,192,195,198-200,202,
205,207,209,213-218,220,222-223,
225-228,238-241,245,249,256,281-283
Heat bath 16,290
Heat capacity see Specific heat
Heat conduction see Fourier law of
heat conduction
Heat current 279
Heat of adsorption see Adsorption
Heat pulse 199-202,205
Heater 199-204
Heating rate 163,199,220
Heisenberg equation 88,103,105,108,
124
Hg 227

Hot wire method 227
Hulthen potential 115
Hybridization 81
in graphite 25,41
Hydrocarbons 224
Hydrodynamics of adsorption 227-281

Image 4,35,286
Infrared spectroscopy 18
InSb 24
Ion stimulated desorption 11
Ionization energy 1-3
Ionization gauge 208
Isothermal desorption 99,142,193,
195-196,198,214
definition 18,190
initial conditions 99,142,190
master equation see Master equation
rate constant see Desorption rate
constant
Isotherms see Adsorption

Jellium 1,34-35,287

K 8,35-36
KCl^- 284
Kinetic equation 208,234,259,277
see also Balance equations; Fokker-
Planck equation; Kramers equation;
Master equation
Knudsen's cosine law see Cosine law
Kr 5,24,132,223,228,245-246,248-
250,256
Kramers equation 89,156,259-260,
266,269,276-277,289
derivation 261-268
generalized 267
radiative boundary conditions 269
Kramers-Kronig relation 24
Kramers-Moyal expansion see Master
equation

Laguerre polynomial 115
Lame constants see Elastic con-
stants
Langevin equation 89,276-277
Langmuir isotherm 108
Laser-induced thermal desorption
220,283
Lateral thermalization see Flash
desorption; Time of flight spectra
Lattice displacement vector see
Displacement vector
Lattice dynamics see Phonon dynam-
ics
Lattice gas 19,108,186

Lattice vibrations see Phonon dyn-
amics
Lennard-Jones interaction
9-3 surface potential 26,42,56,
174
6-12 two-body potential 25,27,29,
41,48,56,78
Level broadening 97
Li 35-36
LiF 24,38-41,118,120,146-149,193,
220,282
Liouville equation 276-277,279
Liouville-von Neumann equation 89,
277
Liouville operator 106
Lippmann-Schwinger equation 123
Local density approximation 46
Local equilibrium 16,187
Local potential 46 see also Sur-
face potential
Localized adsorption see Adsorption
Low energy electron diffraction 18
Lucas mode 78-79

Markov process 89,92,96-97,103,106
definition 91
stationary 93
time homogeneous 93,96,110
Mass spectrometer 18,208,210,224-225
Master equation 11,89-117,92,93,98,
109,118,128,131,142,149,187,
191,195,206,211,213,219,229-230,234-
235,259,274,276,283-284
accommmodation coefficient see
Accommodation coefficient
at finite coverage 107-111
backward (adjoint) equation 92,96
continuum limit 261,269
discrete form 94
for isothermal desorption 99,142
for one-dimensional model of
desorption 113,142,168
for physisorption kinetics 96-103
for three-dimensional model of
desorption 98,160,170
general properties 94-96
generalized 89,106
initial conditions 95,97,99,142,
160
Kramers-Moyal expansion 259-260,
263-264,266-267
matrix diagonalization 94,100,
142-143,151,155-156,161,182,188,
191-192,272
mesoscopic approach 89-96
microscopic approach 89,103-106
Pauli's derivation 103

perturbation theory 100,150-157,
164,212,232
solution 95,100,211,230
sticking coefficient see Sticking
coefficient
symmetrized version 96,135,151
transition probabilities see
Transition probabilities
Maxwell-Boltzmann statistics/distri-
bution 101,107,195,201,203,210,248,
270,273, 289
Mean field theory 34,112,187,283
for multilayer physisorption
43-54
for physisorption kinetics 107,141
see also Hartree-Fock theory
Mean free path
of gas particles 44,191,197,209
of phonons 16
Mg 35-36
Mobile adsorption see Adsorption
Molar volume 16,187
Molecular beam 18,198,223
Moment see Transition probabilities
Momentun accommodation coefficient
228
Momentum conservation 37-39,112,
133,142,159-160,174,206,215
Monolayer 3-4,10,15,43,99,107,183,
187,207,209,225-226
Morse potential 41,57,62,128,143,
145,152,172-173,175-176,179,213,238,
240,254,260-261,265,268
bound states 42,115,120,147,272
matrix elements 116,161
wave functions 115-116,286
Multilayer physisorption 181-187,
283
mean field approach 43-54,181,185
see also Surface potential
Multiphonon processes 191,207,282
coherent 58,122-132,176-181,268-
269
cascades see Cascade model
see also Desorption rate constant

N_2 8,9,226-227
Na 35-36
NaCℓ 8,40,79-81,149,284
NaF 24,40,82,146-147,149-150,175
Ne 24,131-132,164-167,178,223,227-
228,245,248-250
NH_3 8,284
Ni 5,8,225-226
Nichrome 5,161-163,185-186,199,202,
205,218,225
9-3 potential see Lennard-Jones in-
teraction

NiO 40
Noble gas   see Rare gas
Nonequilibrium pressure   see Pressure
Nonlocal potential   see Surface potential
Normal vibrations (modes) 65
  for infinite continuum 58
  for semi-infinite continuum 58,66
N_2O 4,285-286

O_2 9,17,227-228
Occupation probabilities 46-48,
  98-99,105,111,156,161-163,170,181,
  187,191-195,211,229-231,266,273-
  275,284
  definition for a stochastic
    process 93
  master equation 93
  see also Maxwell-Boltzmann
    statistics/distribution;
    Bose-Einstein statistics;
    Fermi-Dirac statistics
One-phonon processes
  58,112-121,149,170,178,205-207,213,
  216,219,240,268,282
One-way equilibrium desorption rate
  104
One-way equilibrium transition rate
  see Transition probability
Onsager coefficients 281
Optical modes   see Phonon dynamics

Pair interaction   see Two-body interaction
Particle induced desorption 15,186
  see also Electron stimulated
  desorption; Ion induced desorption
Partition function 13-15,290
Pd 6,9,24,226
Permeation 198
Phase diagram 10
Phase transition 44,283
Phenomenological rate equation   see
  Coverage
Phonon
  ballistic 202-205,220
  creation and annihilation operators 58,87
  polarization 58,114,204-205
  time evolution 103,109
  see also Phonon dynamics
Phonon dynamics 64-86
  for adsorbate covered solid 83-86
  for infinite elastic continuum
    58,64-77
  for semi-infinite elastic continuum 58,64-77

for semi-infinite lattice 80,82
full harmonic theory 78
Green's function method 82
harmonic theory 77
of infinite solid 80
of slabs 78-86
partial harmonic theory 77
partial quasi-harmonic theory 78
quasi-harmonic theory 78
see also Debye model
Phonon linewidth 181
Phonon mediated desorption   see
  Thermal desorption
Photodesorption 11,15,186,283-285
Physisorption 4
Polarizability 24
Precursor state 226,287
Pre-exponential factor   see Prefactor
Prefactor   see Desorption rate constant
Pressure 12,18,101,108,194,201-202,
  278
  time evolution 191-197   see also
  Flash desorption
Probability density   see Stochastic
  process
Probability distribution   see Stochastic process
Projection operator 105
Prompt accommodation coefficient
  see Accommodation coefficient
Prompt scattering 230
Prompt sticking coefficient   see
  Sticking coefficient
Pt 8,220-224,227,230,242,246,256
Pyridine 284

Quasi-stationary distribution   see
  Steady state occupation

Rainbow effect   see Time of flight
Random phase approximation 106,141
Rare gas 4,22,32,131,223-224,227,
  245,249-250,286
Rate   see Adsorption; Desorption
  rate constant
Rayleigh wave 41,67
  desorption time 158
  in slabs 78
  sticking coefficient 238-239
  surface Debye model 67,70,72-77
  time of flight spectra 216,217
Rb 35
Reaction coordinate 14,287-288
Reaction order
  First order 12,15,96,184-185
  Zero order 15,184-185

Reaction path 14,287-288
Reciprocal lattice
  two-dimensional 26,38-39,59
Recoil 254
Recombination 13,198
Reduced potential see Surface potential
Registered gas 10
Relaxation time 97
  breakdown of relaxation time
    approximation 178
Resonant heating 284
Riemann zeta function see Surface
  potential
Ru 5,8,9,17,285
Rutherford backscattering 18

Sapphire 5,199,202,205
Scattering 37-41
  elastic 37,61
  gas-surface 21,36-41,113,210,223,
    250
  inelastic 11,37-41,82,120,130,
    282-283
  resonant 37
  see also Diffraction
Scattering operator 125
Selective adsorption 37
Self-energy corrections 124
Semi-infinite elastic continuum see
  Debye model
Separable potential see Surface
  potential
SF$_6$ 9
Si 8,223
6-12 Potential see Lennard-Jones
  interaction
Smoluchowski equation 279
Soft cube model 112,133-140,153,
  244-245,258
  definition 133
  see also Sticking coefficient;
    Transition probability
Sound velocity 113
  along the surface 58,66,160
  longitudinal 65,158
  transverse 65,158
Specific heat 16,18,19,43,187-189
Steady-state occupation 153-154,
  230-232
  quasi-equilibrium approximation
    155-156,220,232
Sticking 11,21,223-251,256,283
Sticking coefficient 12,14-15,22,
  37,102,223,232,256-257,261,281,290
  classical theories 235,237,242-
    250
  definition 12,21

energy dependent 233-234,242
experiments 223-226
for mobile adsorption 102,234
full 233,240-241
importance of Rayleigh waves 238-
  239
importance of Umklapp processes
  238-240
low temperature behaviour 237-
  238,256
master equation approach 229-234,
  242,245
prompt 233,236,240-241,245-246
quantum theory 229-242
relation to accommodation coefficient 240-241,246-250,256
role of bound state - continuum
  transitions 229,233
Sticking probability see Sticking
  coefficient
Stochastic process 89,97
  averages 91
  conditional probability 91-92
  definition 90
  joint probability 90-92
  probability density 90,93
  transition probability 92
  see also Master equation; Markov
    process
Stochastic trajectory method 220-
  221,241,257
Stochastic variable 90
  definition 89
  probability distribution 89,90
Stress and strain tensors 64
Surface bond 15
Surface Debye model see Debye model
Surface Debye temperature see
  Debye temperature
Surface modes 78-86,114 see also
  Phonon dynamics; Rayleigh wave;
  Lucas mode
Surface potential 4,10,23-36,42,60,
  87,93,99,102,107,112,115,157,164,
  168-170,198,205,220,222,229,238,
  244-245,260,266,276,282,284-285
  anisotropic harmonic approximation
    128
  attractive part 4,11,23-24,96,
    238,246,250,254
  bare 45,48,50-51,54,110,183
  bound/continuum states 38,40-42,
    51-54,97,101-102,107,110,117-118,
    120,147,169,181,187,205,211,213,
    220,229,251,261,267,269,272
  continuum limit 25,56
  corrugation 27,29,37,39,41,113,
    159,165,168,218,282-283

coverage dependent (mean field)
 43-45,47=48,50-53,110,141
Hulthen 115
ionic solid 24,63
laterally averaged 30,164-165
Lennard-Jones see Lennard-Jones
 interaction
local 122-123
metal 34-35,63
minimum 26-27,29,145,176,286
molecular solid 24,63
Morse potential see Morse
 potential
nonlocal 35,122-123
repulsive part 4,24,96,244,246,
 250,254,257-258
separable 115,122-123,176-177
square well 246,250
sum of two-body interactions
 24-30,63,176
thermally averaged 61-62,128
universal shape (reduced potential) 30-32
zeta form 48

Temperature programmed desorption
 20-21,142,198
Thermal conductivity 16,188-189
Thermal desorption 11,15,63,211,
 213,221,273,279 see also Flash
 desorption; Isothermal desorption;
 Laser induced thermal desorption;
 Temperature programmed desorption
Thermal diffusivity 16,187
Thermalization see Flash desorption
Thermally averaged surface potential
 see Surface potential
Ti 226
Time of flight spectra 18,40,142,
 195,198-222,201,203,207,211,220,
 231,257
 angular dependence 198,205-207,
 211,214-215,220 see also Cosine
 law
 collision induced forward peaking
 207,209,218
 forward peaking 198,205,207,218,
 220
 Maxwell-Boltzmann form 210,215,
 220
 nonequilibrium effects 203-205,
 218
 one-dimensional theory 213,215-
 216
 perturbation theory 212-214
 rainbow effect 206-207,220

three-dimensional theory 213,215-
 217
 with bulk and surface Debye
 models 213,216=218
Transient pressure see Transients;
 Pressure
Transients 275
 in flash desorption 191-197,220
 in physisorption kinetics 100,
 144,147,152,157
Transition probability 63,93,97,99,
 106,110,112-141,148,157,163,165,
 170,179,182,187,191,195,206,213,
 230,235,241,251-252,259
 and energy distribution of
 scattered atoms 131-132,236
 between states in a Morse
 potential 117-121,129,254
 definition 92
 eigenvalues and eigenvectors
 94-95,100,143,147-148,150-152,
 154,156,161,193,230-232
 experiments 22,120,283
 Fermi's golden rule 105,125,152,
 285
 for finite coverage 112
 for localized adsorption 164,171
 for mobile adsorption 112,160,168
 for multiphonon processes
 112,123-132
 for one-dimensional model of
 desorption 112=114,169
 for three-dimensional model of
 desorption 112,160
 Gaussian form 131,140,222,241,
 255-256,259
 in soft cube model 135-140,153,245
 moments 221,241-242,262-265,267,
 270,276
 symmetrized 95,151,154,164
Transition state see Activated complex
Transition state theory 13-14,186,
 289-290
Transmission coefficient 14
Trapping see Sticking; Sticking
 coefficient
Trapping condition 243-244,247
Two-body correlations 45
Two-body interaction 25,30,48,60,
 110,168,278
 attraction 25
 between adsorbed particles 43-45,
 47-49,54
 Gaussian 127
 hard core repulsion 48,107
 Lennard-Jones see Lennard-Jones
 interaction

repulsion 50
  resulting in a Morse surface
    potential 57
Two-dimensional fluid 10
Two-dimensional solid 10,44,50
Two-phonon processes see multipho-
  non processes

Ultraviolet photo spectroscopy 18
Umklapp processes 238,240
Unit surface cell 26-28,48,59,68
Unitary pole approximation 123

van der Waals interaction 4,23-24,
  238,255
  constant C_3 23-24,30,36
Vaporization 44
Vertex correction 124

W 6-8,17,20-21,120-121,131,152-156,
  179-181,223,226-228,239,245,248,
  256,258,263,272-273
Wannier function 127-128

Wigner-Polanyi equation see
  Frenkel-Arrhenius parametrization
WKB approximation 42,127,165,236,
  238
Work function 1-3,35-36

Xe 5-8,20-21,24,86,120,121,132,
  152-156,164-167,180,220-227,
  230-231,242,245-246,248-250,256,
  263,272-273
X-ray absorption fine structure
  spectroscopy 18
X-ray photo spectroscopy 18

Zero order reaction see Reaction
  order
Zero point energy 13
ZnO 8-9,24
Zwanzig model
  accommodation coefficient 257
  sticking coefficient 242-244
Zwanzig's projection operator method
  89,103

Springer Series in Chemical Physics

Editors: V.I. Goldanskii, R. Gomer, F.P. Schäfer, J.P. Toennis

A Selection:

Volume 40
E. Hirota

High-Resolution Spectroscopy of Transient Molecules

With contributions by Y. Endo, K. Kawaguchi, S. Saito, T. Suzuki, C. Yamada
1985. 80 figures. IX, 233 pages
ISBN 3-540-15302-0

Contents: Introduction. – Theoretical Aspects of High-Resolution Molecular Spectra. – Experimental Details. – Individual Molecules. – Applications and Future Prospects. – References. – Subject Index.

Volume 16
V.L. Broude, E.I. Rashba, E.F. Sheka

Spectroscopy of Molecular Excitons

1985. 135 figures. XI, 271 pages
ISBN 3-540-12409-8

Contents: Experimental Background. – Exciton Spectra of Perfect Crystals. – Exciton Spectra of Doped Crystals. – Exciton Spectra of Mixed Crystals. – Band-to-Band Transition Spectra. – Vibronic Spectra of Molecular Crystals. – Conclusion. – Appendix A. – Appendix B. – References. – Subject Index.

Volume 25

Ion Formation from Organic Solids

Proceedings of the Second International Confernce, Münster, Federal Republic of Germany, September 7-9, 1982

Editor: A. Benninghoven
1983. 170 figures. IX, 269 pages
ISBN 3-540-12244-3

Contents: Field Desorption. – ^{252}Cf-Plasma Desorption. – Secondary Ion Mass Spectrometry

(SIMS) Including FAB. – Laser Induced Ion Formation. – Other Ion Formation Processes. – Index of Contributors.

Volume 21

Dynamics of Gas-Surface Interaction

Proceedings of the International School on Material Science and Technology, Erice, Italy, July 1-15, 1981
Editors: G. Benedek, U. Valbusa
1982. 132 figures. XI, 282 pages
ISBN 3-540-11693-1

Contents: Scattering of Atoms from Solid Surfaces. – Characterization of Adsorbed Phases. – Spectroscopy of Surface Optical Excitations. – Surface Phonon Spectroscopy by Atom Scattering. – Index of Contributors.

Volume 15

Vibrational Spectroscopy of Adsorbates

Editor: R.F. Willis
With contributions by B.K. Agrawal, G. Allan, J.A. Creighton, B. Djafari-Rouhani, L. Dobrzynski, B. Feuerbacher, P. Hollins, D.A. King, D.M. Newns, J. Pritchard, N. Sheppard, D.G. Walmsley, R.F. Willis, C.J. Wright
1980. 97 figures, 8 tables. XII, 184 pages
ISBN 3-540-10429-1

Volume 2
M.A. Van Hove, S.Y. Tong

Surface Crystallography by LEED

Theory, Computation and Structural Results
1979. 19 figures, 2 tables. IX, 286 pages
ISBN 3-540-09194-7

Springer-Verlag
Berlin Heidelberg New York Tokyo

Springer

Theory of Chemisorption

Editor: **J. R. Smith**

1980. 116 figures, 8 tables. XI, 240 pages
(Topics in Current Physics, Volume 19)
ISBN 3-540-09891-7

Contents: *J. R. Smith:* Introduction. – *S. C. Ying:* Density Functional
Theory of Chemisorption of Simple Metals. – *J. A. Appelbaum,
D. R. Hamann:* Chemisorption on Semiconductor Surfaces. –
F. J. Arlinghaus, J. G. Gay, J. R. Smith: Chemisorption on d-Band
Metals. – *B. Kunz:* Cluster Chemisorption. – *T. Wolfram, S. Ellial-
tioğlu:* Concepts of Surface States and Chemisorption on d-Band
Perovskites. – *T. L. Einstein, J. A. Hertz, J. R. Schrieffer:* Theoretical
Issues in Chemisorption.

Chemistry and Physics of Solid Surfaces IV

Editors: **R. Vanselow, R. Howe**

1982. 247 figures. XIII, 496 pages. (Springer Series in Chemical
Physics, Volume 20). ISBN 3-540-11397-5

Contents: Development of Photoemission as a Tool for Surface
Science: 1900–1980. – Auger Spectroscopy as a Probe of Valence
Bonds and Bands. – SIMS of Reactive Surfaces. – Chemisorption
Investigated by Ellipsometry. – The Implications for Surface Science
of Doppler-Shift Laser Fluorescence Spectroscopy. – Analytical
Electron Microscopy in Surface Science. – He Diffraction as a Probe
of Semiconductor Surface Structures. – Studies of Adsorption at
Well-Ordered Electrode Surfaces Using Low-Engergy Electron Diff-
raction. – Low-Energy Electron Diffraction Studies of Physically
Adsorbed Films. – Monte Carlo Simulations of Chemisorbed Over-
layers. – Critical Phenomena of Chemisorbed Overlayers. – Struc-
tural Defects in Surfaces and Overlayers. – Some Theoretical
Aspects of Metal Clusters, Surfaces, and Chemisorption. – The
Inelastic Scattering of Low-Energy Electrons by Surface Excitations;
Basic Mechanisms. – Electronic Aspects of Adsorption Rates. –
Thermal Desorption. – Field Desorption and Photon-Induced Field
Desorption. – Segregation and Ordering at Alloy Surfaces Studied by
Low-Energy Ion Scattering. – The Effects of Internal Surface
Chemistry on Metallurgical Properties. – Subject Index.

Chemistry and Physics of Solid Surfaces V

Editors: **R. Vanselow, R. Howe**

1984. 303 figures. XXI, 554 pages. (Springer Series in
Chemical Physics, Volume 35). ISBN 3-540-13315-1

The series **Chemistry and Physics of Solid Surfaces** presents selected
review articles predominantly from the area of solid/gas interfaces.
These articles are written by internationally recognized experts, the
invited speakers of the International Summer Institute in Surface
Science (ISISS).
It is the purpose of this series to bring researchers in academia and
industry up-to-date and to bridge the gap between conventional
surface science textbooks and specialized conference proceedings.
Extensive literature references are provided together with a detailed
subject index.

Springer-Verlag
Berlin Heidelberg
New York Tokyo